273
Current Topics in Microbiology and Immunology

AF785827

Editors

R.W. Compans, Atlanta/Georgia
M.D. Cooper, Birmingham/Alabama
H. Koprowski, Philadelphia/Pennsylvania
F. Melchers, Basel · M.B.A. Oldstone, La Jolla/California
S. Olsnes, Oslo · M. Potter, Bethesda/Maryland
P.K. Vogt, La Jolla/California · H. Wagner, Munich

Springer
Berlin
Heidelberg
New York
Hong Kong
London
Milan
Paris
Tokyo

W. Doerfler · P. Böhm (Eds.)

Adenoviruses: Model and Vectors in Virus-Host Interactions

Immune System, Oncogenesis, Gene Therapy

With 35 Figures and 14 Tables

Springer

Professor Dr. Walter Doerfler
Institut für Klinische und Molekulare Virologie
Universität Erlangen-Nürnberg
91054 Erlangen, Germany
e-mail: doerfler@viro.med.uni-erlangen.de

Petra Böhm
Institut für Genetik
Universität zu Köln
Weyertal 121
50931 Köln, Germany
e-mail: pboehm@scan.genetik.uni-koeln.de

Cover Illustration by Lennart Philipson (this volume):

Binding of 3 receptor IgG domains on the outside of the 3 adenovirus fiber knob polypeptides. The complex is viewed down the fiber shaft.

ISSN 0070-217X
ISBN 978-3-642-05715-1

This work is subject to copyright. All rights are reserved, whether the whole or part of the material is concerned, specifically the rights of translation, reprinting, reuse of illustrations, recitation, broadcasting, reproduction on microfilm or in any other way, and storage in data banks. Duplication of this publication or parts thereof is permitted only under the provisions of the German Copyright Law of September 9, 1965, in its current version, and permission for use must always be obtained from Springer-Verlag. Violations are liable for prosecution under the German Copyright Law.

Springer-Verlag Berlin Heidelberg New York
a member of BertelsmannSpringer Science+Business Media GmbH

http://www.springer.de

© Springer-Verlag Berlin Heidelberg 2004
Softcover reprint of the hardcover 1st edition 2004

The use of general descriptive names, registered names, trademarks, etc. in this publication does not imply, even in the absence of a specific statement, that such names are exempt from the relevant protective laws and regulations and therefore free for general use.

Product liability: The publishers cannot guarantee the accuracy for any information about dosage and application contained in this book. In every individual case the user must check such information by consulting other relevant literature.

Cover Design: Design & Production GmbH, Heidelberg

Printed on acid-free paper

Preface

Science works at its best when new observations are pursued by the persevering investigator to elucidate the molecular mechanisms behind a discovered phenomenon. Time and again, basic research without any, at the time obvious, immediate medical or biological relevance has proved to be the most rewarding endeavor for the promotion of knowledge in science. Biology, with the undisputable advantage of several billions of years in evolutionary experience, is too far ahead for us to be able to anticipate the next important scientific discovery. As in the past, we will have to rely on hard work, analytical scrutiny, and serendipity to guide us to progress. Narrowly focused research aimed at "solving" important medical problems will rarely be worth the time, money, and effort spent on it. The successful design of the Human Genome Project has been based on 50 years of research in molecular biology on a very wide scope, and hence is a case in point rather than an example to the contrary.

In keeping with this long-standing experience, *Adenoviruses: Model and Vectors in Virus–Host Interactions* presents chapters which are devoted to basic research on a group of viruses which have helped unravel many of the basic mechanisms in mammalian molecular biology. While the chapters in the preceding volume 272 of this series dealt with problems of basic molecular virology, we now turn to studies in which adenoviruses have helped to answer more complex problems in biology: How does the virus attach to and penetrate the cell surface? Which mechanisms can the virus avail itself of to circum-vent the immune defenses of the host? Although tumor biology remains as ephemeral and complex as ever, adenovirus oncogenesis offers a resourceful model with the advantage of a high tumor incidence and a short latency period. Different mechanisms have been examined here to better understand cell transformation. The traditional thought that products of the viral E1 region and their interactions with host proteins could play a crucial role in eliciting the oncogenic response has been carefully documented and widely exploited. Moreover, tumorigenesis being a highly complicated biological event, a more generally applicable model has been investigated. Foreign DNA insertion into an established mammalian genome alters patterns of DNA methylation and transcription in the genome of the host. The ensuing structural and functional perturbations are presumed to contribute to the transformation process.

Lastly, much work has been performed adapting adenoviruses as pliable and innocuous gene transfer vectors for human somatic gene therapy. With fatefully bad news from clinical applications of viral gene transfer vectors, we seem to be still very far from this goal. However, basic research on adenovirus vectors and their interactions with host animals, their fate in the living organisms, and the expression of their payload in selected organs, if anything, may improve our knowledge on how to introduce foreign genes into living cells and organisms. Should that aim ever be met, basic research will have led the way. Several of the contributions in this volume point out directions in which solutions to complex biological and medical problems can be sought.

Again, it is a pleasure to thank our colleagues for their excellent chapters and Ms. Clauss of Springer-Verlag in Heidelberg for editorial assistance.

Erlangen/Köln, August 2003

Walter Doerfler
Petra Böhm

List of Contents

List of Contributors

(Their addresses can be found at the beginning of their respective chapters.)

1
Immune System

Evasion of the Immune System by Adenoviruses

G. E. Blair[1] · M. E. Blair-Zajdel[2]

[1] School of Biochemistry and Molecular Biology, University of Leeds, Leeds, LS2 9JT, UK
E-mail: g.e.blair@leeds.ac.uk

[2] Biomedical Research Centre, Division of Biomolecular Sciences, School of Science and Mathematics, Sheffield Hallam University, Sheffield, S1 1WB, UK

Abstract Human adenoviruses (Ads) have the ability to transform primary cells, and certain Ads, the subgenus A adenoviruses such as Ad12, induce tumours in immunocompetent rodents. The oncogenic phenotype of the subgenus A adenoviruses is determined by the viral E1A oncogene. In order to generate tumours, Ad12-transformed cells must evade the cellular immune system of the host. Ad12 E1A gene products mediate transcriptional repression of several genes in the major histocompatibility complex (MHC) involved in antigen processing and presentation, resulting in evasion of cytotoxic T lymphocyte (CTL) killing of transformed cells. In this review, the molecular mechanisms of E1A-mediated transcriptional repression of MHC gene expression are described. In addition, evasion of natural killer (NK) cell killing by Ad-transformed cells is also considered.

1 Introduction

The human adenoviruses (Ads) have proved to be powerful tools in studies on the regulation of the mammalian cell cycle and the mechanisms of cell transformation and oncogenesis (Boulanger and Blair 1991; Moran 1993; Bayley and Mymryk 1994; Gallimore and Turnell 2001). In more recent times, Ads and their gene products have been shown to have potential therapeutic uses. As well as being useful gene therapy vectors (Hitt and Graham 2000), Ads have also been exploited in the selective eradication of human tumours (Hermiston 2000; Heise and Kirn 2000), and expression of the viral E1A gene has been used to increase the chemosensitivity of tumours (Yoo et al. 2001). The host immune response against the adenovirus and its gene products is considered to represent a significant barrier to their therapeutic use (Kafri et al. 1996; Wold et al. 1999). An understanding of the interaction of the adenovirus with the host organism is therefore of fundamental as well as applied importance in aiding the design of gene therapy vectors that do not provoke adverse immunological responses.

There are at least 51 distinct human adenovirus serotypes, which have been classified into six subgenera (A to F) according to various properties, including the oncogenicity of viruses in newborn rodents (Shenk 2001). Viruses of subgenus A (such as Ad12) induce tumours with high frequency and short latency, whereas viruses from subgenus B (such as Ad3 and Ad7) are more weakly oncogenic. Adenoviruses from the remaining subgenera (including the well-studied Ad2 and Ad5 from subgenus C) are non-oncogenic, with the exception of the subgenus D virus Ad9, which forms mammary tumours in the rat Wistar-Furth strain (Javier 1994). All human adenoviruses studied to date can transform primary rodent cells in culture; however, only cells transformed by viruses of subgenus A and B are oncogenic in newborn rodents, paralleling the oncogenic properties of the parental viruses (Raska 1995). Both intact viral DNA and restriction endonuclease fragments can be used to transform primary cells. The viral genes required for transformation are the E1A and E1B genes, which are linked together at the left 11% of the approximately 36-kb linear genome. The E1A gene encodes two major mRNAs, which are generated by alternative RNA splicing and direct synthesis of two proteins (289 and 246 amino acid residues, or R, in Ad2 and Ad5; 266R and 235R in Ad12) that are identical except for an internal peptide segment (Boulanger and Blair 1991; Gallimore and Turnell 2001). Comparison of predicted E1A protein sequences from several viral serotypes led to the identification of three conserved regions, CR1, CR2, and CR3. The CR3 region contains a zinc finger domain and is responsible for transcriptional activation of other viral early promoters (E1B, E2, E3, and E4) as well as the major late promoter (MLP, located at around 16.4% on the viral genome). The MLP directs transcription of late viral gene products, which mainly comprise structural proteins of the viral capsid. E1A proteins do not bind DNA directly, but appear to form heterodimers with basal and ubiquitous transcription factors, thus favouring formation of transcriptional pre-initiation complexes. The CR1 and CR2 regions are involved in cell transformation, by binding the retinoblastoma tumour suppressor gene product, p105-RB and an RB-related protein, p107 (to CR2), and a transcriptional co-activator protein, p300, also termed CBP or CREB binding protein (to the amino-terminal region and CR1). The p105-RB protein is a negative regulator of the transcription factor E2F, which activates genes that are required for the S phase of the cell cycle. E1A binds to hypophosphorylated p105-RB, thus liberating E2F and activating DNA synthesis. The p300/CBP protein possesses histone acetyltransferase (HAT) activity and is involved in chromatin remodelling; E1A binding to p300/CBP activates or represses transcription of target genes (Gallimore and Turnell 2001; Sang et al. 2002). Further, cellular proteins involved in transcriptional regulation of the cell cycle, for example p400 and TRRAP, are targets for binding and modulation by E1A (Deleu et al. 2001; Fuchs et al. 2001). In addition, E1A prevents cell-cycle arrest by cyclin-dependent kinase inhibitor proteins such as $p21^{Cip1/Waf1}$ and $p27^{Kip1}$ (Alivizopoulos et al. 2000), leading to increased cell-cycle progression. E1A proteins therefore have the capacity to act as either transcriptional activators or repressors of target cellular genes (Boulanger and Blair 1991) and can modulate the function of cellular gene products.

The serotypic origin of the *E1A* gene determines the oncogenic phenotype of adenovirus-transformed cells (Raska 1995). It is, therefore, interesting to note that region in oncogenic Ad12, but not in non-oncogenic Ad5 E1A (Williams et al. 1995). A homologous sequence is present in the E1A proteins of the highly oncogenic simian adenovirus 7 (Kimelman et al. 1985). Construction of recombinant viruses and plasmids containing chimaeric Ad5/Ad12 *E1A* genes has suggested that the alanine-rich spacer forms at least one determinant of viral oncogenicity mediated by Ad12 *E1A* (Telling and Williams 1994; Jelinek et al. 1994). Alanine-rich sequences are present in the transcriptional repressor domains of certain transcription factors, such as the *Drosophila* Kruppel factor (Licht et al. 1990). The precise biochemical role of the alanine-rich sequence in Ad12-mediated oncogenesis has yet to be defined, although it is interesting to speculate that it might be involved in transcriptional regulation of genes controlling oncogenesis.

The *E1B* gene encodes two major proteins of 58 kDa and 19 kDa that play important roles in viral infection and transformation. The 19-kDa protein possesses similarities to the bcl-2 family of anti-apoptosis proteins (Chinnadurai 1998). The Ad5 58-kDa protein forms a stable complex with the p53 tumour suppressor gene product in transformed cells (Sarnow et al. 1982) whereas, in contrast, the corresponding Ad12 protein (of 54 kDa) forms a very weak or unstable complex with p53, although p53 is stabilised in Ad12-, as well as Ad5-transformed cells (Grand et al. 1999). The presence of either the Ad5 E1B 58-kDa protein or its Ad12 counterpart is sufficient to abrogate the transcriptional activation properties of p53 (Wienzek et al. 2000). In infected cells, the 58-kDa protein forms complexes with the *E4* ORF3 gene product (Dobner and Kzhyshkowska 2001). Furthermore, the 58-kDa protein also forms a complex with both the E4 ORF 6 gene product and p53, an interaction that recruits a Cullin-containing complex that stimulates ubiquitination of p53 and targets it for degradation by the proteasome (Querido et al. 2001). The half-life of p53 in untransformed cells is short (around 30 min), and p53 is stabilised when cellular DNA is damaged, leading to transcriptional activation of cell-cycle regulatory genes such as $p21^{Cip1/Waf1}$ and either cell-cycle arrest at the G_1 to S phase transition or apoptosis (Woods and Vousden 2001). Thus in both adenovirus transformation and infection, the activity of p53 is regulated by either complex formation or degradation, respectively, to promote cell transformation and prevent apoptosis in viral infection.

The E1A and E4 ORF3 gene products are targeted to nuclear protein complexes termed PML oncogenic domains (PODs) and re-organise these nuclear structures in Ad5- and Ad12-infected cells (Carvalho et al. 1995; Täuber and Dobner 2001; Hösel et al. 2001).

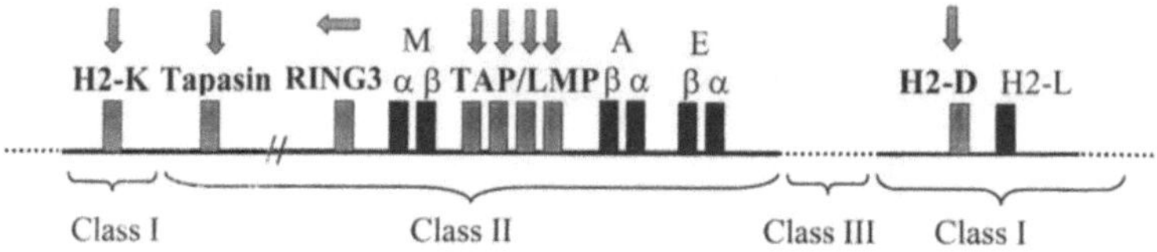

Fig. 1 The organisation of the mouse MHC and regulation of gene expression by oncogenic adenovirus 12 E1A. The locations of MHC genes described in the text are shown in a simplified map of the mouse MHC. Regulation of genes denoted by *dark boxes* has not been examined. *Vertical, downward-pointing arrows* denote the down-regulation of expression of the gene by oncogenic Ad12 E1A, but not by non-oncogenic Ad5 E1A. The *horizontal arrow* denotes a gene, RING 3, whose expression is not altered by Ad transformation

2
Host Cell and Viral Molecules Involved in Antigen Processing and Presentation

Viruses have evolved mechanisms to protect infected and transformed cells from killing by the cellular immune system (Yewdell and Bennink 1999). Infected and transformed cells evade the lytic effects of cytotoxic T lymphocytes (CTLs) and natural killer (NK) cells. The T-cell receptors (TCRs) on the surface of CTLs recognise antigenic peptides bound to class I molecules of the major histocompatibility complex (MHC). Class I molecules are heterodimeric glycoproteins consisting of a polymorphic heavy chain noncovalently linked to an invariant light chain (termed β_2-microglobulin) and they are present on the surface of most somatic cells in the adult. The classical class I heavy chain genes (H-2K, D, and L in the mouse; HLA-A, B, and C in humans) are located within the large 4-Mb MHC gene cluster on mouse chromosome 17 and human chromosome 6 (Fig. 1). Class I molecules function by presenting intracellularly processed peptides on the cell surface, where they are recognised by CD8$^+$ CTLs which can then lyse the presenting cell (Fig. 2).

Transport of class I molecules to the cell surface is dependent on peptide binding to newly synthesised class I polypeptides in the lumen of the endoplasmic reticulum (ER). Antigenic peptides arise by proteolytic cleavage in the cytosol and are pumped into the lumen of the ER by an active transport mechanism involving a protein complex termed the "transporter associated with antigen processing (TAP)" (Lankat-Büttgereit and Tampé 1999). TAP consists of two subunits, TAP1 and

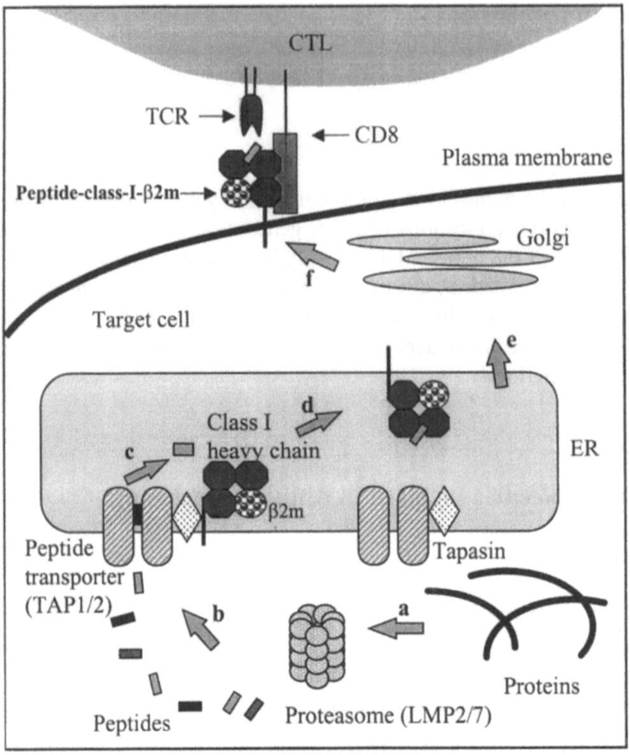

Fig. 2 The MHC class I antigen processing and presentation pathway. (*a*) Endogenous proteins are degraded in the proteasome-dependent degradation pathway, and (*b*) peptides are transported into the ER lumen by TAP dimers. Several molecules in the ER have been implicated in the tightly regulated folding, assembly and loading of MHC class I molecules: calnexin, ERp57, calreticulin and β_2-microglobulin (β_2-m); and (*c*, *d*) tapasin. (*f*) Stable MHC–peptide complexes can leave the ER via the Golgi compartment to the cell surface, potentially for recognition by CTLs

TAP2, encoded by separate genes. Cells in which TAP genes are deleted have very low levels of surface class I molecules, indicating that TAP expression is essential for antigen presentation. Several ER-resident chaperone molecules are required for the assembly of class I molecules, such as calreticulin, calnexin, and ERp57 (Fourie and Yang 1998). Whereas these molecules have general roles in folding and assembly of proteins that transit the ER, an accessory molecule termed tapasin (for TAP-associated glycoprotein) is required for peptide loading of class I molecules. Tapasin has been proposed to facilitate the interaction between the ER lumenal domain of TAP and class I molecules (Li et al. 2000; Grandea III and van Kaer 2001). The genes encoding tapasin, TAP1, and TAP2 are located within the class II region of the MHC gene cluster and are very closely linked to two other genes termed LMP2 and LMP7 (Fig. 2). These genes encode interferon-γ-inducible components of the proteasome, a 26S complex with protease activity that produces peptides that are transported by TAP for assembly with class I molecules (Früh and Yang 1999; Groettrup et al. 1996). Class I heavy chains are terminally glycosylated in the Golgi apparatus, and the mature class I: peptide complex is inserted into the plasma membrane (Fourie and Yang 1998). The amino-terminal domain of Ad5 E1A has been reported to bind to regulatory components of the proteasome, termed Sug1 and S4, and may therefore alter proteolytic activity (Turnell et al. 2000), which could have consequences for antigen processing.

In Ad5-transformed cells, antigenic peptides derived from the carboxy-terminal region of E1A (from residues 232–247 in the 289R protein) are the dominant peptides presented by mouse class I molecules (Kast et al. 1989). Thus, Ad5-transformed cells can be eradicated from the murine host by CTLs. The E1A peptide sequence is not located in the conserved region, and so it could be proposed that Ad12-transformed cells are oncogenic because the corresponding Ad12 E1A peptides are bound to class I molecules inefficiently, or fail to be recognised by CTLs. However, as described below, levels of surface class I molecules are greatly reduced in Ad12-transformed cells and this, as well as possible inefficient viral peptide presentation, may account for the oncogenic phenotype of Ad12-transformed cells.

3
Down-Regulation of Surface MHC Class I Molecules in Adenovirus Infection and Oncogenic Transformation

The E3 region of adenoviruses encodes important regulatory proteins that affect the pathogenicity of adenoviruses in vivo. In particular, a 19-kDa glycoprotein encoded by the E3 region of adenoviruses from subgenera B, C, D, and E has the ability to bind MHC class I molecules within the lumen of the ER, preventing their maturation and traffic to the cell surface (Wold et al. 1999). This constitutes a viral post-translational control on expression of class I molecules, a general strategy noted for many viruses (Yewdell and Bennink 1999), and may be crucial for the establishment of latent and persistent infections by adenoviruses. Not all adenoviruses encode an E3–19-kDa glycoprotein. Sequencing of the Ad12 (from subgenus A) and Ad40 (from subgenus F) genomes failed to provide evidence for an open reading frame corresponding to E3–19-kDa (Sprengel et al. 1994; Davison et al. 1993). Although E3 is not required for transformation or oncogenesis, studies on this gene have provided extensive information on the interaction of adenoviruses with the cellular immune system (see the chapters by Windheim et al. and Fessler et al., this volume). In this chapter, we focus mainly on the evasion of cellular immunity by oncogenic viruses of subgenus A, such as Ad12, which exhibits an alternative strategy for class I down-regulation, namely transcriptional repression of class I heavy chain genes and other genes involved in the MHC class I antigen processing and presentation pathway. We also consider mechanisms whereby adenoviruses evade innate immunity mediated by NK cells.

4
Repression of MHC Class I Gene Expression in Oncogenic Adenovirus 12-Transformed Cells

A number of studies have established that levels of surface MHC class I molecules are down-regulated in rodent and human cells transformed by highly oncogenic Ad12, but not by non-oncogenic Ad2 or Ad5. That this effect has functional significance was demonstrated by the fact that influenza virus-infected Ad12-transformed mouse cells were much more resistant to lysis by influenza-specific CTLs than were infected Ad5-transformed cells (Yewdell et al. 1988). Therefore, greatly diminished levels of surface class I molecules appear to contribute to the ability of Ad12-transformed cells to evade the surveillance of cells of the immune system, in particular $CD8^+$ CTLs. Up-regulation of class I molecules in Ad12-transformed cells by either interferon treatment or transfection of a heterologous class I heavy chain gene appears to diminish oncogenicity. These effects are mediated by expression of Ad12 E1A, which dominates over non-oncogenic Ad5 E1A when both genes are present in the same cell (Ge et al. 1994).

Ad12 E1A proteins transcriptionally repress MHC class I heavy chain gene expression via at least three sites in the 5́-flanking region of the mouse H-$2K^b$ gene:

1. The proximal class I regulatory element (CRE) region at −213 bp to −159 bp, where the start-site for transcription is +1 (Logeat et al. 1991; Kralli et al. 1992; Ge et al. 1992; Meijer et al. 1992; Schouten et al. 1995; Kushner et al. 1996).
2. A 260-bp distal region from −1,440 bp to −1,180 bp (Proffitt et al. 1994).
3. A further upstream 316-bp regulatory sequence located at −1,837 bp to −1,521 bp (Katoh et al. 1990; Ozawa et al. 1993; Tang et al. 1995).

Much research has focused on the regulation of class I heavy chain transcription by factors that bind the conserved enhancer sequence, designated the class I regulatory element (CRE). Katoh et al. (1990) demonstrated that distal elements in the 5́-flanking region of the class I gene repress promoter activity in Ad12-transformed cells, whereas a proximal region, which is part of the CRE, activates it. The CRE consists of three distinct domains, termed Region I, Region II, and Region III (Shirayoshi et al. 1987), and each binds different transcription factors Region I is highly conserved between mouse and human MHC class I genes, and it binds transcription factors of the NF-κB/c-rel family. Region II forms a half-site for the nuclear hormone receptor superfamily of transcription factors, e.g. human RXR-β, mouse H2RIIBP and COUP-TF. Region III contains an NF-κB-like binding site (Blair and Hall 1998).

NF-κB transcription factors are active as heterodimers, consisting of a p65 (relA) and a p50 (NF-κB1-p50) subunit (Karin and Lin 2002). In the unstimulated cell, NF-κB is retained in the cytoplasm, bound to the inhibitory subunit I-κB. The NF-κB1 gene encodes two functionally distinct proteins, the p50 polypeptide subunit of NF-κB and p105, which is a member of the IκB family. The p50 subunit is generated cotranslationally by a proteasome-mediated process that ensures the production of both p50 and p105 (Lin et al. 1998). Upon cellular stimulation, the I-κB p105 protein becomes phosphorylated and is degraded by the 26S proteasome. The p50 subunit is then able to form heterodimers with p65, and this leads to the nuclear localisation of NF-κB. The p50 subunit is also able to dimerise but is transcriptionally inactive in this form.

Regions I and II have both been shown to bind different levels of transcription factors in Ad5- compared to Ad12-transformed cells (Ackrill and Blair 1989; Ge et al. 1992; Meijer et al. 1992). Region I binds several factors, including NF-κB (a p65/p50 heterodimer) and KBF1 (a p50/p50 homodimer), that contribute to regulation of expression of MHC class I genes in vivo (Logeat et al. 1991). It has been consistently observed that both dimers bind their cognate DNA sites to a much lesser extent in Ad12- compared to Ad5-transformed cells (Ackrill and Blair 1989; Ge et al. 1992; Meijer et al. 1992).

In Ad5-transformed cells, the E1A 13S product has been found to direct increased degradation of I-κB and hence stimulate translocation of NF-κB into the nucleus. In addition it is thought to activate the p65 subunit in vivo (Schmitz et al. 1996). The mechanism by which Ad12-E1A 13S reduces the levels of NF-κB DNA binding activity in the nucleus remains unclear. It has been suggested that Ad12-E1A 13S interferes with the proteolytic processing of the 105-kDa product of the NF-κB1 gene, by the 26S proteasome. This interference would alter the p50/p105 ratio in favour of p105. Excess p105 would then dimerise with p50 or p65 and sequester NF-κB in the cytoplasm, resulting in the formation of few p65/p50 or p50/p50 complexes. The low concentrations of the NF-κB heterodimer (p50/p65) and the KBF1 homodimer (p50/p50) would result in less binding to the Region I element and hence reduced expression of MHC class I heavy chain genes. In Ad5 E1A-expressing cells, p50 is more abundant than p105, so p50/p50 and p50/p65 dimers are favoured; NF-κB is, therefore, constitutively active, and increased expression of MHC class I results (Schouten et al. 1995).

Other experiments showed that an alteration in the level of phosphorylation of the p50 subunit is critical for NF-κB binding to Region I (Kushner and Ricciardi 1999). Hypophosphorylation of p50 reduces NF-κB DNA binding, and the level of p50 phosphorylation is greatly reduced in Ad12- compared to Ad5-transformed cells, correlating with the decreased binding activity of NF-κB in Ad12-transformed cells. The mechanism of this altered p50 phosphorylation has not been determined.

In addition to Region I, other elements within the MHC class I heavy chain promoter have been identified as targets for regulation by Ad12 E1A. Ad12-transformed cells appear to exhibit more extensive in vitro binding of nuclear factors to Region II compared to Ad5 E1-transformed cells (Ackrill and Blair 1989). Increased binding of these factors to Region II was reported to account for repression of class I transcription (Ge et al. 1992; Kralli et al. 1992). One factor that binds to Region II in Ad12 E1-transformed cells has been identified as COUP-TFII (Liu et al. 1994; Smirnov et al. 2001). The COUP-TFs (or chicken orphan upstream promoter-transcription factors) comprise members of the large group of so-called orphan steroid hormone receptors, for which no physiological activating ligand has been identified (O'Malley and Conneely 1992). COUP-TFs can dimerise with other steroid hormone receptors, e.g. RXR-β. In Ad12-transformed cells, COUP-TF binds N-CoR, which is a co-repressor of histone deacetylases (HDACs) and thus contributes to transcriptional repression (Smirnov et al. 2000). Inhibition of HDACs in Ad12-transformed cells by the specific inhibitor trichostatin A results in up-regulation of surface class I molecules (Hou et al. 2002). It therefore appears that there is potentially important 'cross-talk' between Regions I and II in Ad12-transformed cells leading to reduced MHC class I transcription, although there is still much to learn about the precise mechanism of transcriptional modulation.

It has been reported that binding of an as yet unidentified factor to a CAA repeat element located between −1837 bp and −1521 bp upstream of the cap site contributes to repression of the H-2K^{b} gene in Ad12 E1-transformed 3Y1 cells (Katoh et al. 1990). The H-2K^{bm1} gene has repeated CAA sequences in regions −1,736 bp to −1,689 bp and −1,616 bp to −1,535 bp, and both are necessary for full negative regulation of this gene by the 13S product of E1A (Ozawa et al. 1993). In addition, a TATA-like sequence (TATAA) in the far upstream region of the 5′-flanking sequence (−1,773 bp to −1,767 bp) appears to be a key element in the negative regulation of expression of the H-2K^{bm1} gene by Ad12 E1A, in conjunction with either upstream or downstream CAA repeats (Tang et al. 1995). This TATA-like element is sensitive to orientation. Tang et al. (1995) observed that Ad12 E1A-dependent negative regulation is dependent upon the presence of the specific sequence TAT(A/T)AA at −1,773 bp to −1,767 bp. This negative regulatory region may well extend back to −1,811 bp and contains several elements capable of binding proteins that participate in the down-regulation of MHC class I by E1A (Katoh et al. 1990). Finally, a 260-bp sequence from −1,180 bp to −1,440 bp has been shown to play a major role in Ad12 E1A-mediated down-regulation of the H-2K^{b} promoter (Proffitt et al. 1994). More than one element is targeted over this region, and these elements function co-operatively to achieve maximal down-regulation. The precise mechanism of class I repression via this region is not known, but cellular factors are probably involved either by induction or activation by Ad12 E1A. However, analysis of DNA-protein complexes binding to this region has shown that Ad12 E1A, but not Ad5 E1A proteins, form a component of the complex that binds this region of the H-2K^{b} gene (N. Philpott and

G. Blair, unpublished data). The significance of this region in regulation of the class I heavy chain promoter may extend wider than the oncogenic adenoviruses, since the high-risk human papillomavirus E7 proteins of HPV 16 and 18 also target this region and repress the class I promoter (Georgopoulos et al. 2000).

5 Repression of Other Genes in the MHC Complex in Oncogenic Adenovirus 12-Transformed Cells

Ad12 E1A has been shown to reduce expression of other MHC genes at the level of transcription. TAP 1 and 2, as well as LMP 2 and 7 are all down-regulated in Ad12-, but not in Ad5-transformed cells (Rotem-Yehudar et al. 1994; 1996). Proffitt and Blair (1997) reported that the bidirectional promoter, shared by the TAP1 and LMP2 genes, is repressed in Ad12-, but not Ad5-transformed cells. Interestingly the TAP1-LMP2 bidirectional promoter contains an NF-κB site which is conserved between the mouse, rat, and human genes (Proffitt and Blair 1997). Therefore, these genes could potentially be repressed by Ad12 E1A via a mechanism similar to that which regulates the NF-κB element of Region I of the CRE in MHC class I genes (see above). Pleiotropic down-regulation of all four genes would cause a reduction in cell surface expression of peptide-containing MHC class I molecules. Although this is an attractive hypothesis, it is not supported by analysis of the mouse TAP2 promoter, which failed to demonstrate the presence of a functional NF-κB element or an NF-κB-like element that might be conserved between the mouse and human TAP2 promoters (Arons et al. 2001).

Genes that are linked to the MHC but are not involved in the antigen presentation pathway are also modulated in Ad 12-transformed cells. Eyler et al. (1997) observed that the gene product of *Waf*-1, which encodes the $p21^{Cip1/Waf1}$ protein that is involved in cell-cycle control and the complement factors C3 and C2, are all down-regulated in Ad 12-transformed cells. These genes are all located on mouse chromosome 17 in a region that includes the MHC, but it is not yet known whether expression of these genes is directly regulated by the E1A gene products of Ad 12. This suggested that a very large segment of the chromosome containing the MHC and flanking genes might be repressed in Ad12-transformed cells. However, expression of the RING 3 gene is not differentially regulated in Ad12- compared to Ad5-transformed cells, implying a model of selective repression of genes in the MHC complex (L. Stanbridge and G.E. Blair, unpublished data).

6 The Role of Ad12 E1A Gene Expression in Repression of MHC Class I Transcription

Mapping of the domains on Ad12 E1A involved in oncogenesis and in class I down-regulation would be desirable. In particular, if one domain was implicated in both oncogenesis and in class I down-regulation, this would provide a possible connection between the two processes. At present no single domain on Ad12 E1A has been shown to be involved in both oncogenesis and class I down-regulation. Most studies have pointed to the importance of the CR3 transactivating domain, or the carboxy-terminal portion of CR3, in down-regulation of both surface class I molecules (Huvent et al. 1996) and MHC class I promoter constructs (Meijer et al. 1989) in Ad12-transformed cells. In contrast, tumorigenicity of Ad12-transformed cells is governed by Ad12 E1A exon 1 sequences, excluding the transactivating domain (Telling and Williams 1994; Jelinek et al. 1994). Modulation of levels of transcription factors binding to the NF-κB and Region II sites of the CRE in Ad12-transformed cells is also associated with exon 1 sequences, suggesting that the processes of altered transcription and tumorigenesis could be linked (Kushner et al. 1996). The alanine-rich region is an intriguing potential transcriptional regulator, as discussed above, based on its presence in oncogenic adenoviruses, its role in tumorigenicity of Ad12-transformed cells and virions and the existence of alanine-rich sequences in transcriptional repressors. However, present indications are that point mutagenesis of the alanine-rich region does not alter levels of surface class I molecules in Ad12-transformed cells (Williams 1995; Pereira et al. 1995). Therefore, both the function of the alanine-rich region and the identification of domain(s) on Ad12 E1A that regulate class I expression require further analysis. Ad12 also expresses a unique 52R E1A protein that has been reported to interfere with c-jun mediated up-regulation of the class I promoter (Brockmann et al. 1994; 1996). It remains to be established whether this protein mediates the oncogenic phenotype of Ad12 E1A.

7
Evasion of Natural Killer Cell-Mediated Immunity

Natural killer (NK) cells are a sub-population of lymphocytes that mediate natural immunity to pathogen-infected and tumour cells, by their spontaneous (NK) and lymphokine activated (LAK) killer functions. They are also capable of antibody-dependent cell-mediated cytotoxicity (ADCC) that is triggered when antibody bound to the surface of a cell interacts with Fc receptor on the NK cells. Cytotoxicity without prior priming is the defining characteristic of NK cells. Direct cell-mediated killing is their fundamental mechanism of action. However NK cells also produce cytokines (e.g. interferon-γ) and therefore can participate not only in innate but also in shaping the adaptive immune response. Recognition of targets by NK cells involves receptors with either activating or inhibitory activity. Multiple receptors are involved in this process, and the final outcome of receptor cross-linking is determined by a balance between positive and negative signals transduced by them. A number of receptor-ligand systems involved in NK activation have been identified. NK cells are currently known to utilize at least three families of receptors, KIR, NKG2, and NCR, to identify cells undergoing infection or oncogenic transformation (Long 2002).

The KIR group (killer cell immunoglobulin-like receptors), members of which contain several extracellular immunoglobulin-like domains, recognise classical MHC class I molecules on target cells. The inhibitory KIRs are characterised by an immunoreceptor tyrosine-based inhibitory motif (ITIM) in their cytoplasmic tail, that upon phosphorylation, binds tyrosine phosphatases SHP-1 and SHP-2 that inactivate the signal molecules required for cytotoxicity. The lectin-like NK receptors, the NKG2 group, recognise, for example, MICA and MICB antigens which are induced by stress (NKG2D) and HLA-E (NKG2A). Members of the NKG2 and KIR families exist in inhibitory and activating forms. NK cells also express natural cytotoxicity receptors (NCRs) such as NKp30, NKp44, and NKp46 (Moretta et al. 2001). They recognise non-MHC molecules. Integrins play a role in the recognition process by facilitating adhesion of NKs to target cells. In contrast to cytotoxic T lymphocytes, NK cells typically target cells that do not express MHC class I molecules and lyse them according to the 'missing self' principle. Target cells that do not express (or express low levels) of MHC class I molecules trigger NK cell-activating receptors without engaging their inhibitory receptors and are therefore sensitive to NK cell lysis. However, the complete loss of MHC class I molecules is not a necessary requirement for lysis by NK cells. The profile of activating and inhibitory receptors on NK cells, including the clonal distribution of inhibitory receptors, allows cytolysis of targets that have lost a single HLA-C and (in some cases) single HLA-B allotypes.

Natural killer cells have the ability to eliminate their targets by two major mechanisms (Djeu et al. 2002). One mechanism is calcium-dependent and involves the directed exocytosis of cytotoxic granules. The major granule components are perforin and serine proteases termed granzymes (specifically granzyme B). Another mechanism occurs through death receptor pathways such as FasL-Fas, tumour necrosis factor-alpha (TNF-α) and TNF-related apoptosis-inducing ligand (TRAIL) systems (Screpanti et al. 2001).

In the adenovirus system, oncogenic potential has been correlated not only with the ability of the virus to transform and immortalise cells, but also with its ability to induce apoptosis and to effect the evasion of immune surveillance. Adenoviruses modulate cell susceptibility to both CTLs and NK cells. Human and rodent cells transformed with non-oncogenic subgenus C adenoviruses (Ad2/Ad5) are highly susceptible to killing by NK cells and activated macrophages (Cook et al. 1989). However, cells transformed by oncogenic Ad12 (belonging to subgenus A of adenoviruses) are resistant to the cytotoxic effects of NK cells (Sawada et al.1985). The mechanism of these differing cytolytic susceptibilities has not been fully elucidated. It has, however, been shown that for subgenus C adenoviruses it is not a necrotic, but an apoptotic process which is independent of p53 and is not inhibited by the E1B-19K protein (Cook et al. 1999a,b). Expression of high levels of E1A protein, higher than those required for immortalisation of cells, appears to be important. Both direct effects of E1A on cytolytic susceptibility of target cells as well as an indirect influence through modulation of gene expression have been proposed. Cells transformed with Ad2/5 have the same pattern and at least the same levels of expression of surface MHC class I molecules as their parental counterparts, and therefore, their enhanced susceptibility to NK killing cannot be explained by the 'missing self' principle. Moreover, when mouse MCA-102-E1A cells (H-2^b) that express E1A and are sensitive to NK cell lysis were transfected with H-$2D^d$ (the ligand for the NK inhibitory receptor Ly 49A^+), they were protected from lysis by Ly

49A$^+$ NK cells (Routes et al. 2001). This indicates that E1A protein does not interfere with signal transduction by inhibitory NK receptors.

Evidence has been presented to suggest that the post-recognition (post-triggering) stages are mainly involved in E1A sensitisation (Cook et al. 1996b). Comparison of the interaction of NK cells with E1A-expressing Rat 1 cells or the parental Rat1 cell line revealed identical levels of binding (Klefstrom et al. 1999). Moreover, treatment with IL-2 resulted in similar increases in binding of NK cells to both targets. In addition, fibroblasts expressing high levels of E1A were more susceptible to the cytolytic effects of isolated NK cell-derived granules. Mutational analysis has shown that amino acids 4–18 in the N-terminal region and 46–68 (CR1) located in the first exon of E1A are required for the induction of cytolytic susceptibility. These regions contain binding sites for the p300 family of transcriptional co-activators, suggesting a requirement for E1A-p300 complex formation (Cook et al. 1996a). Expression of the second exon is also necessary for E1A-induced susceptibility to NK cell lysis (Krantz et al. 1996). Studies with E1A mutants containing deletions in the second exon have identified two non-overlapping regions that seem to contribute equally to the induction of susceptibility to NK cell lysis, and either of those regions is required to cooperate with the first exon in this function. One of those regions spans the first 12 amino terminal residues of exon 2, and the second is located within the 221–289 aa segment of this exon. It is noteworthy that Ad5 mutants lacking E1A (and also the E3 region) have been shown to be able to cause cell death in the liver of mice (Liu et al. 2000).

Expression of E1A of sub-genus C adenoviruses blocks IFN-γ-induced resistance of cells to NK lysis (Routes et al. 1996). This allows discrimination between infected and uninfected cells. Binding of p300 transcriptional co-activators by E1A is required for this function. It has, therefore, been proposed that the capacity of E1A to block IFN-γ-induced cytolytic resistance is probably secondary to the ability of E1A to inhibit IFN-γ-stimulated gene expression.

Interestingly, rodent cells infected with subgenus C adenoviruses are susceptible to NK cytolysis, whereas human cells are not (Routes and Cook 1989, 1995). It can be postulated that the difference in cytolytic susceptibility of human adenovirus Ad2/Ad5-transformed and -infected cells to NK cells, regardless of comparable levels of E1A expression, is determined by inhibitory effects of the products of certain viral genes that are expressed during the infectious cycle but are absent in transformed cells. The E3 14.7-kDa protein has, however, not been shown to be responsible for this process.

Adenovirus 12, belonging to subgenus A, efficiently induces tumours after injection into newborn rodents. Tumorigenic rat cells transformed with adenovirus 12 are significantly more resistant to NK cell lysis than syngeneic cells transformed with non-oncogenic adenovirus 2. The cytolytic resistance of cells transformed in vitro with recombinant Ad5/Ad12 to NK cells suggests the involvement of the E1A region of highly oncogenic Ad12 (Kenyon et al. 1991). Cells transformed with Ad12 have significantly lower levels of surface MHC class I molecules than their parental counterparts. However, their behaviour in the reaction with NK cells does not obey the 'missing self' principle, and their cytolytic susceptibility to NK cells is not increased. The mechanism of this process remains to be elucidated. Interestingly, cells transformed with Ad12 E1A are not resistant to lysis by LAK cells (Kenyon et al. 1986).

Death receptor pathways are important in NK cytolysis. They have also been shown to be affected by E1A in other systems. Expression of Ad2 or Ad5 E1A proteins sensitises cells to TNF-α-mediated cytolysis (Duerksen-Hughes et al. 1989). This effect is, however, cell type-dependent (Vanhaesebroeck et al. 1990). It has been reported, for example, that in primary cells and normal rat kidney cells, there was no correlation between E1A expression and sensitivity to TNF-α-mediated killing. Expression of E1A was, however, found to enhance the sensitivity of NIH3T3 cells to TNF-α. Clonal variation in susceptibility of E1A-expressing NIH3T3 cells also occurs at a high rate. These differences may explain discrepancies in reports on the involvement of the TNF-α pathway in E1A-induced susceptibility to NK cytolysis. Apparently conflicting data on the role of Ad12 E1A proteins in response to TNF-α have been reported. Whereas the induction of sensitivity to TNF-α was shown in one report, other investigators observed a broad range of sensitivity of cell lines to TNF-α that did not correlate with either their susceptibility to NK/LAK cytolysis or their tumorigenic potential. Proteins specified by other genes of sub-genus C adenoviruses, e.g. E1B and E3, can protect cells against TNF-α cytolysis. The E3 14,700-M_r protein (14.7-kDa protein) protects mouse cells against TNF-α. Whereas this protein acts autonomously, other E3 encoded proteins (E3 10.4 kDa and E3 14.5 kDa) form a complex with similar activity (Gooding et al. 1991b). The E1B 19,000-M_r protein (19-kDa protein) prevents TNF-α cytolysis of human (but not mouse) cells (Gooding et al. 1991a).

E1A enhances the lysis of stably transfected human melanoma (A2058) and fibrosarcoma (H4) cells by TRAIL (TNF-related apoptosis-inducing ligand) (Routes et al. 2000). However, Ad5-infected cells expressing equally high levels of E1A proteins remain resistant to TRAIL. E3 gene products and, to a lesser extent, E1B-19K have been shown to inhibit E1A-induced sensitisation to TRAIL. Cell lysis effected by another member of the TNF receptor superfamily, Fas (CD95), which is also involved in NK killing, is also controlled by the adenovirus E3–10.4K/14.5K complex. This complex mediates the loss of cell surface Fas, without altering intracellular levels of Fas or its synthesis (Shisler et al. 1997). Interestingly, this complex does not induce the loss of TNF-α receptors.

In the adenovirus system, two important factors which determine the tumorigenicity of transformed cells are their intrinsic susceptibility to immune-mediated injury and the immunological status of the host. Expression of E1A confers different tumorigenic potential on cells, dependent on its sub-genus origin. Certain cell lines transformed with sub-genus C adenoviruses also express the E3 region, which encodes proteins that possess inhibitory activity against immune-mediated injury. The tumorigenicity of these cells is predicted to be directly related to the overall effect of the adenovirus-encoded gene products. Adult rodents possess normal T-cell (thymus-dependent) and NK cell responses, whereas nude rodents lack T-cell function (due to their athymic status), but possess functional NK cells. NK cells derived from nude rodents differ in their ability to lyse xenogeneic cells expressing E1A of sub-genus C adenoviruses (Cook and Routes 2001). Nude mice, in contrast to nude rats, have defective NK cytolytic activity against Ad5 E1A-expressing hamster and rat cells. Intact NK cell responses considerably reduce the tumorigenicity of cells expressing E1A of sub-genus C adenoviruses. In contrast, the expression of E1A of sub-genus A adenoviruses results in increased tumorigenic potential, due to evasion of immune surveillance by CTLs and NK cells.

8 Concluding Remarks

Much has been learned about the transcriptional regulation of MHC gene expression by oncogenic adenoviruses over the past decade. Although many target genes and transcription factors have been identified, intriguing new ones are emerging. The MHC gene cluster has been shown to be specifically associated with PML bodies (or POD domains, described above; Shiels et al. 2001). Furthermore the PML proto-oncogene has been reported to regulate multiple genes in the MHC class I pathway, including LMP2 and 7, TAP1 and 2 (Zheng et al. 1998). Since E1A proteins are targeted to PODs (Carvalho et al. 1994), this may provide a new avenue of study of MHC regulation. The p53 tumour suppressor gene product has also been reported to induce the expression of TAP1 mRNA and protein (Zhu et al. 1999), and the alteration of p53 transcriptional activity by E1B-58 kDa might suggest a further level of regulation of the MHC in Ad-transformed cells. Although the mechanism(s) of down-regulation of multiple MHC genes by Ad12 E1A has yet to be established, it is interesting to note that similar transcriptional alterations in MHC gene expression also exist in several mouse and human tumours (Garrido et al. 1995; Johnson et al. 1998; Delp et al. 2000; Seliger et al. 2000). This might raise the possible existence of E1A-like molecules in naturally occurring tumours and also reinforce the usefulness of Ad-transformed cells in the study of molecular interactions in the immune system. It may also be expected that cDNA microarray analysis of gene expression will reveal more information on the nature and groups of genes regulated by E1A, although progress to date has been limited (Kiemer et al. 2000; Vertegaal et al. 2000). This might also be coupled with studies on transgenic mouse models of E1A and E1B gene expression (Duncan et al. 2000; Chen et al. 2002).

Acknowledgements. Work in the authors' laboratories was supported by Yorkshire Cancer Research and the Medical Research Council. The help of Diane Baldwin in preparation of the manuscript is gratefully acknowledged. The authors also thank Dr. Graham Cook for comments on the manuscript.

References

Ackrill AM, Blair GE (1989) Nuclear proteins binding to an enhancer element of the major histocompatibility class I promoter: differences between highly oncogenic and nononcogenic adenovirus-transformed rat cells. Virology 172:643–646

Alevizopoulos K, Sanchez B, Amati B (2000) Conserved region 2 of adenovirus E1A has a function distinct from pRb binding required to prevent cell cycle arrest by $p16^{INK4a}$ or $p27^{Kip1}$. Oncogene 19:2067–2074

Arons E, Kunin V, Schechter C, Ehrlich R (2001) Organisation and functional analysis of the mouse transporter associated with antigen processing 2 promoter. J Immunol 166:3942–3951

Bayley, ST, Mymryk, JS (1994) Adenovirus E1A proteins and transformation. Int J Oncol 5:425–444

Blair GE, Hall KT (1998) Human adenoviruses: evading detection by cytotoxic T lymphocytes. Sem Virol 8:387–397

Boulanger PA, Blair GE (1991) Expression and interactions of human adenovirus oncoproteins. Biochem J 275:281–299

Brockmann D, Feng L, Kroner G, Tries B, Esche H (1994) Adenovirus type 12 early region 1A expresses a 52R protein repressing the trans-activating activity of transcription factor c-jun/AP-1. Virology 198:717–723.

Brockmann D, Schäfer D, Kirch H-C, Esche H (1996) Repression of c-Jun-induced mouse major histocompatibility class I promoter (H-$2K^b$) activity by the adenovirus type 12-unique 52R E1A protein. Oncogene 12:1715–1725

Carvalho T, Seeler J-S, Õhman K, Jordan P, Pettersson U, Akusjärvi G, Carmo-Fonseca M, Dejean A (1995) Targetting of adenovirus E1A and E4-ORF3 proteins to nuclear matrix-associated PML bodies. J Cell Biol 131:45–56

Chen Q, Ash JD, Branton P, Fromm L, Overbeek P (2002) Inhibition of crystallin expression and induction of apoptosis by lens-specific E1A expression in transgenic mice. Oncogene 21:1028–1037

Chinnadurai G (1998) Control of apoptosis by human adenovirus genes. Seminars in Virology 8:399–408

Cook JL, May DL, Wilson BA, Holskin B, Chen M-JY, Shalloway D, Walker TA (1989) Role of tumor necrosis factor-α in E1A oncogene-induced susceptibility of neoplastic cells to lysis by natural killer cells and activated macrophages. J Immunol 152:4527–4534

Cook JL, Routes BA, Walker TA, Colvin KL, Routes JM (1999a) E1A oncogene induction of cellular susceptibility to killing by cytolytic lymphocytes through target cell sensitization to apoptotic injury. Exp Cell Res 251:414–423

Cook JL, Routes BA, Leu CY, Walker TA, Colvin KL (1999b) E1A oncogene-induced cellular sensitisation to immune-mediated apoptosis is independent of p53 and resistant to blockade by E1B 19 kDa protein. Exp Cell Res 252:199–210

Cook JL, Krantz CK, Routes BA (1996a) Role of p300-family proteins in E1A oncogene induction of cytolytic susceptibility and tumor cell rejection. Proc Natl Acad Sci USA 93:13985–13990

Cook JL, Potter TA, Bellgrau D, Routes BA (1996b) E1A oncogene expression in target cells induces cytolytic susceptibility at a post-recognition stage in the interaction with killer lymphocytes. Oncogene 13:833–842

Cook JL, Routes JM (2001) Role of the innate immune response in determining the tumorigenicity of neoplastic cells. Dev Biol. Basel, Karger 106:99–108

Davison AJ, Telford EAR, Watson MM, McBride K, Mautner V (1993) The DNA sequence of adenovirus type 40. J Mol Biol 234:1308–1316

Delp K, Momburg F, Hilmes C, Huber C, Seliger B (2000) Functional deficiencies of components of the MHC class I antigen pathway in human tumours of epithelial origin. Bone Marrow Transplant 25:S88-S95

Deleu L, Shellard S, Alevizopoulos K, Amati B, Land H (2001) Recruitment of TR-RAP required for oncogenic transformation by E1A. Oncogene 20:8270–8275

Djeu JY, Jian K, Wei S (2002) A view to a kill: Signals triggering cytotoxicity. Clin Cancer Res 8:636–640

Dobner T, Kzhyshkowska J (2001) Nuclear export of adenovirus RNA. Curr Topics in Microbiol and Immunol 259:25–54

Duncan MD, Tihan T, Donovan DM, Phung QH, Rowley DL, Harmon JW, Gearhart PJ, Duncan KLK (2000) Esophagogastric adenocarcinoma in an E1A/E1B transgenic model involves p53 disruption. J Gastrointest Surg 4:290–297

Duerksen-Hughes P, Wold WSM, Gooding LR (1989) Adenovirus E1A renders infected cells sensitive to cytolysis by tumor necrosis factor. J Immunol 143:4193–4200

Eyler YL, Siwarski DF, Huppi KE, Lewis AM Jr (1997) Down-regulation of Waf1, C2, C3 and major histocompatibility class I loci within an 18 cM region of chromosome 17 in adenovirus-transformed mouse cells. Mol Carcinogenesis 18:213–220

Fourie AM, Yang Y (1998) Molecular requirements for assembly and intracellular transport of class I major histocompatibility complex molecules. Curr Topics in Microbiol & Immunol 232:49–74

Früh K, Yang Y (1999) Antigen presentation by MHC class I and its regulation by interferon γ. Curr Opin Immunol 11:76–81

Fuchs M, Gerber J, Drapkin R, Sif S, Ikura T, Ogryzko V, Lane WS, Nakatani Y, Livingston DM (2001) The p400 complex is an essential E1A transformation target. Cell 106:297–307

Gallimore PH, Turnell AS (2001) Adenovirus E1A: remodelling the host cell, a life or death experience. Oncogene 20:7824–7835

Garrido F, Cabrera T, Lopez Nevot MA, Ruiz Cabello F (1995) HLA class I antigens in human tumors. Adv Cancer Res 67:155–195

Ge R, Kralli A, Weinmann R, Ricciardi RP (1992) Down-regulation of the major histocompatibility class I enhancer in adenovirus 12-transformed cells is accompanied by an increase in factor binding. J Virol 66:6969–6978

Ge R, Liu X, Ricciardi RP (1994) E1A oncogene of adenovirus 12 mediates trans-repression of MHC class I transcription in Ad5/Ad12 somatic hybrid transformed cells. Virology 203:389–392

Georgopoulos NT, Proffitt JL, Blair GE (2000) Transcriptional regulation of the major histocompatibility complex (MHC) class I heavy chain, TAP1 and LMP2 genes by the human papillomavirus (HPV) type 6b, 16 and 18 E7 oncoproteins. Oncogene 19:4930–4935

Gooding LR, Aquino L, Duerksen-Hughes PJ, Day D, Horton TM, Yei S, Wold, WSM (1991a) The E1B 19,000-molecular weight protein of group C adenoviruses prevents tumor necrosis factor cytolysis of human cells but not of mouse cells. J Virol 65:30893–3094

Gooding LR, Ranheim TS, Tollefson AE, Aquino L, Duerksen-Hughes P, Horton TM, Wold WSM (1991b) The 10,400- and 14,500-dalton proteins encoded by region E3 of adenovirus function together to protect many but not all mouse cell lines against lysis by tumor necrosis factor. J Virol 65:4114–4123

Grand RJA, Parkhill J, Szestak T, Rookes SM, Roberts S, Gallimore PH (1999) Definition of a major p53 binding site in Ad2E1B58 K protein and a possible nuclear localization signal on the Ad12E1B54K protein. Oncogene 18:955–965

Grandea AG III, Kaer LV (2001) Tapasin: An ER chaperone that controls MHC class I assembly with peptide. Trends Immunol 22 194–1999

Groettrup M, Soza A, Kuckelkorn U, Kloetzel P M (1996) Peptide antigen production by the proteasome: complexity provides efficiency. Immunol Today 17:429–435

Heise C, Kirn DH (2000) Replication-selective adenoviruses as oncolytic agents. J. Clin. Invest. 105:847–851

Hermiston T (2000) Gene delivery from replication-selective viruses: arming guided missiles in the war against cancer. J. Clin. Invest. 105:1169–1172

Hitt MM, Graham FL (2000) Adenovirus vectors for human gene therapy. Advances in Virus Research, 55:479–505

Hösel M, Schröer J, Webb D, Jaroshevskaja E, Doerfler W (2001) Cellular and early viral factors in the interaction of adenovirus type 12 with hamster cells: the abortive response. Virus Research 81:1–16

Hou S, Guan H, Ricciardi RP (2002) In adenovirus type 12 tumorigenic cells, major histocompatibility complex Class 1 transcription shutoff is overcome by induction of NF-κB and relief of COUP-TFII repression. J Virol 76:3212–3220.

Huvent I, Cousin C, Kiss A, Bernard C, D'Halluin JC (1996) Susceptibility to natural killer cells and down-regulation of MHC class I expression in adenovirus 12 transformed cells are regulated by different E1A domains. Virus Res 45:123–134

Javier RT (1994) Adenovirus type 9 E4 open reading frame 1 encodes a transforming protein required for the production of mammary tumors. J. Virol.68:3917–3924

Jelinek T, Pereira DS, Graham FL (1994) Tumorigenicity of adenovirus-transformed rodent cells is influenced by at least two regions of adenovirus type 12 E1A. J Virol 68:888–896

Johnsen A, France J, Sy M-S, Harding CV (1998) Down-regulation of the transporter for antigen presentation, proteasome subunits and class I major histocompatibility complex in tumor cell lines. Cancer Res 58:3660–3667

Kafri T, Morgan D, Krahl T (1996) Cellular immune response against adenoviral infected cells does not require de novo viral gene expression: implications for gene therapy. Proc. Natl Acad. Sci. USA 95:314–317

Karin M, Lin A (2002) NF-κB at the crossroads of life and death. Nature Immunol 3:221–227

Kast, WM, Offringa, R, Peters PJ, Voordonw AC, Meloen RH, van der Eb, AJ, Melief CJM (1989) Eradication of adenovirus E1-induced tumors by E1A-specific cytotoxic T lymphocytes. Cell 59:603–614

Katoh S, Ozawa K, Kondoh S, Soeda E, Israel A, Shiroki K, Fujinaga K, Itakura K, Gachelin G, Yokoyama K (1990) Identification of sequences responsible for positive and negative regulation by E1A in the promoter of H-2K^{bm1} class I MHC gene. EMBO J 9:127–135

Kenyon DJ, Dougherty J, Raska K (1991) Tumorigenicity of adenovirus-transformed cells and their sensitivity to tumor necrosis factor α and NK/LAK cell cytolysis. Virology 180:818–821

Kenyon DJ, Raska K (1986) Region E1A of highly oncogenic adenovirus 12 in transformed cells protects against NK but not LAK cytolysis. Virology 155:644–654

Kiemer AK, Takeuchi K, Quinlan MP (2001) Identification of genes involved in epithelial-mesenchymal transition and tumor progression. Oncogene 20:6679–6688

Kimelman D, Miller JS, Porter D, Roberts, BE (1985) E1A regions of the human adenoviruses and of the highly oncogenic simian adenovirus 7 are closely related. J Virol 53:399–409

Klefstrom J, Kovanen PE, Somersalo K, Hueber A, Littlewood T, Evan GI, Greenberg AH, Saksela E, Timonen T, Alitalo K (1999) c-Myc and E1A induced cellular sen-

sitivity to activated NK cells involves cytotoxic granules as death effectors. Oncogene 18:218–2188
Kralli A, Ge R, Graeven U, Ricciardi RP, Weinmann R (1992) Negative regulation of the major histocompatibility class I enhancer in adenovirus type 12-transformed cells via a retinoic acid response element. J Virol 66:6979–6988
Krantz CK, Routes BA, Quinlan MP, Cook JL (1996) E1A second exon requirements for induction of target cell susceptibility to lysis by natural killer cells: implications for the mechanism of action. Virology 217:23–32
Kushner DB, Pereira DS, Liu X, Graham FL, Ricciardi RP (1996) The first exon of Ad12 E1A excluding the transactivation domain mediates differential binding of COUP-TF and NF-κB to the MHC class I enhancer in transformed cells. Oncogene 12:143–151
Kushner DB, Ricciardi RP (1999) Reduced phosphorylation of p50 is responsible for diminished NF-κB binding to the major histocompatibility complex class I enhancer in adenovirus type 12-transformed cells. Mol Cell Biol 19:2169–2179
Lankat-Büttgereit B, Tampé R (1999) The transporter associated with antigen processing TAP: structure and function. FEBS Letts 464:108–112
Li S, Paulsson KM, Chen S, Sjögren H-O, Wang P (2000) Tapasin is required for efficient peptide binding to transporter associated with antigen processing. J Biol Chem 275:1581–1586
Licht JD, Grossel MJ, Figge, J, Hansen, UM (1990) *Drosophila* kruppel protein is a transcriptional repressor. Nature 346:76–79
Liu X, Ge R, Westmoreland S, Cooney AJ, Tsai SY, Tsai M-J, Ricciardi RP (1994) Negative regulation by the R2 element of the MHC class I enhancer in adenovirus 12 transformed cells correlates with high levels of COUP-TF binding. Oncogene 9:2183–2190
Liu X, Govindarajan S, Okamoto S, Dennert G (2000) NK cells cause liver injury and facilitate the induction of T cell-mediated immunity to a viral liver infection. J Immunol 164:6480–6486
Logeat F, Israel N, Ten R, Blank V, Le Bail, Kourilsky P, Israel A (1991) Inhibition of transcription factors belonging to the rel/NF-kappaB family by a transdominant negative mutant. EMBO J 10:1827–1832
Long EO (2002) Tumour cell recognition by natural killer cells (2002) Sem Cancer Biol 12:57–61
Meijer I, Jochemsen AG, de Wit CM, Bos JL, Morello D, van der Eb AJ (1989) Adenovirus type 12 E1A down-regulates expression of a transgene under control of a major histocompatibility complex class I promoter: evidence for transcriptional control. J Virol 63:4039–4042
Meijer I, van Dam H, Boot AJM, Bos JL, Zantema A, van der Eb AJ (1991) Co-regulated expression of *jun*B and MHC class I genes in adenovirus-transformed cells. Oncogene 6:911–916
Meijer I, Boot AJM, Mahibin G, Zantema A, van der Eb AJ (1992) Reduced binding activity of the transcription factor NF-κB accounts for the MHC class I repression in adenovirus 12 E1-transformed cells. Cell Immunol 145:56–65
Moran E (1993) DNA tumor virus transforming proteins and the cell cycle. Curr Opin Gen Dev 3:63–70

Moretta A, Bottino C, Vitale M, Pende D, Cantoni C, Mingari MC, Biassoni R, Moretta L (2001) Activating receptors and coreceptors involved in human natural killer cell-mediated cytolysis. Annu Rev Immunol 19:197–223

O'Malley BW, Conneely OM (1992) Orphan receptors: In search of a unifying hypothesis for activation. Mol Endocrinol 92:1359–1361

Ozawa K, Hagiwara H, Tang X, Saka F, Kitabayashi I. Shiroki K, Israël A, Gachelin G, Yokoyama K (1993) Negative regulation of the gene for H-2K^b class I antigen by adenovirus 12 E1A is mediated by a CAA repeated element. J Biol Chem 268:27258–27268

Pereira DS, Rosenthal, KL, Graham, FL (1995) Identification of adenovirus E1A regions which affect MHC class I expression and susceptibility to cytotoxic T lymphocytes. Virology 211:268–277

Proffitt JA, Blair GE (1997) The MHC-encoded TAP1/LMP2 bidirectional promoter is down-regulated in highly oncogenic adenovirus type 12 transformed cells. FEBS Lett 400:141–144

Proffitt JL, Sharma E, Blair GE (1994) Adenovirus 12-mediated down-regulation of the major histocompatibility (MHC) class I promoter: Identification of a negative regulatory element responsive to Ad12 E1A. Nucleic Acids Res 22:4779–4788

Querido E, Blanchette P, Yan Q, Kamura T, Morrison M, Boivin D, Kaelin WG, Conaway RC, Conaway JW, Branton PE (2001) Degradation of p53 by adenovirus E4orf6 and E1B55 K proteins occurs via a novel mechanism involving a Cullin-containing complex. Genes Dev 15:3104–3117

Raska Jr K (1995) Functional domains of adenovirus E1A oncogenes which control interactions with effectors of cellular immunity. Curr. Topics in Microbiol. & Immunol 199 (part III):131–148

Rotem-Yehudar R, Groettrup M, Soza A, Kloetzel PM, Ehrlich R (1996) LMP-associated proteolytic activities and TAP-dependent peptide transport for class I MHC molecules are suppressed in cell lines transformed by the highly oncogenic adenovirus 12. J Exp Med 183:499–514

Rotem-Yehudar R, Winograd S, Sela S, Coligan JE, Ehrlich R (1994) Down-regulation of peptide transporter genes in cell lines transformed with the highly oncogenic adenovirus 12. J Exp Med 180:477–488

Routes JM, Cook JL (1989) Defective E1A gene product targeting of infected cells for elimination by natural killer cells. J Immunol 141:4022–4026

Routes JM, Cook JL (1995) E1A gene expression induces susceptibility to killing by NK cells following immortalization but not adenovirus infection of human cells. Virology 210:421–428

Routes JM, Li H, Bayley ST, Ryan S, Klemm, DJ (1996) Inhibition of IFN-stimulated gene expression and IFN induction of a cytolytic resistance to natural killer cell lysis correlate with E1A-p300 binding. J Immunol 156:1055–1061

Routes JM, Ryan S, Clase A, Miura T, Kuhl A, Potter TA, Cook JL (2000) Adenovirus E1A oncogene expression in tumor cells enhances killing by TNF-related apoptosis-inducing ligand (TRAIL). J Immunol 165:4522–4527

Routes JM, Ryan JC, Ryan S, Nakamura M (2001) MHC class I molecules on adenovirus E1A-expressing tumor cells inhibit NK cell killing, but not NK cell-mediated tumor rejection. International Immunology 13:1301–1307

Sang N, Caro J, Giordano A (2002) Adenoviral E1A: everlasting tool, versatile applications, continuous contributions and new hypotheses. Frontiers in Bioscience 7:407–413

Sarnow P, Ho Y, Williams J, Levine A (1982) Adenovirus E1b-58kd tumor antigen and SV40 large tumor antigen are physically associated with the same 54 kd cellular protein in transformed cells. Cell 28:387–394

Sawada Y, Föhring B, Shenk TE, Raska K (1985) Tumorigenicity of adenovirus-transformed cells: region E1A of adenovirus 12 confers resistance to natural killer cells. Virology 147:413–421

Screpanti V, Wallin RPA, Ljunggren H-G, Grandien A (2001) A central role for death receptor-mediated apoptosis in the rejection of tumours by NK cells. J Immunol 167:2068–2073

Seliger B, Wollsheid U, Momburg F, Blankenstein T, Huber C (2000) Coordinate downregulation of multiple MHC class I antigen processing genes in chemical-induced murine tumor cell lines of distinct origin. Tissue Antigens 56:327–336

Shiels C, Islam SA, Vatcheva R, Sasieni P, Sternberg MJE, Freemont PS, Sheer D (2001) PML bodies associate specifically with the MHC gene cluster in interphase nuclei. J Cell Sci 114:3705–3716

Shisler J, Yang C, Walter B, Ware CF, Gooding LR (1997) The adenovirus E3-10.4 K/14.5 K complex mediates loss of cell surface fas (CD95) and resistance to fas-induced apoptosis. J Virol 71:8299–8306

Schmitz ML, Indorf A, Limbourg FP, Städtler H, Traeckner EB-M, Baeurle PA (1996) The dual effect of adenovirus type 5 E1A 13S protein on NF-κB activation is antagonized by E1B 19K. Mol Cell Biol 15:4052–4063

Schouten GJ, van der Eb, Zantema A (1995) Down-regulation of MHC class I expression due to interference with p105-NF-κB1 processing by Ad12 E1A. EMBO J 14:1498–1507

Schrier PI, Bernards R, Vaessen, RTMJ, Houweling A, van der Eb AJ (1983) Expression of class I major histocompatibility antigens is switched off by highly oncogenic adenovirus 12 in transformed rat cells. Nature 305:771–775

Shenk T (2001) Adenoviridae: the viruses and their replication. In: Knipe D, Howley PM (eds) Fields' Virology, 4th edition, Lippincott Williams and Wilkins, Philadelphia, pp 2265–2300

Shirayoshi Y, Miyazaki J, Burke P, Hamada K, Appella E, Ozato K (1987) Binding of multiple nuclear factors to the 5' upstream regulatory element of the murine major histocompatibility class I gene. Mol. Cell. Biol. 7:4542–4548

Smirnov DA, Hou S, Ricciardi R (2000) Association of histone deacetylase with COUP-TF in tumorigenic Ad12-transformed cells and its potential role in shut-off of MHC Class 1 transcription. Virology 268:319–328

Smirnov DA, Hou S, Liu X, Claudio E, Siebenlist UK, Ricciardi RC (2001) COUP-TFII is up-regulated in adenovirus type 12 tumorigenic cells and is a repressor of MHC Class I transcription. Virology 284:13–19

Sprengel J, Schmitz B, Hens-Neitgel D, Zock C, Doerfler W (1994) Nucleotide sequence of human adenovirus 12: comparative functional analysis. J Virol 68:379–389

Tang X, Li H-O, Sakatsume O, Ohta T, Tsutsui H, Smit AFA, Horikoshi M, Kourilsky P, Israël A, Gachelin G, Yokayama K (1995) Cooperativity between an upstream

TATA-like sequence and a CAA repeated element mediates E1A-dependent negative repression of the H-2K^{b} class I gene. J Biol Chem 270:2327–2336
Täuber B, Dobner T (2001) Adenovirus early E4 genes in viral oncogenesis. Oncogene 20:7847–7854
Telling GC, Williams J (1994) Constructing chimeric type 12/type 5 adenovirus E1A genes and using them to identify an oncogenic determinant of adenovirus type 12. J Virol 68:877–887
Turnell AS, Grand RJ, Gorbea C, Zhang X, Mymryk JS, Gallimore PH (2000) Regulation of the 26S proteasome by adenovirus E1A. EMBO J 19:4759–4773
Vanhaesebroek H, Timmers HTM, Pronk GJ, Van Roy F, Van der eb AJ, Fiers W (1990) Modulation of cellular susceptibility to the cytotoxic/cytostatic action of tumour necrosis factor by adenovirus E1 gene expression is cell type-dependent. Virology 176:362–368
Vertegaal AC, Kuiperij HB, van Laar T, Scharnhorst V, van der Eb AJ, Zantema A (2000) cDNA micro array identification of a gene differentially expressed in adenovirus type 5- versus type 12-transformed cells. FEBS Lett 487:151–155
Wienzek S, Roth J, Dobbelstein M (2000) E1B 55-kilodalton oncoproteins of adenovirus types 5 and 12 inactivate and relocalize p53, but not p51 or p73, and cooperate with E4orf6 proteins to destabilize p53. J Virol 74:193–202
Williams J, Williams M, Liu C, Telling G (1995) Assessing the role of E1A in the differential oncogenicity of group A and group C human adenoviruses. Curr Top Microbiol Immunol 199 (part III):149–175
Wold WSM, Doronin K, Toth K, Kuppuswamy M, Lichtensein DL, Tollefson A (1999) Immune responses to adenoviruses: viral evasion mechanisms and their implications for the clinic. Current Opinion in Immunology 11:380–386
Woods DB, Vousden KH (2001) Regulation of p53 function. Experimental Cell Research 264:56–66
Yewdell JW, Bennink JR, Eager KB, Ricciardi RP (1988) CTL recognition of adenovirus-transformed cells infected with influenza virus: lysis by anti-influenza CTL parallels adenovirus 12-induced suppression of class I molecules. Virology 162:236–238
Yewdell JW, Bennink JR (1999) Mechanisms of viral interference with MHC class I antigen processing and presentation. Annu Rev Cell Dev Biol 15:579–606
Yoo GH, Hung MC, Lopez-Berestein G, LaFollette S, Ensley JF, Carey M, Batson E, Reynolds TC, Murray JL (2001) Phase 1 trial of intratumoral liposome E1A gene therapy in patients with recurrent head and neck cancer. Clin. Cancer Res 7:1237–1245
Zheng P, Guo Y, Niu Q, Levy DE, Dyck JA, Lu S, Shelman A, Liu Y (1998) Proto-oncogene PML controls genes devoted to MHC class I antigen presentation. Nature 396:373
Zhu K, Wang J, Zhu J, Jiang J, Shou J, Chen X (1999) p53 induces TAP1 and enhances the transport of MHC class I peptides. Oncogene 18:7740–7747

Immune Evasion by Adenovirus E3 Proteins: Exploitation of Intracellular Trafficking Pathways

M. Windheim[1] · A. Hilgendorf[2] · H.-G. Burgert[3]

[1] Aventis, DG Metabolic Diseases, 65926 Frankfurt, Germany

[2] Gene Centre, Ludwig-Maximilians University, Feodor Lynen Str. 25, 81377 Munich, Germany

[3] Department of Biological Sciences, The University of Warwick, Coventry, CV4 7AL, UK

E-mail: H-G.Burgert@warwick.ac.uk

Abstract Adenoviruses (Ads) are nonenveloped viruses which replicate and assemble in the nucleus. Therefore, viral membrane proteins are not directly required for their multiplication. Yet, all human Ads encode integral membrane proteins in the early transcription unit 3 (E3). Previous studies on subgenus C Ads demonstrated that most E3 proteins exhibit immunomodulatory functions. In this review we focus on the E3 membrane proteins, which appear to be primarily devoted to remove critical recognition structures for the host immune system from the cell surface. The molecular mechanism for removal depends on the E3 protein involved: E3/19K prevents expression of newly synthesized MHC molecules by inhibition of ER export, whereas E3/10.4-14.5K down-regulate apoptosis receptors by rerouting them into lysosomes. The viral proteins mediating these processes contain typical transport motifs, such as KKXX, YXXΦ, or LL. E3/49K, another recently discovered E3 protein, may require such motifs to reach a processing compartment essential for its presumed immunomodulatory activity. Thus, E3 membrane proteins exploit the intracellular trafficking machinery for immune evasion. Conspicuously, many E3 membrane proteins from Ads other than subgenus C also contain putative transport motifs. Close inspection of surrounding amino acids suggests that many of these are likely to be functional. Therefore, Ads might harbor more E3 proteins that exploit intracellular trafficking pathways as a means to manipulate immunologically important key molecules. Differential expression of such functions by Ads of different subgenera may contribute to their differential pathogenesis. Thus, an unexpected link emerges between viral manipulation of intracellular transport pathways and immune evasion.

1 Introduction

Adenoviruses (Ads) are nonenveloped viruses with a double-stranded DNA genome which is encapsidated in icosahedral virions of 70–90 nm in diameter. Fifty-one different human Ad serotypes have been described that are classified into subgenera A–F mainly by serological criteria (Shenk 1996; De Jong et al. 1999). Ads gain entrance into the cells after attaching to specific receptors, such as CAR and sialic acid (Bergelson et al. 1997; Tomko et al. 1997; Arnberg et al. 2000; Nemerow 2000), and this initial step is followed by internalization of virions into an endosomal compartment (Shenk 1996). During this process, the stepwise release of virion proteins destabilizes the capsid, which penetrates the lipid bilayer of the endosomes, reaches the cytoplasm, and migrates along microtubules to the nuclear pore complex (Suomalainen et al. 1999). After disassembly of the capsid, the DNA genome and associated proteins are imported into the nucleus, where early gene transcription starts (Russell 2000; Whittaker et al. 2000). The early Ad proteins induce cell-cycle progression without triggering the intrinsic apoptosis pathways, prepare the cell for DNA replication, and provide multiple protective measures against the attack by the immune system. In the late phase of infection, mainly structural proteins are being synthesized, and new virions are assembled in the nucleus. Eventually, viral particles are released from lysed cells (Shenk 1996; Russell 2000). Thus, the Ad replication cycle involves only some of the compartments of a eukaryotic cell: the plasma membrane and the endosomal system during uptake, and the cytoplasm and the nucleus for assembly and replication. In addition, the integrity of the mitochondria is preserved, possibly through E1B/19K-mediated inhibition of apoptotic mitochondrial pathways (White 1998). Other membrane enwrapped compartments, the endoplasmic reticulum (ER), the Golgi/*trans*-Golgi network (TGN), peroxisomes, and lysosomes are not directly required for Ad entry, replication, and release. Accordingly, Ads do not code for any membrane proteins that are essential for virus replication. However, transmembrane proteins are encoded in the early transcription unit 3 (E3), a region of the Ad genome that is dispensable for Ad replication in cell culture. This region varies in size and gene composition between the different subgenera but is present in all human Ads (Burgert and Blusch 2000). Examining the E3 genes of Ads from the various subgenera for the presence of potential signal sequences and transmembrane segments, which are typical features of integral membrane proteins, reveals that the great majority of the Ad E3 proteins are indeed transmembrane proteins (Fig. 1, and Table 3). Frequently, these proteins are glycosylated and localize to the ER, Golgi/TGN, plasma, or nuclear membrane. The only exceptions are the Ad2 12.5K and 14.7K proteins and their homologues in subgenera A–E and A–F, respectively. The 6.3K open reading frame (ORF) of Ad4 may also encode a cytosolic protein, but its expression has not yet been confirmed. Most E3 proteins are predicted to be type Ia transmembrane proteins with an N-terminal signal sequence (Table 3). Only the subgenus C-specific E3 proteins, 11.6K and 6.7K, and the 9K ORF present in subgenus B1 lack such an N-terminal sequence, and instead contain a

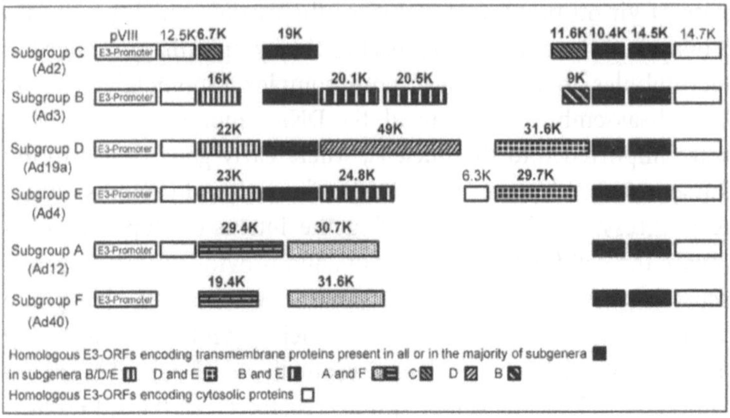

Fig. 1 E3 region coding capacity of representative members of Ad subgenera A–F. Open reading frames (ORFs) are indicated as *bars* and are drawn to scale. The shading code is depicted *below* the figure. ORFs encoding transmembrane proteins are indicated by *filled boxes* and *bold characters*, and ORFs encoding cytosolic proteins are represented by *open boxes*. Significant overall homology (similarity >25%) is indicated by *identical shading*. Homology to a portion of a protein was neglected. The size/name of common ORFs is only given once. Gaps between ORFs have been artificially introduced to align homologous ORFs. The E3 promoter overlaps with part of the pVIII ORF, which is not an E3 gene

signal-anchor domain (type Ib transmembrane proteins). As the E3 region is not required for Ad replication in vitro, but is present in all human Ads, it is thought to play a crucial role for efficient virus propagation in the face of the host immune response in vivo and thus for Ad pathogenesis and persistence. This was experimentally confirmed by the discovery that E3 proteins exhibit multiple immunomodulatory functions (Wold et al. 1995; Mahr and Gooding 1999; Burgert and Blusch 2000; Burgert et al. 2002).

By studying the E3 proteins of Ad2 and Ad5 of subgenus C, novel viral mechanisms were discovered, aimed at evading the recognition and elimination by the host immune system: E3/19K retains MHC class I molecules in the ER and thereby prevents presentation of viral peptides and the subsequent lysis of infected cells by cytotoxic T-lymphocytes (CTLs) (Burgert and Kvist 1985; Burgert 1996). The E3/14.7K protein can protect infected cells from tumor necrosis factor-α (TNFα)-induced cell death and is implicated in the inhibition of the inflammatory response (Mahr and Gooding 1999; Wold et al. 1999; Horwitz 2001). A complex of two E3 proteins, 10.4K and 14.5K, may also confer resistance against TNFα-mediated cytolysis and down-regulates the epidermal growth factor receptor (EGFR) from the cell surface (Mahr and Gooding 1999; Wold et al. 1999; Burgert and Blusch 2000). Moreover, the same protein complex (also called RID, for receptor internalization and degradation) dramatically reduces the expression of the apoptosis receptor Fas on the cell surface of infected cells by inducing its internalization and degradation in lysosomes (Shisler et al. 1997; Elsing and Burgert 1998; Tollefson et al. 1998). Similarly, the 10.4–14.5K complex down-regulates TNF-related apoptosis-inducing ligand (TRAIL) receptor 1 and 2 (TRAIL-R1, TRAIL-R2), the latter possibly in concert with E3/6.7K (Benedict et al. 2001; Tollefson et al. 2001). Thus, Ad E3 proteins counteract antiviral immune responses, such as recognition and lysis by CTLs and natural killer (NK) cells (Harty et al. 2000; Biron and Brossay 2001; Burgert et al. 2002). Therefore, these E3 functions are proposed to contribute profoundly to immune evasion and to the establishment of persistent infections in vivo (Mahr and Gooding 1999; Wold et al. 1999; Burgert and Blusch 2000; Burgert et al. 2002). Interestingly, most of these immunosubversive E3 functions exploit the membranous organelles of the secretory pathway and the endosomal/lysosomal system, which are not essential for replication in vitro. Collectively, E3 proteins target important host recognition molecules that are removed from the cell surface, either by preventing expression of newly synthesized molecules on the cell surface or by rerouting them into lysosomes. Both mechanisms are brought about by viral exploitation of the cellular protein-sorting machinery. In this review, we focus on the manipulation of trafficking pathways by integral membrane proteins of the E3 region. After giving an overview about the protein-sorting machinery and cellular-sorting signals, we describe in more detail how Ad E3 proteins counter immune responses by exploiting various cellular transport systems. Only a fraction of Ad E3 proteins has been functionally characterized to date. Close inspection of all E3 sequences reveals that many E3 proteins contain putative transport motifs. To evaluate their potential significance, we examined the sequences surrounding these motifs for additional features characteristic of functional trafficking signals.

2
The Cellular Protein-Sorting Machinery

Eukaryotic cells contain a number of membranous compartments, composed of a distinct set of proteins. Many of these organelles exchange material by vesicular transport, while some may also be connected by a tubular network (Lippincott-Schwartz et al. 2000). Despite a continuous flux of protein transport through these compartments, their identity, as defined by the distinct protein and lipid composition, is maintained. This is achieved by bidirectional trafficking whereby molecules of the donor membrane are retrieved again from the acceptor compartment (Mellman and Warren 2000). For the Golgi apparatus and the lysosomes also maturation models have been proposed to explain the differentiation into distinct organelles or subcompartments (*cis*- vs. *trans*-Golgi; early to late endosomes and lysosomes). Numerous components involved in sorting proteins to their specific destinations have already been characterized (Kirchhausen 2000b). Three major types of coat complexes have been identified that play a role in vesicular transport: Clathrin, COPI, and COPII (Barlowe 2000; Kirchhausen 2000a; Mellman and Warren 2000). A striated coat visualized in specialized regions/invaginations of the plasma membrane enriched for caveolin, the caveolae, also contributes to trafficking between the plasma membrane and the TGN (Anderson 1998). All these coats are derived from soluble, cytosolic precursors that are recruited to the ER and Golgi membranes upon activation of GTPases and the binding of cargo receptors that recognize special features in the lumenal portion of proteins. This initial step is energy-dependent and requires GTP. Intrinsic sorting signals in the cytoplasmic tail (C-tail) may be either directly recognized by the coat (COPI and COPII) or may be bound by so-called adaptor complexes, such as AP-1 or AP-2, which link the selected receptors to the coat components such as clathrin. Growth of the coat leads to membrane deformation; SNAREs (soluble *N*-ethylmaleimide-sensitive factor attachment protein receptors) are incorporated into the budding vesicle, and completion of the coat results in scission at the neck of the invagination. Before fusion with a specific acceptor membrane, which is thought to be controlled by the pairing of v-SNAREs on vesicles and t-SNARES on target membranes (Chen and Scheller 2001), the coat is removed. Small GTP-binding proteins play a crucial role in the regulation of these processes: Rab proteins and the Sar1p/ADP ribosylation factors (ARFs) control the budding process of vesicles as well as the subsequent release of the coat proteins from the vesicle (uncoating). Rab proteins also coordinate tethering/docking/fusion of the vesicles by regulating the formation of SNARE complexes (Kirchhausen 2000b; Chen and Scheller 2001; Zerial and Mc-Bride 2001). In addition, Rab effectors, e.g., p115 in the intra-Golgi transport and EEA-1 in endosomal fusion, act as so-called tethering factors upstream of SNAREs and confer targeting specificity to the vesicles (Pfeffer 1999).

Cargo proteins may contain positively selecting export signals or may be transported by a default pathway ("bulk flow"). As a consequence, resident proteins of a particular compartment should either contain retention signals or retrieval signals which mediate the return of proteins from the acceptor compartment to the donor membrane. As many E3 proteins contain targeting signals, either for the ER or the Golgi/TGN and the endocytic pathway (early/late endosomes, lysosomes), we will describe sorting events within these compartments in more detail below (Fig. 2).

2.1 Sorting in the ER, ERGIC, and Golgi Complex

To enter the secretory pathway in mammalian cells, most proteins contain an N-terminal signal sequence that is recognized by the signal recognition particle (SRP) when it emerges from the ribosome (Brodsky 1998). Upon binding of the SRP-ribosome complex to the SRP receptor on the ER membrane, proteins are translocated cotranslationally into the ER, either as soluble proteins or anchored in the lipid bilayer by their transmembrane domain (TMD). Some proteins are also targeted to the ER posttranslationally. Trafficking between the ER, ER-Golgi intermediate compartment (ERGIC), and the Golgi/TGN complex is accomplished by COPI and COPII coats, composed of eight and five subunits, respectively (Barlowe 2000; Kirchhausen 2000b; Klumperman 2000; Lippincott-Schwartz et al. 2000; Mellman and Warren 2000). COPII vesicles are utilized for the anterograde transport between the ER and the ERGIC, also known as VTC (vesicular tubular clusters or transitional elements), whereas COPI-coated vesicles function in the anterograde transport of proteins from the ERGIC to the Golgi, the transport between Golgi cisternae, and the retrograde transport from the Golgi to the ERGIC and the ER (Fig. 2). Evidence was also provided for a retro-

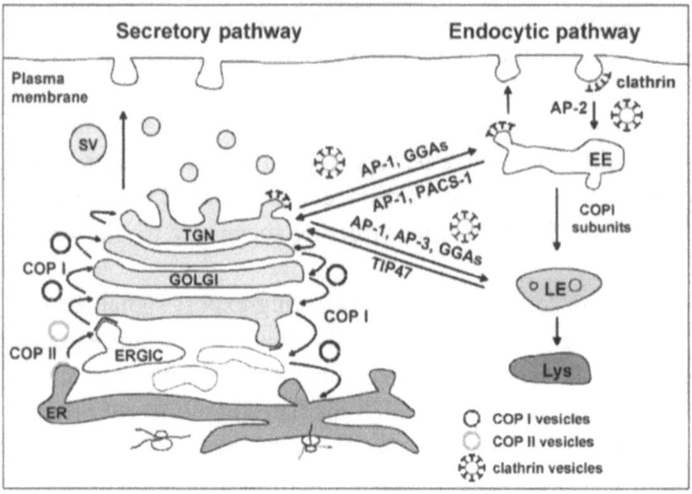

Fig. 2 Trafficking in the secretory and endocytic pathway. Export from the ER to the ERGIC occurs via COPII-coated vesicles, whereas transport from the ERGIC to the *cis*-Golgi, between the Golgi cisternae and retrograde to the ER is mediated by COPI-coated vesicles. From the TGN, proteins can be transported (1) to the plasma membrane via secretory vesicles or (2) to early (*EE*) or late (*LE*) endosomes by clathrin-coated vesicles involving the clathrin adaptor proteins AP-1 and AP-3 and the GGAs. In addition, early endosomes can be reached via endocytosis from the plasma membrane, which depends on the clathrin adaptor complex AP-2 and clathrin. From early endosomes, three trafficking routes are possible: (1) Recycling to the cell surface, (2) recycling to the TGN involving clathrin-coated vesicles and the clathrin adaptors AP-1 and PACS-1, (3) transport to late endosomes requiring COPI subunits. From late endosomes, proteins are either delivered to lysosomes (*Lys*) to be degraded or to fulfill their function as lysosomal residents, or they are recycled to the TGN. The latter pathway seems to involve TIP47. For details and references see text

grade pathway independent of COPI (Klumperman 2000). Budding of COPII vesicles at specific ER exit sites requires the small GTPase Sar1p and depends on GTP, whereas COPI is recruited to Golgi membranes by the small GTP-binding protein ARF1 that initiates the formation of COPI-coated vesicles (Barlowe 2000).

Sorting begins already at the level of the ER and may involve the recognition of specific ER export signals, retention/retrieval motifs in resident ER proteins (Table 1), and the subsequent segregation of certain sets of proteins, e.g., GPI-anchored proteins, from others (Muniz et al. 2001). Most newly synthesized proteins will leave the ER at defined,

Table 1 ER/Golgi sorting signals

Comp.	Signal	Location in protein	Protein example (s)	Function	References
ER	KDEL	COOH-terminus, lumenal	Protein disulphide isomerase	Retrieval of lumenal ER proteins from the Golgi	Teasdale and Jackson 1996
ER	KKXX, KXKXX	COOH-terminus, cytoplasm	Glucose-6-phosphatase, oligosaccharyl-transferase	Retrieval of ER membrane proteins from the Golgi	Teasdale and Jackson 1996
ER	XXRR	N-terminus, cytoplasm	MHC II invariant chain	Retrieval of ER membrane proteins from the Golgi	Teasdale and Jackson 1996
ER	Charged, hydrophilic aa	TMD	TCR-α, membrane IgM	ER retention (degradation)	Ellgaard et al. 1999
ER	DXE	COOH-terminus, cytoplasm	VSV G protein	ER export	Klumperman 2000
ER	FF	COOH-terminus, cytoplasm	p24, ERGIC-53	ER export	Klumperman 2000
Golgi	Short TMD of about ~17 aa	TMD	Glycosyl-transferases, t-SNARES	Localization to certain membrane regions or aggregation	Munro 1998
Golgi	GRIP domain	C-terminus of peripheral Golgi proteins	Golgin-245, golgin-97	Peripheral association with the Golgi	Barr 1999; Kjer-Nielsen et al. 1999; Munro and Nichols 1999

For abbreviations and details see text.

highly organized membrane domains, called ER exit sites. There is increasing evidence that cargo is actively partitioned into these sites, and for some proteins, specific export signals have been defined. For example, the cargo receptors p24 and ERGIC-53 (Table 1), seem to be concentrated in COPII vesicles via direct interaction of two C-terminal phenylalanines (acting as specific ER export signals) with COPII (Hauri et al. 2000; Klumperman 2000).

A variety of sorting signals have been characterized that are critical to maintain high steady-state levels of proteins in the ER (Table 1). Lumenal residents of the ER characteristically have the sequence KDEL at their carboxy-terminus (C-terminus) and are retrieved from the *cis*-Golgi by the KDEL receptor (Lewis et al. 1990). A class of ER membrane proteins is retrieved by means of cytosolically exposed peptides containing two basic residues. In type I transmembrane proteins, two critical lysines are found at position −3 and −4 or −3 and −5 from the C-terminus (KKXX or KXKXX), whereas type II transmembrane proteins contain an XXRR motif at the N-terminus (with K and R representing lysine and arginine, respectively, and X any amino acid) (Teasdale and Jackson 1996). The position of the basic amino acids (aa) is absolutely critical for functional activity, but the local sequence context also profoundly influences the efficacy of retrieval. A recent investigation using a combinatorial approach in vivo showed that different KKXX motifs will give rise to the entire spectrum of predominantly ER to predominantly surface localization (Zerangue et al. 2001). Amino acids favoring ER localization were in +1: Y, F, T and R, and in +2: L, V, N, Q. However, each combination appears to have its intrinsic efficacy. Originally, it was proposed that all proteins containing di-lysine motifs reach the *cis*-Golgi and are retrieved back to the ER by direct binding to COPI (Cosson and Letourneur 1994). In some cases, KKXX signals may in addition to ER retrieval also mediate ER retention (Andersson et al. 1999).

ER retention can be achieved by the presence of single basic or acidic aa in the TMDs of many multimeric complexes (e.g., the T-cell receptor–CD3 complex). Proper pairing of the subunits neutralizes the charges and allows efficient export. In this case, ER retention serves as a quality control mechanism for oligomeric proteins. In addition, ER retention can also be mediated by stretches of hydrophilic serine and threonine residues (Ellgaard et al. 1999). The major quality control system in the ER is based on proper folding facilitated by the interaction with ER chaperones, such as calnexin or calreticulin (Ellgaard et al. 1999). These chaperones contain ER retrieval or retention motifs and keep proteins in the ER until they assume the "correct" structure. Consequently, ER localization can also occur by attachment of a protein without ER retention or retrieval signal to one bearing such signals.

The steady-state localization of the cargo receptor ERGIC-53 in the ER-Golgi intermediate compartment is not maintained by a specific ERGIC localization signal but is due to a dynamic process relying on the presence of both ER export (FF; see Table 1) as well as KKXX retention and retrieval signals (Hauri et al. 2000). ERGIC-53 is a lectin thought to function as a cargo receptor for some glycoproteins to mediate their transport from the ER to the ERGIC. The ERGIC is thought to be a major site for cargo concentration (Martinez-Menarguez et al. 1999) and is considered a relay station, sorting out proteins for anterograde transport to the Golgi from those that must be returned to the ER (Hauri et al. 2000; Klumperman 2000; Lippincott-Schwartz et al. 2000).

From the ERGIC, proteins are delivered to the *cis* face of the Golgi complex. Transit through the Golgi complex involves the sequential passage through 3–5 cisternae (*cis*, *medium*, *trans*). The state of the Golgi/TGN complex is still controversial. Two major models have been postulated: cisternal maturation versus vesicular transport between static cisternae (Glick and Malhotra 1998; Mellman and Warren 2000). The former model proposes that new Golgi cisternae are permanently created. Consequently, cargo proteins travel via maturation of the cisternae, whereas resident Golgi proteins must be transported in a retrograde fashion. Thus, Golgi residents in this model would require retrieval signals. In the second model, resident Golgi proteins are retained in static cisternae and cargo proteins are transported in anterograde direction by default ("bulk flow") or by active selection. A concept was proposed that reconciles the two models, suggesting that both processes may co-exist in a cell: slow transport may be achieved via maturation and fast transport through vesicles. Utilization of each of these pathways may depend on the cell type or the differentiation state (Mellman and Warren 2000; Pelham and Rothman 2000).

Golgi residents contain signals to specify their Golgi localization. In contrast to the ER, no soluble Golgi residents appear to exist; instead, all Golgi proteins are either integral membrane proteins or peripheral membrane proteins attaching to the Golgi from the cytoplasmic face (Munro 1998). For some of the latter, the so-called GRIP domain seems to mediate Golgi targeting (Barr 1999; Kjer-Nielsen et al. 1999; Munro and Nichols 1999). For transmembrane proteins, sorting signals in either the TMD or the C-tail have been implicated. Glycosyltransferases, SNARE proteins, and some viral proteins are targeted to the Golgi by their TMD. The mechanism as to how the TMD mediates Golgi localization is still ill-defined, but based on the vesicular model for Golgi transport, two hypotheses have been proposed: an oligomerization ("kin-recognition") model and a "bilayer-thickness" model. The kin-recognition model proposes that Golgi proteins form large complexes that are excluded from anterograde transport vesicles, whereas in the latter model the TMD of Golgi proteins, which is on average five residues shorter than that of plasma membrane proteins, directs the proteins to specialized membrane areas that are excluded from sphingolipid/cholesterol-rich membranes used for export of proteins (Munro 1998). In other proteins, the determinant for Golgi localization has been mapped within the cytoplasmic tail (Locker et al. 1994; Andersson et al. 1997). The steady-state localization of some *cis*-Golgi proteins, e.g., the KDEL receptor Erd2p, is maintained by a dynamic process involving export from and recycling to the ER. Sorting at the TGN will be discussed in the following chapter, because the TGN seems to be closely connected to the endosomal/lysosomal system, which utilizes similar targeting motifs.

2.2 Sorting at the TGN, the Plasma Membrane, and in the Endosomal/Lysosomal System

2.2.1 Sorting Motifs, Adaptor Complexes, and Coat Proteins

The TGN serves as a sorting station where proteins destined for the different post-Golgi compartments, the specialized secretory vesicles, the plasma membrane, and the endosomal/lysosomal system are separated (Traub and Kornfeld 1997). In polarized cells, proteins are additionally sorted to the apical or basolateral membrane by determinants interpreted at the TGN and/or in sorting endosomes (Heilker et al. 1999; Keller et al. 2001). In general, transport from the TGN to the cell surface is thought to be mediated by secretory vesicles without involving any intermediate organelle, but a few plasma membrane proteins may take an indirect route via an endosomal compartment (Traub and Kornfeld 1997). No coat protein of secretory vesicles or for apical transport has been identified yet. Most other trafficking pathways between TGN, endosomes/lysosomes, and the plasma membrane (including basolateral transport) seem to be mediated by clathrin-coated vesicles. In general, clathrin coats do not directly bind to cargo but are recruited to cargo proteins by different members of the AP complexes. These AP complexes recognize di-leucine and YXXΦ motifs (see below) in the cytoplasmic tail of cargo proteins and have the additional capacity to bind clathrin. They share a similar architecture. All four AP complexes are tetramers with a molecular mass of about 250–300 kDa, consisting of homologous but distinct subunits (AP-1: γ, $\beta1$, $\sigma1$, $\mu1$; AP-2: α, $\beta2$, $\sigma2$, $\mu2$; AP-3: δ, $\beta3$, $\sigma3$, $\mu3$; AP-4: ε, $\beta4$, $\sigma4$, $\mu4$) (Kirchhausen 1999; Robinson and Bonifacino 2001). Several of the subunits exist as different isoforms. AP complexes are associated with different cellular compartments and, therefore, they appear to be involved in distinct sorting events. AP-1 localizes to the TGN and has been implicated in sorting between the TGN and the endosomal/lysosomal system and the basolateral membrane (Fig. 2). AP-2 localizes to the plasma membrane and directs endocytosis of a number of transmembrane proteins, including the receptors for transferrin, low-density lipoprotein and the EGF. Upon ligand binding, these receptors are rapidly internalized into early endosomes, from which they are either recycled to the cell surface or transferred to the lysosome for degradation (Fig. 2) (Kirchhausen 1999; Robinson and Bonifacino 2001). AP-2 might also participate in the budding of clathrin-coated vesicles in a retrograde trafficking pathway out of the lysosomal compartment (Heilker et al. 1999). AP-3 is associated with the TGN and peripheral endosomal structures and seems to be involved in the transport of a set of proteins from the TGN to lysosomes (Le Borgne et al. 1998; Dell'Angelica et al. 1999). AP-4 is a component of a non-clathrin coat associated with the TGN (Robinson and Bonifacino 2001). For AP-1 and AP-2, the association with clathrin-coated vesicles was demonstrated, whereas for AP-3 clathrin association is controversial and AP-4 has not yet been found in clathrin coats. Particularly important for clathrin binding seem to be the β subunits of the AP complexes. The contact site for clathrin has been mapped to the hinge region, which contains a clathrin-binding consensus sequence, the so-called "clathrin-box" (L, L/I, D/E/N, L/F, D/E), which is also found in other clathrin-interacting proteins. The clathrin-box is present in the β subunits of AP-1, AP-2, and AP-3, but not in AP-4, consistent with the view that AP-4 does not bind to clathrin (Kirchhausen 2000a; Robinson and Bonifacino 2001). A novel class of adaptors was recently implicated in TGN-endosome targeting, the Golgi-associated, γ ear containing, ARF-binding proteins (GGAs) (Black and Pelham 2001). Three GGAs are expressed in mammals and localize to the TGN. GGAs were shown to bind a set of cargo proteins, e.g., mannose-6-phosphate receptors (MPRs) and sortilin, by a motif consisting of an acidic stretch and an LL motif to mediate TGN export (Nielsen et al. 2001; Puertollano et al. 2001; Zhu et al. 2001). At present, it is unclear whether AP-1 and GGAs have overlapping functions. One possibility is that AP-1 is involved in one of the different recycling pathways between early/recycling endosomes or late endosomes and the TGN (Meyer et al. 2000; Rohn et al. 2000) (Fig. 2). In accord with this notion, AP-1 has been detected on endosomal membranes and was found to bind to immature secretory granules. AP complexes are recruited to specific membranes through clathrin-independent docking mechanisms. The small GTP-binding protein ARF1 is essential for the association of AP-1 and GGA-3 to the TGN membrane as well as for endosomal membrane association of AP-3 (Robinson and Bonifacino 2001). No related small GTP-binding protein has been found for the docking of AP-2 to the plasma membrane. In this case, docking might be initiated by phosphoinositides. Many other adaptor-associated proteins have been described that seem to regulate clathrin-coated vesicle formation. A particularly complex network of interactions seems to control vesicle formation at the plasma membrane, involving proteins, such as amphiphysin or dynamin. For instance, amphiphysin binds AP-2 and clathrin and acts as a receptor for dynamin, a GTPase required for fission of the coated vesicle from the plasma membrane (Hinshaw 2000; Kirchhausen 2000b).

Cargo selection by AP complexes is mediated by direct binding of some of their subunits to sorting signals in the cytoplasmic tails of transmembrane proteins. Three major motifs for sorting in the TGN and the endosomal/lysosomal system have been identified: NPXY, YXXΦ, and LL (Table 2), where Φ represents a bulky hydrophobic aa (L, V, I, F, M) and X any aa (Bonifacino and Dell-Angelica 1999; Heilker et al. 1999; Kirchhausen 1999). Most di-leucine (LL) motifs consist of a pair of leucines, but one leucine might also be substituted for another bulky hydrophobic aa (M, V, I, F). In this review, we will use the term di-leucine motif for all of these cases. Many proteins contain more than one motif and frequently combinations of different signals, e.g., the MPRs (Table 2). These most frequent sorting signals will be discussed in more detail in Sect. 2.2.2. Certain proteins achieve a steady-state localization in the

Table 2 TGN and endosomal/lysosomal sorting signals

Signal (s)	Protein	Destination			Adaptor binding	References
NPXY motifs						
NFD**NPVY**	LDL receptor	E	EE	B	AP-2, Clathrin	Heilker et al. 1999; Kirchhausen 1999
YXXΦ *motifs*						
GGEPLS**YTRF**	Transferrin receptor	E	EE	B	AP-2	Heilker et al. 1999
KRSHAG**YQTI**	Lamp-1	E	L	B	AP-1, -2, -3	Heilker et al. 1999
RAGHSS**YTPL**	HLA-DM*β*	E	MIIC		AP-1, -2	Heilker et al. 1999
RPKASD**YQRL**	TGN38/46	E	TGN	B	AP-1, -2	Banting and Ponnambalam 1997
LL *motifs*						
EAENTITYS**LL**KH	FcRII-B2 (mouse)	E		B		Hunziker and Fumey 1994
EGTADERAP**LI**RT	LimpII	E	L	B	AP-3	Heilker et al. 1999
MDDQRD**LI**SN LISNNEQLP**ML**GR	Invariant chain	E	MIIC	B	AP-1, -2	Heilker et al. 1999
VTSEPDKHS**LL**VG	Menkes disease protein	E	TGN			Petris et al. 1998; Francis et al. 1999
Sorting motifs in MPRs						
SNVSYK**YSKV**	CI-MPR (MPR300)	E	TGN/EE	B	AP-2	Heilker et al. 1999
SFHDDSDED**LL**HI	CI-MPR (MPR300)	E	TGN/EE	B	AP-1, GGA1, 2, 3	Heilker et al. 1999; Puertollano et al. 2001; Zhu et al. 2001
RNVPAA**YRGV**	CD-MPR (MPR46)	E	TGN/EE	B	AP-2, -3	Heilker et al. 1999; Storch and Braulke 2001
EESEERDDH**LL**PM	CD-MPR (MPR46)	E	TGN/EE	B	AP-3, GGA1, (3)	Heilker et al. 1999; Puertollano et al. 2001; Zhu et al. 2001; Storch and Braulke 2001

Table 2 (continued)

Signal (s)	Protein	Destination		Adaptor binding	References
Other sorting signals					
Acidic cluster	e.g. HIV-1 Nef, Furin	E	TGN	PACS-1	Voorhees et al. 1995; Piguet et al. 2000
Ubiquitin	e.g. EGFR	E			Hicke 1999
KKFF	ERGIC-53	E			Hauri et al. 2000
KRFY	VIP36				
FW	CI- and CD-MPR		TGN/EE	TIP47	Carroll et al. 2001
Palmitoylation	CD-MPR				Schweizer et al. 1996
Man-6-P	Lysosomal hydrolases		L		Roberts et al. 1998

E, endocytosis; EE, early endosomes; B, basolateral membrane; L, lysosomes; MIIC, MHC II compartment. For other abbreviations and details see text.

TGN utilizing these sorting signals. For instance, TGN localization of TGN38 and furin was shown to depend on YXXΦ motifs (see below) in their cytoplasmic tails which mediate their retrieval from the cell surface (Table 2). The retention of TGN38 in the TGN is mediated by its TMD (Banting and Ponnambalam 1997), whereas an acidic cluster motif with a casein kinase II phosphorylation site is required for furin localization in the Golgi (Molloy et al. 1999). Several other sorting signals have been suggested to play a role in post-Golgi trafficking pathways, and the most prominent ones are listed in Table 2. Moreover, a variety of additional proteins, e.g., TIP47 and PACS-1 (Fig. 2) have been implicated in those pathways. TIP47 was recently shown to interact on endosomal membranes with Rab9-GTP, a protein which seems to regulate the generation of transport vesicles in endosomes destined for the TGN (Carroll et al. 2001). PACS-1 forms ternary complexes with cargo proteins containing acidic clusters and AP-1 or AP-3, but does not bind to AP-2. Overexpression of a dominant negative PACS-1 mutant leads to a redistribution of furin and the cation-independent MPR (CI-MPR) from the TGN to a vesicular, most likely endosomal compartment, suggesting a role of PACS-1 in endosome-TGN transport (Crump et al. 2001). COPI subunits have also been implicated in trafficking within the endosomal/lysosomal system (Gu and Gruenberg 1999) and have been proposed to be crucial for the biogenesis of so-called endocytic carrier vesicles, which are thought to transport proteins from early to late endosomes (Fig. 2). Not included in Table 2 are the various motifs implicated in targeting proteins to the apical membrane in polarized cells, such as N-glycans, O-glycans, TMDs, cytoplasmic peptides, and GPI-anchors (Rodriguez-Boulan and Gonzalez 1999).

2.2.2
YXXΦ and Di-Leucine Motifs

Sorting of numerous transmembrane proteins in the late secretory and endocytic pathways depends on YXXΦ and di-leucine (LL) motifs in their cytoplasmic tails, and these motifs are also present in many adenovirus E3 proteins (see below). YXXΦ motifs interact with the μ1-μ4 subunits of the AP complexes (Ohno et al. 1998; Bonifacino and Dell'Angelica 1999), whereas it is controversial as to whether the LL motif is bound by the *β* or the μ subunit of AP complexes (Heilker et al. 1999; Kirchhausen 1999). The crystal structures of the μ2 binding domain complexed to a peptide containing the YXXΦ motifs of TGN38 or the EGF receptor revealed that the peptide was bound in an extended conformation and not in a tight turn structure as previously suggested (Owen and Evans 1998). The NPXY motif (Table 2) was proposed to directly interact with clathrin, although it is unclear whether these low-affinity interactions are specific (Kirchhausen 1999).

Strikingly, both YXXΦ and LL motifs in the C-tails of transmembrane proteins mediate sorting at multiple sites: They serve as signals for endocytosis and for transport to the basolateral surface of polarized cells, or mediate sorting at the TGN and within the endocytic pathway (Table 2). Therefore, apart from the basic YXXΦ and LL motifs, additional information must determine the distinct trafficking and localization of individual proteins. How exquisite the interpretation of this distinct sorting information in each protein must be is perhaps best illustrated considering sorting at the TGN. Here, proteins with YXXΦ and LL motifs interacting with AP-1 and AP-3 for direct targeting to endosomal/lysosomal compartments must be discriminated from other YXXΦ-containing proteins destined for the plasma membrane, where their YXXΦ motifs are recognized by AP-2 and thereby induce endocytosis. For example, the transferrin receptor contains a YTRF sorting motif that interacts with μ2 but not with μ1, whereas the LL motifs of LIMP-II and tyrosinase bind in vitro with a higher affinity to AP-3 than to AP-1 and AP-2 (Höning et al. 1998; Bonifacino and Dell'Angelica 1999). At present, it is not clear what determines the fine-tuning of YXXΦ and LL motifs; however, it is reasonable to assume that residues surrounding the motifs are of particular importance. These aa may fulfill two functions; (1) they might create a structure enabling correct presentation and recognition of the motif, and (2) they might directly participate in specific interactions with the adaptor proteins. In an attempt to extract additional information, we screened 48 functional YXXΦ motifs in 45 proteins and 35 LL motifs in 31 proteins for the frequency of different aa residues in positions −6 to +3 relative to the tyrosine (Y) and −9 to +3 relative to the first leucine, respectively. This value was compared with the statistical frequency of each aa in 2,912 representative proteins (White 1994). A residue was considered as favored if its frequency was twofold higher than its statistical frequency. Highly overrepresented residues with a frequency of threefold or higher than their statistical frequency are written below in bold. Residues overrepresented in the vicinity of YXXΦ motifs are at positions −6: Y (2.7), K (2.6), R (2.3); −5: **R (3.4)**, N (2.5); −4: K

(2.5); −3: **H (3.9)**, R (2.2), P (2.2); −2: Q (2.7), P (2.2); −1: G (2.8); +1: **Q (4.3)**, R (2.9), S (2.4); +2: P (2.5), T (2.2). Interestingly, at positions −3, −2, −1, +1, and +2, a preference of residues frequently found in turn structures (G >N, P, D, S) is noted. This supports earlier suggestions (Collawn et al. 1990) that the YXXΦ motif may be part of a turn structure, or may lack a defined secondary structure and might bind adaptors in an extended conformation, as suggested by the crystal structure (Owen and Evans 1998). Although the relative enrichment of basic aa in positions −4 to −6 might be biased due to the close vicinity of some motifs to the transmembrane domains often followed by a cluster of basic amino acids, this argument cannot explain the overrepresentation of these aa in position −3 and +1. The bulky hydrophobic aa (Φ) in position +3 of YXXΦ motifs was most frequently leucine (54.2%), followed by valine (12.5%), isoleucine (10.4%), phenylalanine (8.3%), and methionine (6.3%). The presence of these different aa did not correlate with specific sorting events (Bonifacino and Dell-Angelica 1999; Heilker et al. 1999; Kirchhausen 1999).

The sequence requirements of YXXΦ motifs for interaction with AP complexes have been investigated using combinatorial peptide libraries or the yeast-two-hybrid system. For the interaction with μ2, a YXRL consensus motif was found in a combinatorial library approach (Boll et al. 1996). No preference for surrounding residues was observed. Applying the yeast-two-hybrid system, it was shown that mutations in the SDYQRL sequence of TGN38 affected its interactions with μ1 and μ2 differentially (Ohno et al. 1996), indicating that flanking residues may contribute to the selective interaction between YXXΦ motifs and different AP complexes. In this motif, a free hydroxyl group (S, T) at position −2 seems to be crucial for correct trafficking from endosomes to the TGN (Stephens et al. 1997; Roquemore and Banting 1998). Similar and divergent sequence requirements were revealed for the interactions of a combinatorial XXXYXXΦ library with μ1, μ2, μ3, and μ4 (Ohno et al. 1998; Aguilar et al. 2001). This indicates that AP complexes recognize similar but not identical motifs, and these differences are possibly the basis for their involvement in different trafficking pathways. In accord with this concept, some studies show that motifs that specify different functions, e.g., internalization versus basolateral localization, overlap but are not identical (Lin et al. 1997; Heilker et al. 1999).

Screening the sequence context of the 35 LL motifs revealed preferences for E (2.5), W (2.2) in position −9; Y (2.7), S (2.6), E (2.5), R (2.0) in position −8; **R (3.6)**, H (2.7), S (2.2) in position −7; E (2.4), M (2.2) in position −6; **D (4.4), E (3.3)** in position −5; **E (4.3)**, S (2.9), D (2.8) in position −4; K (2.9), N (2.6), T (2.0) in position −3; **Q (5.9)**, H (2.6), R (2.0) in position −2; **P (4.6)**, H (2.6) in position −1; **H (4.1)**, P (2.5), S (2.2), R (2.1), E (2.0) in position +1 and N (2.9), P (2.6), M (2.5), E (2.2) in position +2. Interestingly, as seen for the YXXΦ motifs, aa favoring turn structures were enriched in positions −1, +1, and +2. Particularly impressive was the presence of proline in position −1 in ~23% of the proteins. This might indicate that LL motifs also adopt a turn or an extended conformation. Most remarkable is the enrichment of negatively charged aa or serine from positions −8 to −4, with a particular high frequency at positions −4 and −5. Sixty-eight percent of all motifs contained either a D or an E at one of these positions. This was noted previously, and a consensus motif was proposed with a negatively charged aa (E, D or phosphoserine) between −3 and −5 from the first leucine (Kirchhausen 1999). The importance of surrounding aa for specific functional activities of LL motifs was underscored by plasmon resonance and immunofluorescence studies showing that D and E in positions −5 and −4, respectively, are important for the interaction of a peptide containing the LI sorting motif of LimpII with AP3 (Höning et al. 1998), and positively charged residues in these positions drastically changed the trafficking of LimpII (Sandoval et al. 2000). Although the glutamic acid residue at position −4 is critical for efficient intracellular sorting of LIMPII to lysosomes, it is dispensable for its internalization from the cell surface. However, for the di-leucine motif-based internalization of the invariant chain, an acidic residue in −4 (E, D) was shown to be important (Pond et al. 1995). In LL motifs, leucine was preferred in both positions (L1/L2) with 85.7% and 82.9%, respectively. Among the proteins with LL motifs, 5.7% contained methionine in the L1 position, whereas valine, isoleucine, and phenylalanine were found in one protein only (2.86%). In the L2 position, isoleucine (11.4%) was more frequent than methionine or valine (2.86% each). Taken together, these data provide evidence for the existence of distinct YXXΦ- and leucine-based motifs which are differentially interpreted at different cellular sites.

Moreover, the accessibility of YXXΦ and LL motifs can be influenced by phosphorylation. Recognition of some LL motifs seems to be positively regulated by serine phosphorylation, e.g., in CD4 or CD3γ, whereas the ability of the YXXΦ motif of CTLA-4 to bind the µ2 subunit of AP-2 is abolished by tyrosine phosphorylation (Heilker et al. 1999). Upstream serine residues were also shown to be important for the function of the LL motif of the IL-6 signal transducer gp130, which might indicate regulation by phosphorylation (Dittrich et al. 1996). Sorting of CI- and CD-MPRs depends on a casein kinase II phosphorylation site, but it is unclear whether phosphorylation is indeed important for sorting (Mauxion et al. 1996; Chen et al. 1997). Consistent with a putative regulatory role of phosphoserine, our analysis reveals an enrichment of serine in positions −8, −7, −4 and +1.

The distance of YXXΦ and LL motifs to the transmembrane region of proteins varies considerably. Accordingly, rather large deletions in the cytoplasmic tail moving the YXXΦ motif of the transferrin receptor (a type II transmembrane protein) closer to the TMD were tolerated without affecting the internalization rate. However, a minimal distance of at least 7 aa between the F of the YTRF motif and the transmembrane segment was required (Collawn et al. 1990). Similarly, a 7-aa spacing between the TMD and the Y of the YXXΦ motif of Lamp-1 was essential for transport to lysosomes. Whereas alteration of the distance to 6 or 12 aa changed the internalization rate only modestly, lysosomal targeting was dramatically impaired (Rohrer et al. 1996). A similar minimal distance seems to be required for the functionality of LL motifs whereby slight differences are observed depending on whether the motif is regulated by phosphorylation. For the function of the phosphorylation-dependent LL motif (SDKQTLL) of CD3γ, a minimum of 7 aa between the phosphorylated S and the TMD is required, whereas for the constitutively active, phosphorylation-independent LL motif, DKQTLL, a minimal spacing of six residues between the TMD and the acidic residue was required for optimal activity (Geisler et al. 1998). Thus, a minimal distance between a transport motif and the lipid bilayer must be retained, whereas other changes in the distance between a motif and the TMD seem to be differentially tolerated by different proteins.

In many cases the trafficking characteristics of a protein, e.g., of tyrosinase (Calvo et al. 1999), HLA-DM (Copier et al. 1996), or CD3γ, could be transferred by transplanting the cytoplasmic tail containing YXXΦ or LL motif to the ectodomain and TMD of a bona-fide plasma membrane protein, such as CD8 or the IL-2α receptor (Letourneur and Klausner 1992). However, there are other examples in which the transfer of the signal motif alone was not sufficient to switch the trafficking characteristics of the recipient. For instance, the SDYQDL motif of TGN38 transferred to the transferrin receptor could mediate internalization but not transport to the TGN (Johnson et al. 1996). Generally, the requirements seem less stringent for endocytosis than for subsequent sorting events. Also, the functionality of YXXΦ motifs artificially introduced into hemagglutinin and glycophorin A by site-directed mutagenesis seemed to depend on the surrounding residues rather than on the position in the cytoplasmic tail (Zwart et al. 1996).

In summary, sequence and positional requirements seem to exist for the function of YXXΦ and LL motifs in different trafficking pathways, yet a general correlation between the presence of certain aa at certain positions and a particular trafficking route has not emerged. This might be due to the presence of additional sorting information in proteins or might relate to the complexity of the transport pathway, since many destinations require several sorting decisions to be made. Despite this, the integration of the observed aa frequencies may help to predict the general functionality of a motif.

3
E3/19K-Mediated Inhibition of Antigen Presentation by ER Retention and ER Retrieval

One major mechanism of the immune system to control viral infections is based on recognition and elimination of infected cells by $CD8^+$ CTLs. CTLs recognize viral peptides presented by MHC class I molecules on the cell surface of infected cells (Pamer and Cresswell 1998). Recognition of the MHC antigen complex by the T-cell receptor of CTLs triggers the fusion of intracellular granules with the plasma membrane, releasing cytotoxicity-promoting factors, such as perforin and granzymes A and B. The latter are serine proteases that are able to induce apoptosis in a caspase-dependent and -independent manner (Trapani et al. 2000). Apart from this granule-mediated cell death, apoptosis of infected cells by CTLs and NK cells can also be triggered by the expression of proapoptotic factors, such as Fas ligand, TNF, and TRAIL (Nagata 1997; Degli-Esposti 1999). To ensure a continuous replication in the presence of these host mechanisms, viruses have evolved strategies to prevent the presentation of viral peptides on the cell surface and their recognition by CTLs (Burgert 1996; Pamer and Cresswell 1998; Yewdell and Bennink 1999; Alcami and Koszinowski 2000; Tortorella et al. 2000).

Antigenic peptides are predominantly generated by proteasomes in the cytosol and are translocated across the ER membrane by the trans-

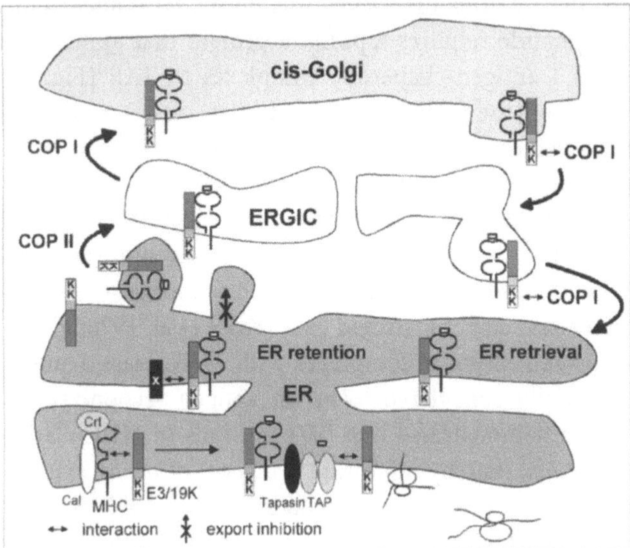

Fig. 3 Role of ER retention and retrieval for E3/19-mediated inhibition of antigen presentation. ER localization of E3/19K and E3/19K-MHC class I complexes seems to be achieved by both ER retention and retrieval mechanisms. E3/19K binds MHC class I molecules early after translocation, presumably during folding which is facilitated by chaperones such as calnexin (*Cal*) or calreticulin (*Crt*). For clarity, not all proteins interacting with MHC molecules during assembly (e.g., APLP2) are shown. ER retention involves the transmembrane domain, but the mechanism of how this domain contributes to ER retention is not yet understood. One possibility is that the TMD of E3/19K directly interacts with an ER-located factor X and thereby is retained and prevented from being packaged into ER export vesicles. Critical for retrieval from the ERGIC and *cis*-Golgi to the ER is a di-lysine motif (*KK*) at the carboxy-terminus of E3/19K which mediates the packaging of E3/19K into COPI-coated retrograde vesicles by direct interaction with COPI. TAP, transporter associated with antigen processing

porter associated with antigen processing (TAP) (Pamer and Cresswell 1998). In the ER lumen, these peptides might be further trimmed or even generated de novo by N-peptidases and the signal peptidase, respectively. Binding of the peptides to the MHC class I heavy chain/ β_2-microglobulin complex in the ER is assisted by a number of chaperones (Fig. 3), such as calnexin, calreticulin, and ERp57 (Yewdell and Bennink 1999; Burgert et al. 2002). Calnexin may facilitate the initial folding of the heavy chain, enabling β_2-microglobulin binding, whereas others are involved in correct intramolecular disulfide bond formation or folding of the $\alpha1$ and $\alpha2$ domains. Depending on the HLA allele, efficient transfer of the peptide requires tapasin, a protein that appears to tether the MHC class I antigen-chaperone complexes to TAP (Fig. 3). Peptide binding completes the folding process of MHC class I molecules and triggers their dissociation from TAP. Subsequently, they seem to be selectively recruited to ER exit sites (Spiliotis et al. 2000).

Ads invented the E3/19K protein to target the antigen presentation pathway. E3/19K was first described as a membrane glycoprotein that could be detected by nonspecific surface labels such as iodination on the cell surface of transformed and transfected cells (Kvist et al. 1978; Pääbo et al. 1983). As it formed physical complexes with MHC class I molecules, E3/19K was initially considered to be an antigen recognized by CTLs. Only later it was demonstrated that E3/19K binds to MHC class I antigens already in the ER, thereby inhibiting transport of newly synthesized MHC molecules to the cell surface (Andersson et al. 1985; Burgert and Kvist 1985). As a result, CTL recognition in vitro is drastically suppressed (Andersson et al. 1987; Burgert et al. 1987; Flomenberg et al. 1996). In vivo data generated in animal models strongly support an immunomodulatory function of the E3 region as a whole and E3/19K in particular. Lungs of cotton rats infected with wild-type Ad show a less severe immunopathology than those infected with a mutant Ad lacking E3/19K (Ginsberg et al. 1989). Moreover, E3/19K in combination with the other E3 proteins can prevent allograft rejection of transplanted pancreatic islets, and remarkably, suppresses virus-induced diabetes in a murine model (Efrat et al. 1995; von Herrath et al. 1997). Depending on the mouse strain and the experimental system used, integration of the Ad E3 region or E3/19K alone in Ad-based gene therapy vectors can prolong transgene expression (Bruder et al. 1997; Ilan et al. 1997; Schowalter et al. 1997).

With the exception of Ads from subgenus A and F, all human Ads express an E3/19K-like protein (Fig. 1). E3/19K-like proteins are type Ia transmembrane proteins with an N-terminal signal sequence (Table 3) (Burgert 1996). Depending on the number of N-linked glycans, the apparent molecular mass varies from 25 to 35 kDa (Deryckere and Burgert 1996). Although their function seems to be conserved (Pääbo et al. 1986a; Deryckere and Burgert 1996), the aa sequence homology of E3/19K proteins is surprisingly low (Burgert and Blusch 2000; Burgert et al. 2002). Only 20 of the 139–151 residues within the mature E3/19K proteins of subgenera B–E are strictly conserved. The capacity of E3/19K to

Table 3 Putative sorting motifs in the cytoplasmic tails of Ad E3 transmembrane proteins

Protein	Subgenus	Serotype	Signal seq.[a]	TMD[b]	TM protein type[c]	Potential sorting motif[d]
19K	B	Ad3	+	+	Ia	TMD 9 EKEKMP
	C	Ad2	+	+	Ia	TMD 9 DEKKMP
	D	Ad19a	+	+	Ia	TMD 9 EKEKLL
	E	Ad4	+	+	Ia	TMD 9 EKEKQP
16K	B	Ad3	+	+	Ia	TMD YTPCCAYLVILCC 8
	B	Ad7	+	+	Ia	TMD YIPCCAYLVILRC 8
	B	Ad35	+	+	Ia	TMD YTPCFTFLVVLWY 8
22K	D	Ad19a	+	+	Ia	TMD QLPCWVEIKIFIC 9
23K	E	Ad4	+	+	Ia	TMD YLPCFSYLVVLCC 9
6.7K	C	Ad2	–	+	Ib	TMD 5 CTHVCTYCQL 9
20.1K	B	Ad3	+	+	Ia	TMD 3 KKFQHKQDPLLNFNI
20.5K	B	Ad3	+	+	Ia	TMD 3 RKHEQKGDALLNFDI
24.8K	E	Ad4	+	+	Ia	–
49K	D	Ad19a	+	+	Ia	TMD 2 RKRPRAYNHMVDPLLSFSY
11.6K	C	Ad2	–	+	Ib	TMD 6 RARPPIYRPI 26
	C	Ad5	–	+	Ib	TMD 6 RARPPIYSPI 25
	C	Ad2	–	+	Ib	TMD 27 RLDGLKPCSLLLQYD
	C	Ad5	–	+	Ib	TMD 27 RLDGLKHMFFSLTV
9K	B1	Ad3	-	+	Ib	TMD 10 SKRRPIYSPM 11
	B1	Ad7	-	+	Ib	–
31.6K	D	Ad19a	+	+	Ia	TMD 4 KSRRPIYRPV 23
29.7K	E	Ad4	+	+	Ia	TMD 3 PNAKPVYKPI 24
31.6K	D	Ad19a	+	+	Ia	TMD 20 LQVEGGLRNLLFS 4
29.7K	E	Ad4	+	+	Ia	TMD 20 LQVEGGLRNLLFS 4
29.4K	A	Ad12	+	+	Ia	–
19.4K	F	Ad40	+	+	Ia	–
30.7K	A	Ad12	+	+	Ia	TMD 3 SLCVSKTEPLMPIPY
31.6K	F	Ad40	+	+	Ia	–

Table 3 (continued)

Protein	Subgenus	Serotype	Signal seq.[a]	TMD[b]	TM protein type[c]	Potential sorting motif[d]
10.4K	A	Ad12	+	+	Ia	TMD 18 RYQNPQIAA**LL**QLQP
	B	Ad3	+	+	Ia	TMD 18 EYRNQNVAA**LL**R**LI**
	C	Ad2	+	+	Ia	TMD 18 QYRDRTIAD**LL**R**IL**
	D	Ad19a	+	+	Ia	TMD 18 EYRDKNVAR**IL**R**LI**
	E	Ad4	+	+	Ia	TMD 18 QYRDQRVAQ**LL**R**LI**
	F	Ad40	+	+	Ia	TMD 18 QYRNHEVAT**LL**CLS
14.5K	A	Ad12	+	+	Ia	TMD **YARLNF** 45
	B	Ad3	+	+	Ia	TMD **YPTFNF** 54
	C	Ad2	+	+	Ia	TMD **YPYLDI** 51
	D	Ad19a	+	+	Ia	TMD **YPCFDL** 59
	E	Ad4	+	+	Ia	TMD **YPRFDF** 64
	F	Ad40	+	+	Ia	TMD **YGCLHL** 44
	A	Ad12	+	+	Ia	TMD 36 PPSVVS**Y**FK**F** 5
	B	Ad3	+	+	Ia	TMD 45 LPPAIS**Y**FN**L** 5
	C	Ad2	+	+	Ia	TMD 42 TPTEIS**Y**FN**L** 5
	D	Ad19a	+	+	Ia	TMD 50 PPSTVS**Y**FH**I** 5
	E	Ad4	+	+	Ia	TMD 55 MLPAIS**Y**FN**L** 5
	F	Ad40	+	+	Ia	TMD 35 APSVIS**Y**FH**L** 5
14.7K	A-F		–	–	–	
12.5K	A-E		–	–	–	
6.3K	E	Ad4	–	–	–	

[a] Signal peptide prediction was performed with the SignalP V1.1 program (Nielsen et al. 1997).

[b, c] Transmembrane domains and their orientation were predicted with the TMHMM program (http://www.cbs.dtu.dk/services/TMHMM), and proteins were classified according to the nomenclature of Singer (1990).

[d] Potential sorting motifs and flanking sequences are shown, and their distance (aa) to the transmembrane domain (TMD) and the carboxy-terminus are indicated by underlined numbers. Residues supposed or proven to be critical for function of a potential sorting motif are highlighted by bold characters.

bind MHC class I antigens resides primarily in the lumenal portion of the protein (Pääbo et al. 1986b; Hermiston et al. 1993; Sester and Burgert 1994). This portion of the Ad2 E3/19K protein forms two intramolecular disulfide bonds that are absolutely critical for HLA binding (Sester and Burgert 1994). All cysteines contributing to these disulfide bonds are strictly conserved. Other conserved aa are dispersed throughout the lumenal domain, and several were shown to be essential for E3/19K function (Deryckere and Burgert 1996; Burgert and Blusch 2000; Burgert et al. 2002; Sester et al., in preparation).

The Ad2 and Ad5 E3/19K proteins, and presumably also their homologues in Ads of other subgenera, bind the majority if not all human HLA alleles, albeit with different affinity (Beier et al. 1994; Körner and Burgert 1994; Burgert 1996). They are also capable of interacting with MHC alleles of other species; however, some murine MHC alleles (e.g., K^k, D^d) do not bind E3/19K and hence are not susceptible to its transport inhibition function (Burgert and Kvist 1987; Cox et al. 1990). Using hybrid MHC molecules containing domains from E3/19K-binding and nonbinding MHC alleles, the polymorphic $\alpha 1$ and $\alpha 2$ domains of MHC molecules comprising the peptide binding pocket have been identified as being essential for complex formation with E3/19K (Burgert and Kvist 1987; Beier et al. 1994; Feuerbach et al. 1994). Further characterization of the critical structural determinants using site-directed mutagenesis and antibody binding suggests that the contact site is formed, or at least influenced, by aa within the carboxy-terminal part of the $\alpha 2$ helix at the junction to $\alpha 3$ and the amino-terminal part of the $\alpha 1$ helix (Feuerbach et al. 1994; Flomenberg et al. 1994). Despite the vicinity of the proposed E3/19K attachment site to the peptide binding groove, there is no evidence as yet that E3/19K interferes with peptide binding (Cox et al. 1991; H.-G. Burgert et al., unpublished data). Based on the broad reactivity of E3/19K, the critical structure of HLA is thought to be rather conserved. Interestingly, the cocrystallization of HLA-A2 with the CMV protein US2, that also interferes with antigen presentation, revealed a similar or identical site for attachment at the junction of the peptide-binding region and the $\alpha 3$ domain (Gewurz et al. 2001). As E3/19K can be coprecipitated with calnexin and MHC class I heavy chains, the interaction between E3/19K and MHC molecules seems to occur soon after translocation, perhaps during binding of MHC molecules to calnexin (Sester et al. 2000). E3/19K can also bind to TAP independently of classical MHC molecules and therefore might affect efficient peptide loading of tapasin-dependent MHC alleles (Bennett et al. 1999). At present, it is unclear whether E3/19K also interferes with TAP function (Fig. 3). Interestingly, E3/19K increases complex formation of the MHC K^d molecule and possibly other alleles with the amyloid precursor-like protein 2, a protein previously implicated in peptide transfer to K^d (Feuerbach and Burgert 1993; Sester et al. 2000). Taken together, these finding suggest that E3/19K does not seem to grossly alter the assembly of MHC class I molecules but rather abrogates the export of completely assembled MHC antigens out of the ER/*cis*-Golgi compartment.

The mechanism of how E3/19K retains MHC molecules is incompletely understood, but seems to be based on its ability to localize to the ER. Deletion analysis of the cytoplasmic tail of E3/19K suggested that the last 6–8 aa are crucial for ER localization and E3/19K function (Nilsson et al. 1989; Cox et al. 1991). A fusion protein containing the lumenal portion and the TMD of the bona-fide plasma membrane protein CD8 and the C-tail of E3/19K localizes to the ER. Mutational analysis of this chimera revealed a di-lysine motif, KKXX or KXKXX, at positions −3 and −4 or −3 and −5 from the C-terminus, respectively, which appeared to act as a retrieval signal mediating retrograde transport from the ERGIC/*cis*-Golgi to the ER (Fig. 3) (Jackson et al. 1990, 1993). Monitoring mainly the effect on HLA transport also, a more complex structure within the C-tail of E3/19K was proposed to be responsible for ER retrieval (Gabathuler and Kvist 1990). Since this first description of an ER retrieval signal, many more have been identified in typical cellular ER proteins (Teasdale and Jackson 1996). Expression in mammalian cells of combinatorial libraries based on KKXX or KXKXX motifs gave rise to the whole range of predominantly ER- to predominantly plasma membrane-localized proteins (Zerangue et al. 2001). Thus, the efficiency of ER retrieval dramatically depends on the sequence context and strongly correlated with the efficiency of αCOP binding (see also Sect. 2.1).

All E3/19K-like proteins contain di-lysine-based ER retrieval signals (Burgert and Blusch 2000; Burgert et al. 2002). Interestingly, E3/19K proteins of subgenus C Ads display a KKXX motif, whereas those of subgenus B, D, and E Ads bear KXKXX motifs (Table 3). According to the data derived from the combinatorial approach, the efficacy for ER retrieval of the latter should generally be lower than that of KKXX-based motifs. Surprisingly, the combinations of aa that conferred ER localization most efficiently in this system are not found in the motifs of E3/19K-like proteins. Thus, either other features within E3/19K may contribute to this property, or efficient ER retrieval is not decisive for E3/19K function. As ERGIC-53, containing a KKFF motif, is not processed by *cis*-Golgi-specific enzymes when the overlapping ER export signal FF (Table 1) is mutated to AA, it was suggested that di-lysine motifs may exhibit in addition an ER retention function (Andersson et al. 1999). Therefore, the motifs found in Ad 19K-like proteins might mediate retrieval as well as retention. However, we recently observed that although mutation of the KK motif in the Ad2 E3/19K protein allows cell surface expression, the great majority of the protein and associated HLA remains in the ER. Only upon the additional replacement of the TMD of E3/19K by that of a bona-fide plasma membrane protein, is efficient cell surface expression observed. Thus, the transmembrane segment of E3/19K profoundly contributes to ER retention and thereby adds to the efficiency of HLA interaction and retention (Fig. 3; M. Sester, Z. Ruzsics, and H.-G. Burgert, manuscript in preparation). As discussed above, ER retention of transmembrane proteins can be achieved by charged aa or stretches of polar aa. The E3/19K TMDs contain an unusually high number of polar residues (T, S, C, Y), and this feature is well conserved among E3/19K proteins, suggesting functional importance (Burgert et al. 2002). Therefore, it seems likely that polar residues in the TMD of E3/19K mediate ER retention of the molecule. Although the detailed mechanism of ER retention of proteins with hydrophilic or charged aa remains unknown, it is striking that a viral protein has adopted a cellular ER quality control signal for retaining unassembled proteins of multimeric complexes in the ER to exert its immunomodulatory function.

In summary, the structure of E3/19K combines two modules, one for recruiting HLA molecules and a second one for localization in the ER. This latter feature appears to consist of two elements; (1) an ER retention signal contained in the transmembrane segment, and (2) an ER retrieval signal in the cytoplasmic tail which may act as a back-up system to increase the efficiency of retention.

4 Inhibition of Apoptosis by the E3/10.4–14.5K Complex: Rerouting of Plasma Membrane Receptors to Lysosomes for Degradation

Over the last few years it became clear that the adenovirus E3/10.4–14.5K proteins specifically modulate a set of plasma membrane receptors involved in apoptosis and growth control. The story began with the observation that the epidermal growth factor receptor (EGFR) was down-regulated from the cell surface during Ad infection (Carlin et al. 1989). The same group of investigators also reported that members of the class I and class II family of receptor tyrosine kinases, such as insulin growth factor-1 and insulin receptors, are affected albeit with reduced efficiency (Kuivinen et al. 1993). Initially, this effect was ascribed solely to the E3/10.4K protein (Carlin et al. 1989). Since then, several studies demonstrated that both the E3/10.4K and the 14.5K proteins are required for this modulation (Tollefson et al. 1991; Wold et al. 1995; Elsing and Burgert 1998). Presumably, the former data were misinterpreted due to the use of incompletely characterized virus insertion/deletion mutants which exhibited altered splicing of E3 RNAs and consequently altered expression of the E3/10.4K and 14.5K proteins.

In addition, several groups independently reported that a complex of 10.4–14.5K prevents apoptosis triggered by FasL or agonist Fas antibodies. This phenomenon is accompanied by down-modulation of the apoptosis receptor Fas from the cell surface of Ad-infected and E3-transfected cells (Shisler et al. 1997; Elsing and Burgert 1998; Tollefson et al. 1998). Fas mRNA levels are not significantly affected by 10.4–14.5K (Shisler et al. 1997), indicating that 10.4–14.5K interfere with a posttranscriptional process. Loss of cell surface Fas is accompanied by a similar decrease of total Fas levels. By contrast, in the presence of lysosomotropic agents, Fas accumulates in late endosomal/lysosomal vesicles (Elsing and Burgert 1998). This suggested that 10.4–14.5K induce internalization of the Fas receptor and its subsequent degradation in endosomes/lysosomes.

A similar mechanism was recently reported to be responsible for 10.4–14.5K-dependent down-modulation of TRAIL-R1/DR4, rendering Ad-infected cells resistant to TRAIL-induced apoptosis (Routes et al. 2000; Tollefson et al. 2001). In addition, 14.7K was implicated in inhibition of TRAIL-R1-induced apoptosis. By contrast, E3/10.4–14.5K do not suffice to modulate expression levels of the related TRAIL-R2/DR5 (S. Obermeier, unpublished data). Evidence has been presented that this process requires a third E3 protein, 6.7K (Benedict et al. 2001). Interestingly, two other members of the TNF/nerve growth factor receptor superfamily, murine TNFRI and human CD40, are not significantly modulated by subgenus C Ads (Shisler et al. 1997; Elsing and Burgert 1998). Thus, the 10.4–14.5K complex affects members of this superfamily differentially. The term down-regulation was originally coined to describe the process of ligand-induced internalization of activated receptors via coated pits and sorting in early endosomes which is followed by receptor degradation in lysosomes (Sorkin and Waters 1993). This process is thought to be a mechanism to attenuate or modulate ligand-induced signaling. Therefore, 10.4–14.5K-receptor interaction might be considered a surrogate ligand-receptor interaction aimed at blocking signaling of cell surface receptors by causing their removal from the cell surface and their degradation. The molecular basis of this interesting activity of 10.4–14.5K awaits further investigation.

The 10.4K and 14.5K proteins are both transmembrane proteins and are encoded by all serotypes of human adenoviruses (Fig. 1; Table 3). The sequence homology of 10.4K proteins of different subgenera is high (35%–72%, average 47.5%), and hydrophobic amino acids constitute about 50% of the 10.4K sequence. With the exception of the Ad40 and Ad41 homologues (subgenus F), 10.4K proteins have a length of 91 aa, and 15 aa are strictly conserved (Burgert and Blusch 2000; Burgert et al. 2002). Two 10.4K isoforms are expressed, one with the signal peptide sequence cleaved, whereas in the other it remains attached and seems to function as a second membrane anchor (Wold et al. 1995). The two 10.4K species are linked by a disulfide bond between the strictly conserved cysteine at position 31. One or both isoforms may associate noncovalently with the 14.5K protein (Tollefson et al. 1991; Stewart et al. 1995). 14.5K is a type Ia transmembrane protein (Table 3). 14.5K proteins consist of a short extracellular domain of variable length (20–40 aa), a transmembrane segment, and a cytoplasmic tail of 45–66 amino acids. The sequence homology between 14.5K proteins of different subgenera is comparably low (21–50%, average ~30%). In the mature protein, only nine (of 91–127) amino acids are strictly conserved (Burgert and Blusch 2000; Burgert et al. 2002).

To search for structural features with potential functional relevance, we sequenced 10.4K and 14.5K proteins of subgenera D and E, thereby allowing the sequence comparison of proteins from all subgenera. In doing so, we noticed that both proteins contain sequence elements in their cytoplasmic tails which conform to consensus YXXΦ and LL transport motifs, in 14.5K and 10.4K, respectively. In 14.5K, a YXXΦ motif consisting of a strictly conserved tyrosine and a bulky hydrophobic aa in position +3, either leucine (YXXL) in subgenera A, C, and F or F (YXXF) in subgenera B, D, and E, is conserved among all subgenera (Table 3, upper panel). However, according to the predicted membrane topology of

14.5K, the tyrosine may be located adjacent to or even within the lipid bilayer, which argues against a role of this motif in trafficking. Partially overlapping with this sequence, a second YXXΦ is found in subgenus C sequences only (Table 3, Ad2). Apart from the isoleucine in position +3, its surrounding amino acids are not typical for transport motifs and with a very short distance if any to the lipid bilayer, it should not be functional. This conclusion is supported by its restricted presence in Ads of a single subgenus. A third YXXΦ motif with the Y in position −9 from the C-terminus is strictly conserved and consists of the sequence YXXL in subgenera B, C, E, and F, YXXI in subgenus D, and YXXF in Ad12 (Table 3, lower panel). Remarkably, amino acids flanking the tyrosine are strictly conserved: a serine residue in position −1 and phenylalanine in +1. No preference for a phenylalanine at this position has been noted in our analysis, but serine in position −1 is somewhat enriched (1.5) in functional YXXΦ motifs (see Table 2 and references therein). The position close to the C-terminal end of the cytoplasmic tail should make it easily accessible for recognition by adaptor protein complexes and suggests that this motif might be functional.

Remarkably, the interaction partner of 14.5K, 10.4K, also contains potential sorting motifs in its cytoplasmic tail. Interestingly, these conform to the consensus of the aforementioned second large class of transport signals, the di-leucine motifs. Two leucines (LL), or an IL pair in 10.4K proteins of subgenus D, are present in position −4 and −5 from the C-terminus. If isoleucine is part of functional di-leucine motifs in mammalian proteins, it is more frequently found in position L2 than in L1; nonetheless, functionality of an IL transport signal was recently reported for the type II TGF-β receptor (Ehrlich et al. 2001). Comparing the sequence composition flanking the LL element of 10.4K with the aa environment of functional LL motifs, a number of preferred aa conserved in 10.4K proteins of subgenera B, C, D, and E are noted: R in +1 (2.1), A in −2 (1.8), R in −7 (**3.6**), and Y in −8 (2.7). Although Ad12 (subgenus A) lacks two of these favorable residues, it is still able to reduce Fas and the EGFR expression (H.-G. Burgert, unpublished data). The conserved position (−4/−5 from the C-terminus) of a potential endosomal/lysosomal sorting signal among 10.4K proteins of all subgenera suggests an important role for 10.4K function. Interestingly, the LLRIL sequence found in 10.4K proteins of subgenus C is also present in a target molecule of 10.4–14.5K, the EGFR. This sequence is located on the cytosolic portion of the EGFR adjacent to the TMD and has been described as being necessary for efficient sorting of ligand-receptor complexes in early endosomes en route to lysosomes (Kil et al. 1999; Kil and Carlin 2000). It was proposed that interaction of 10.4K with the EGFR might lead to oligomerization of di-leucine-type signals (Crooks et al. 2000), and this might be important for regulating the sorting activity of these signals in different compartments (Arneson and Miller 1995). In subgenera B–E, the last two aa of 10.4K (IL or LI) may also constitute a sorting motif. As leucines are rarely found preceding functional di-leucine motifs, the C-terminal aa are unlikely to represent functional transport motifs. A possible mechanistic picture becomes even more complicated, when we assume the involvement of a third E3 protein, E3/6.7K, in down-regulation of TRAIL-R2 by subgenus C Ads. The 6.7K protein lacks a classical amino-terminal signal sequence, but contains an internal hydrophobic region that may act as a signal-anchor domain (Wilson-Rawls and Wold 1993). Therefore, the N-terminus of 6.7K is located in the ER lumen, and the protein can be classified as a type Ib transmembrane protein (Table 3). Based on differential extraction into the detergent phase of Triton-X 114 and immunofluorescence staining, it was proposed that 6.7K is an ER transmembrane protein. Cell surface expression of a tagged 6.7K version that forms a complex with 10.4–14.5K indicates that the natural 6.7K might also be expressed at the cell surface (Benedict et al. 2001). In its short cytoplasmic tail, 6.7K contains the sequence element YXXL, conforming to the YXXΦ consensus (Table 3). Its surrounding sequence with a high content of Cys residues does not resemble any of the known functional transport motifs. It remains to be investigated why the modulation of TRAIL-R2 requires 6.7K in addition to 10.4–14.5K, whereas modulation of the closely related TRAIL-R1 or Fas and the EGFR are not dependent on the presence of 6.7K.

Recognition of transport motifs in the cytoplasmic tails of membrane proteins by adaptor protein complexes such as AP-1 or AP-2 (Kirchhausen 1999; Robinson and Bonifacino 2001) is a key element for sorting proteins into endosomes/lysosomes. Indeed, we recently showed that C-terminal peptides of the Ad2 14.5K and 10.4K proteins bind to AP-1 and AP-2 complexes in vitro. Binding of the adaptor proteins is abrogated by substituting alanine for the Y of the C-terminal YXXΦ motif in 14.5K and the LL in the tail peptide of 10.4K, respectively (A. Hilgendorf et al., manuscript in preparation). These data are underscored by expressing mutant proteins in vivo. Remarkably, both mutations abolish the capacity of the 10.4–14.5K complex to modulate surface expression levels of Fas and the EGFR. Concomitantly, cell surface expression levels of 14.5K were modified by either mutation, indicating that both motifs play a decisive role in regulating trafficking of the two proteins (A. Hilgendorf et al., manuscript in preparation).

Inhibition of cell surface transport by brefeldin A treatment of uninfected cells results in a steady decline of Fas from the cell surface. Comparing this kinetic with that seen upon infection shows that the loss of cell surface Fas upon Ad2 infection is much more rapid (Elsing and Burgert 1998). This suggests that 10.4–14.5K target Fas molecules on the cell surface or in a kinetically closely linked compartment such as early endosomes, and that the viral proteins do not deviate newly synthesized receptors directly from the TGN to the lysosome. By analogy, we assume a similar targeting mechanism for the EGFR and TRAIL-receptors. Consistent with the proposed mechanism, 10.4–14.5K are known to be expressed on the plasma membrane (Stewart et al. 1995). Down-regulation kinetics of the receptors in Ad-infected cells suggest that 10.4–14.5K either trigger receptor internalization or act subsequently to their autonomous internalization.

The model currently envisaged for 10.4–14.5K-mediated modulation of Fas is depicted in Fig. 4. We propose that a complex of E3/10.4–14.5K binds to Fas at the cell surface. Thus far, no evidence has been presented for a direct interaction, indicating that the interaction might be short-lived or possibly mediated by other cellular proteins ("x" in Fig. 4). The Fas-associated 14.5K protein, in complex with 10.4K, may then be recognized via its tyrosine-based sorting signal by the AP-2 adaptor, and the complex is recruited into coated pits and internalized into endosomes (1 in Fig. 4). Alternatively, the 10.4–14.5 complex might be internalized independent of its target molecules and interacts with these first in early endosomes (EE, 2). In support of this model, we found that mutation of the YXXΦ motif in 14.5K interferes with endocytosis of the 10.4–14.5K complex. Subsequently, 10.4–14.5K dissociate and may be recycled to the cell surface to bind newly synthesized Fas molecules, whereas Fas is targeted to lysosomes where it is degraded. Treating infected cells with lysosomotropic agents, such as chloroquine, ammonium chloride, or bafilomycin A1, an inhibitor of the vacuolar ATPase (Elsing and Burgert 1998; Tollefson et al. 1998), prevents degradation of Fas and leads to its accumulation predominantly in vesicles staining for lysosomal-associated membrane protein 2. Under these conditions, 14.5K exhibits only a limited colocalization with Fas (A. Hilgendorf et al., unpublished data).

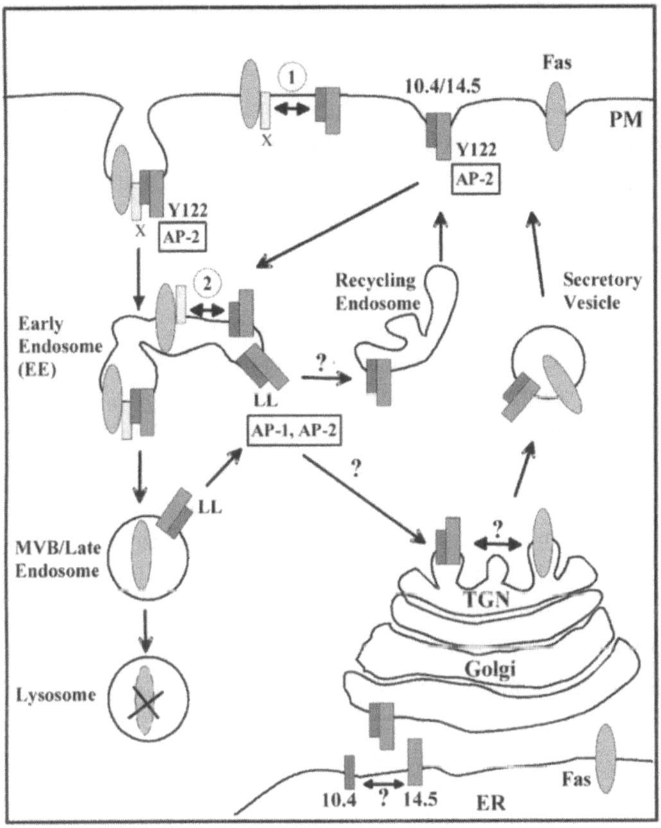

Fig. 4 Proposed mechanism for 10.4–14.5K-mediated down regulation of Fas: After synthesis, 10.4K and 14.5K associate with each other and are transported as a complex to the plasma membrane (PM), where they directly or indirectly, via an undefined cellular partner X, interact with Fas and induce endocytosis of Fas (1). Alternatively, interaction occurs first in early endosomes (2). Efficient down-modulation requires an intact $Y_{122}XX\Phi$ motif in 14.5K which may be bound by the clathrin adaptor AP-2. Interaction with 10.4–14.5K targets Fas for degradation in late endosomes/lysosomes, whereas 10.4–14.5K recycle to the PM via EE, the recycling endosome or the TGN. This process may depend on the LL motif in 10.4K and may involve AP-1

Therefore, a mechanism may exist for 10.4–14.5K to escape from the endocytic compartments and to be rescued from degradation in lysosomes. We think that the di-leucine motif in 10.4K might provide this function by sorting the 10.4–14.5K complex into a recycling pathway. At present, it is unclear whether 10.4–14.5K recycle from early or late endosomes to the TGN or to the plasma membrane. The dependence of such recycling pathways on the presence of transport motifs has been demonstrated (Fabbri et al. 1999). Among the multiple sorting signals characterized in MPR46, a di-leucine motif in the cytoplasmic tail is not only capable of mediating endocytosis but also fulfills a role in sorting the receptor from endosomes back to the TGN (Tikkanen et al. 2000). The two adaptor complexes, AP-1 and AP-2, that recognize the di-leucine motif in 10.4K, can be recruited to early endosomes and lysosomes by interacting with tyrosine and di-leucine motifs of multimeric receptor complexes (Marks et al. 1996; Sorkina et al. 1999). Therefore, the di-leucine motif in 10.4K might be necessary for the rescue of 10.4–14.5K from lysosomes or early/recycling or late endosomes.

The 10.4–14.5K-induced down-regulation of the EGFR occurs with the same kinetics as that of the constitutively internalized unoccupied EGFR (Hoffman and Carlin 1994). These data indicate that down-regulation of EGFR is caused by enhanced degradation of constitutively internalized receptors, which instead of being recycled accumulate in a prelysosomal compartment. In support of our model, cell fractionation and immunocytochemistry show that the 10.4K protein is associated with the EGFR in early endosomes. Whereas EGFRs proceed to lysosomes for degradation, 10.4K is retained on limiting membranes of endosome-to-lysosome transport intermediate compartments (Crooks et al. 2000).

5
Transport of a Novel Ad E3-Encoded Lymphocyte-Binding Factor to a Site of Proteolytic Processing Required for Secretion

We identified a novel ORF of 49K in the E3 region of Ad19a, a subgenus D Ad associated with epidemic keratoconjunctivitis (EKC) (Deryckere and Burgert 1996). Subsequently, we showed that the corresponding 49K protein is expressed by all subgenus D adenoviruses and is not restricted to EKC-causing serotypes of subgenus D (Blusch et al. 2002). Therefore, 49K may contribute to subgenus D-specific diseases. 49K is predicted to be a type Ia transmembrane protein with an N-terminal signal sequence (Table 3). This prediction was substantiated by experimental data. Ad19a 49K is highly glycosylated, containing N- as well as O-linked glycans, and migrates with a molecular mass of 80–100 kDa (Windheim and Burgert 2002). Synthesis of the 49K protein begins in the early phase of infection and continues throughout the infection cycle. Immunofluorescence studies localized the protein primarily in the Golgi/TGN, but the protein is also detected in early endosomes and on the plasma membrane. After treatment with lysosomotropic agents, 49K accumulated in swollen endosomal vesicles. Remarkably, the 49K protein is proteolytically processed and the N-terminal ectodomain is secreted (M. Windheim et al., manuscript in preparation). This is a completely novel phenomenon for Ad proteins. Based on the localization of the 49K ORF in the Ad E3 region, it is proposed that the protein has an immunomodulatory function. In support of this hypothesis, we found that the purified secreted protein binds to NK cells, but not to fibroblasts (M. Windheim, C. Falk, E. Kremmer, H.-G. Burgert, manuscript in preparation). Interestingly, the lumenal part of 49K contains an immunoglobulin-like domain. As these domains are frequently involved in protein-protein interactions within the immune system (Barclay et al. 1997), the presence of such a domain in 49K may explain the NK binding activity. Taking these findings together, it seems that we have identified for the first time a soluble Ad E3 encoded factor that may not act on infected cells but rather on cells of the immune system.

Interestingly, the 49K protein displays an internal repeat structure. Three imperfect repeats (Rs), comprising about 80 aa each, with a homology of 47% (R1/R2), 36% (R1/R3), and 29% (R2/R3), form the backbone of the lumenal/ectodomain (Deryckere and Burgert 1996). These repeats share significant homology to an 80 aa stretch within the lumenal domains (CR1=conserved region 1) of the Ad3 20.1K and 20.5K proteins. The homology of 20.1K CR1 to R1–3 is even higher (34%/27%/33%) than to 20.5K CR1 (22%). Another region of homology between these proteins comprises the TMD and flanking aa (CR2). The short cytoplasmic tail of 49K contains two putative transport motifs, a membrane-proximal YXXΦ motif, and a more distal LL motif. A similar LL motif is also found in the cytoplasmic tails of 20.1K and 20.5K. Thus, 49K shares some structural features with these subgenus B proteins; however, the overall homology is below 25%.

In the neighborhood of the 49K YXXΦ motif, residues overrepresented in functional motifs are found in positions −3 (P, 2.2) and −6 (R, 2.3). The LL motif is flanked by highly favored residues in positions −1 **(P, 4.6)** and +1 (S, 2.2). Notably the proline in −1 was observed in 23% of the functional LL motifs examined, suggesting that this motif might serve a sorting function. In support of this hypothesis, we showed by

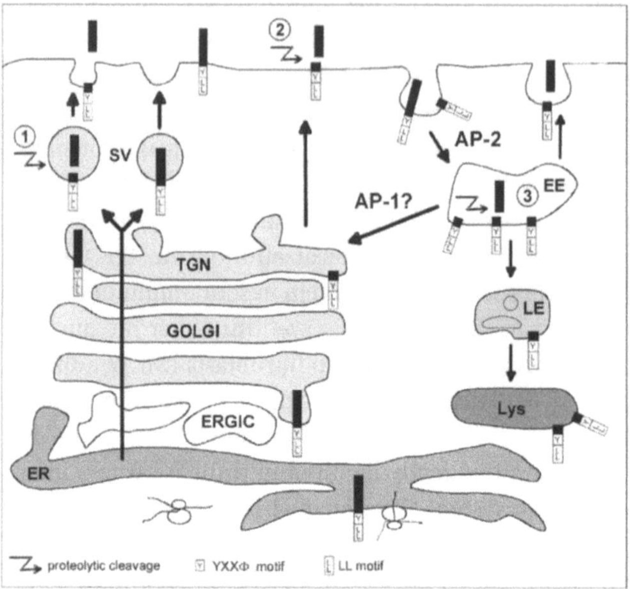

Fig. 5 Model of E3/49K trafficking and proteolytic processing. At steady state, E3/49K is found in the Golgi/TGN, at the plasma membrane, and in early endosomes. After synthesis, E3/49K is extensively modified with N- and O-glycans and intramolecular disulfide bonds. At the TGN, E3/49K is packaged into secretory vesicles. Proteolytic cleavage releasing the N-terminal ectodomain most likely occurs at the cell surface (*2*), but cleavage in secretory vesicles (*1*) or endosomes (*3*) cannot be excluded. The secreted N-terminal ectodomain can bind to NK cells. The cytoplasmic tail and uncleaved full-length E3/49K might be endocytosed by virtue of the LL and the YXXΦ motifs in the cytoplasmic tail which can interact with AP-2. One or both sorting motifs might also participate in a recycling pathway from early endosomes (*EE*) to the TGN, possibly by interacting with AP-1. A minor pathway may direct 49K and/or C-tail to late endosomes (*LE*) and lysosomes (*Lys*)

surface plasmon resonance analysis that a peptide comprising the entire cytoplasmic tail of 49K was able to interact with both AP-1 and AP-2 (M. Windheim, S. Hoening, H.-G. Burgert, unpublished observation). Whereas mutations of either one of the motifs reduced the affinity, binding was abolished after elimination of both motifs. These in vitro studies indicated that both potential sorting motifs in the C-tail of 49K may be functional. The current model for Ad19a E3/49K trafficking is depicted in Fig. 5. E3/49K enters the secretory pathway and is extensively post-translationally modified by formation of disulfide bonds and the addition of N- and O-glycans in the ER and the Golgi. At the TGN, it is packaged into secretory vesicles destined for the plasma membrane. Proteolytic processing of E3/49K seems to take place in a post-Golgi compartment, e.g., in secretory vesicles (1), at the plasma membrane (2), or in endosomes (3), as this step is inhibited by brefeldin A (BFA). The cleaved N-terminal domain of 49K cannot be detected in significant amounts intracellularly, suggesting that the time gap between cleavage and secretion is very short and that cleavage may occur at the plasma membrane (M. Windheim, unpublished observation). The C-terminal fragment seems to be internalized via the sorting motifs and may be delivered to late endosomes/lysosomes or may recycle to the TGN. Thus, the typical Golgi distribution might be based on trafficking dynamics rather than on active retention. Some data also indicate the existence of a minor pathway to lysosomes. Whereas early in infection, no colocalization with late endosomal/lysosomal markers was found, late in infection colocalization was detected. This phenomenon might be explained by an accumulation of C-tail in this compartment during the time course of infection, because its half-life is significantly longer than that of full-length 49K protein (Windheim and Burgert 2002).

6
E3 Transmembrane Proteins with Unknown Functions Containing Putative Trafficking Motifs in Their Cytoplasmic Tails

Until now, only functional activities for E3 proteins of subgenus C Ads have been described (see Sects. 1, 3, and 4). By screening all E3 transmembrane proteins for the presence of transport motifs (Table 3) and by incorporating the knowledge about the preferred sequence context, we attempted to obtain a very first hint regarding whether motifs present in E3 proteins of other subgenera may be of functional relevance.

6.1
The 20.1K and 20.5K Proteins of Subgenus B and the 24.8K Protein of Subgenus E

The 20.1K and 20.5K proteins are unique to subgenus B Ads (Signäs et al. 1986; Mei and Wadell 1992; Basler et al. 1996). Interestingly, the 20.5K proteins exhibit a homology to the 20.1K proteins of ~28%, suggesting that these proteins arose by gene duplication. The Ad4 24.8K protein also seems to belong to this family, with a homology to the Ad3 20.1K/20.5K proteins of 26% and 28%, respectively. As discussed above, there is also a significant homology to the 80-amino acid repeat units present in Ad19a 49K of subgenus D; however, the overall homology (17%) drops below the threshold considered to be significant. All family members are predicted to be type Ia transmembrane proteins containing a signal sequence (Table 3). For 20.5K this prediction was substantiated by experimental data (Hawkins et al. 1995; Hawkins and Wold 1995). The protein is expressed in the early and late phase of Ad3 infection and is N- and O-glycosylated. The intracellular localization and function are currently unknown. However, the cytoplasmic tails of 20.1K and 20.5K contain LL motifs which are conserved among subgenus B Ads. A prediction as to their functional relevance is difficult based on the residues surrounding the motifs. The Ad3 20.1K sequence has the favored proline in position −1 (**4.6**), whereas it is substituted by leucine in Ad11 and Ad35. Interestingly, sequences of the latter viruses contain a favored arginine in position +1 (2.1). For the 20.5K sequences the situation is similar; the B1 serotypes such as Ad3 show no favored aa, whereas the subgenus B2 Ads have the favored proline and arginine in positions −1 and +1, respectively. The aspartic acid in −2 fits the consensus proposed by Kirchhausen (1999). These findings and the fact that three (20.1K) and two (20.5K) of the four aa flanking the motif are identical to residues surrounding the functional LL motif in 49K may indicate that the motifs in the 20.1K/20.5K proteins are indeed functional. This is also supported by their distance (12–15 aa) to the TMD. Surprisingly, no such motif is found in the Ad4 24.8K, suggesting that these proteins may exhibit differential activities.

6.2
The 11.6K Protein of Subgenus C

The 11.6K protein is an atypical E3 protein in that it is only scarcely expressed early in infection but its expression increases dramatically (~400-fold) late in infection. This massive expression is achieved by splicing of the 11.6K coding region to a major late promoter-derived transcript (Tollefson et al. 1992). 11.6K proteins of subgenus C Ads are predicted to be type Ib transmembrane proteins with an internal hydrophobic stretch of amino acids that may function as a signal-anchor domain (Table 3). This prediction was supported by experimental data.

The Ad2 11.6K protein is modified at one site with a complex-type N-linked oligosaccharide (Scaria et al. 1992; Wold et al. 1995) and also contains O-linked sugars. Furthermore, 11.6K is palmitoylated involving two cysteine residues adjacent to the transmembrane domain (Hausmann et al. 1998). The role of this modification for 11.6K trafficking is unknown. Palmitoylation of the CD-MPR seems to be important for preventing its transport to and degradation in lysosomes (Table 3; Schweizer et al. 1996). 11.6K localizes to the ER, the Golgi, and the nuclear membrane (Scaria et al. 1992) and promotes cell death and the release of Ads from infected cells late in infection (Tollefson et al. 1996a,b). Therefore, 11.6K was named the "adenovirus death protein" (ADP). The 11.6K-mediated death of infected cells did not show the characteristic signs of apoptosis. As 11.6K is a subgenus C-specific E3 protein, this function, which is supposedly critical for efficient spread of the virus, is either lacking in other Ads or is represented by an unrelated protein. Based on its cell death-promoting properties, 11.6K might be applied in cancer gene therapy (Doronin et al. 2001).

In the cytoplasmic tail of 11.6K of Ad2, a YXXΦ and an LL motif can be recognized (Table 3). The YXXI motif exhibits remarkable similarities to sequences found in other functional YXXΦ motifs. Favored residues are located in positions −6 (R, 2.3), −3 (P, 2.2), −2 (P, 2.2), +1 (R, only in Ad2 and Ad6, 2.9), and +2 (P, 2.5). Thus, the YXXΦ motif is expected to represent a functional motif, although it is relatively close to the TMD (distance between TMD and Y=12 aa). It is not obvious at which step the YXXΦ motif may function to bring about the observed localization of 11.6K in the ER, Golgi, and the nuclear membrane, although this steady-state distribution may not reflect the trafficking dynamics of 11.6K. Therefore, it is not excluded that the 11.6K motif is required for some post-Golgi sorting step. Interestingly, a YXXΦ motif with a virtually identical sequence context is found in subgenus D 31.6K proteins and Ad3 9K (subgenus B1). The LL residues are not conserved among subgenus C viruses, and the surrounding sequence does not seem to resemble functional LL motifs.

6.3
The 9K Protein of Subgenus B1

The 9K protein of Ad3 (subgenus B) is like 11.6K of subgenus C predicted to be a type Ib transmembrane protein, but exhibits apart from a small stretch of 20 aa no significant homology to 11.6K (Fig. 1; Table 3). Subgenus B adenoviruses have been subdivided into two DNA homology clusters, B1 (Ad3, 7, 16, 21) and B2 (Ad11, 14, 34, 35) (Wadell 1984). Sequence analysis of the Ad11 and Ad35 E3 regions did not reveal an equivalent 9K ORF. Therefore, the 9K ORF seems to be characteristic for subgenus B1 Ads. The cytoplasmic tail of Ad3 but not that of Ad7 contains a YXXΦ motif that shares surrounding aa with those seen in subgenus C 11.6K (Table 3). Thus, the YXXΦ motif of Ad3 9K might be functional; however, no evidence that the gene is expressed has been provided to date.

6.4 The 30K Protein Family of Subgenera D and E

A compilation of 31K protein sequences of seven subgenus D serotypes (e.g., Ad19a 31.6K) and of the Ad4 29.7K protein (subgenus E) reveals a clear phylogenetic relationship, exhibiting an amino acid homology of 33%. The C-terminal 18 aa are even identical. Within a short stretch of amino acids in the cytoplasmic tail, there is also a very patchy homology to the 11.6K proteins, but its significance is unknown. The 31.6K proteins and Ad4 29.7K contain an N-terminal signal sequence and are predicted to be type Ia transmembrane proteins (Table 3). Both proteins are predicted to be heavily glycosylated, containing 9–10 (31K) and seven sites (Ad4 29.7K), respectively, for N-linked glycosylation and multiple sites for O-glycosylation. Experimentally the presence of high-mannose and/or hybrid-type carbohydrates has been verified (Li and Wold 2000). Similar to Ad2 11.6K, synthesis is low in the early phase and significantly higher in the late phase. Upon transfection of an expression vector, 29.7K is predominantly located in the ER and the nuclear membrane. Its localization in the viral context remains unknown. No data have been presented to date for the 31K proteins of subgenus D.

Like 49K, both proteins contain two potential sorting motifs in the cytoplasmic tail, one YXXΦ motif and an LL motif. The sequence context of the YXXΦ motifs resembles that of the YXXΦ motifs of subgenus C 11.6K and Ad3 9K (Table 3). However, the bulky aa in position +3 is not conserved but is isoleucine (29.7K, 11.6K), valine (31.6K), or methionine (9K). The functional relevance of these differences is unclear. In the subgenus D 31K proteins, residues overrepresented in functional YXXΦ motifs are found in positions −6 (K, 2.6), −3 (R, 2.2), −2 (P, 2.2), +1 (R,

2.9), and +2 (P, 2.5), and in the subgenus E 29.7K protein in positions −5 (N, 2.5), -2 (P, 2.2), and +2 (P, 2.5). The LL motifs in subgenus D 31K and subgenus E 29.7K are not flanked by favored aa and, therefore, they seem not to be functional.

6.5 The 16K–23K Protein Family of Subgenera B, D, and E

The subgenus B 16K proteins display a significant homology to the E3/22K proteins of subgenus D (26.7%) and the E3/23K protein of subgenus E (34.2%) (Fig. 1). The relevant proteins of subgenus B and E also show a weak homology to the C-terminal part of the 6.7K protein of subgenus C Ads (22–24.6% overall). All family members are predicted to be type Ia transmembrane proteins with an N-terminal signal sequence (Table 3). The 16K proteins of subgenus B Ads (Ad3, Ad7) have been characterized biochemically (Hawkins et al. 1995). Some evidence was provided that the putative signal sequence of 16K is not cleaved. Therefore, 16K is probably anchored in the membrane by two transmembrane regions. Although 16K contains two potential sites for N-glycosylation, no N-linked oligosaccharides could be detected. Ad7 16K is posttranslationally modified, although the nature of the modification is unknown. The 22K proteins of subgenus D and E contain five to eight potential sites for N-glycosylation.

In the 16K proteins of subgenus B and the 23K protein of subgenus E, a stretch of five aa is found which contains four overlapping putative sorting motifs, one YXXΦ motif and three LL motifs. Their position in close vicinity to or even within the TMD precludes a function in transport. This conclusion is confirmed after inspection of the surrounding aa which consist of bulky hydrophobic aa, such as leucine or valines. These types of aa are generally not found in positions +1 and +2 of YXXΦ motifs or in position −1, −2, +1, and +2 of LL motifs. Consistent with their proposed nonfunctional features, none of these motifs is found in the subgenus D 22K proteins.

6.6 The Subgenus A and F-Specific E3 Proteins

The E3 region of Ad12 of subgenus A encodes two proteins, 29.4K and 30.7K, which show similarities to the 19.4K (28.9%) and 30.4K (30.9%)

proteins of Ad40/41 of subgenus F, respectively, but not to any other E3 protein (Fig. 1). The A/F type of proteins are predicted to contain a signal sequence and to be type Ia transmembrane proteins (Table 3). Only in the cytoplasmic tail of Ad12 30.7K is a potential LL motif (leucine/methionine) found. The position in the cytoplasmic tail four aa from the C-terminus is identical to that of the LL motifs of subgenus B 20.1K and 20.5K and subgenus D 49K proteins. Also, the flanking aa support the notion that the motif is functional. Favored residues are found in positions −3 (T, 2.0), −1 (**P, 4.6**), and +1 (P, 2.5). Remarkable again is the presence of the highly favored proline in position −1.

In summary, from the 22 types of E3 proteins (19K, 10.4K, and 14.5K proteins are only counted once), 19 (86%) have all features of integral membrane proteins. Of those, 14 contain putative transport motifs, and 11 (50%) are predicted to have at least one functional motif (all except 6.7K and the 16–23K family). For four of these proteins (19K, 10.4K, 14.5K, and 49K), functionality has been experimentally confirmed. We hypothesize that (1) more E3 proteins will be discovered that exploit the intracellular trafficking machinery, and (2) some of these proteins may utilize their motifs to reroute their cellular targets into a degradative compartment for destruction.

7
Concluding Remarks and Perspectives

In the absence of an appropriate in vivo model and significant sequence homologies to known cellular proteins which would allow some functional predictions, the analysis of Ad E3 proteins in cell culture has proven to be a very valuable approach to elucidate their function. With this approach, much has been learned about the activities of Ad E3 proteins of subgenus C. New immunomodulatory functions were discovered: inhibition of antigen presentation through inhibition of HLA transport by E3/19K, and the removal of death receptor family members from the cell surface by 10.4–14.5K (+6.7K). These functions corroborated the initial concept that E3 proteins are generally devoted to subvert the host immune response. While the mechanistic details of these viral functions are being worked out, the search is now opened to also characterize E3 proteins of other subgenera and to identify new cellular target proteins. To this end, a new method has been developed that allows the easy manipulation of E3 genes from essentially all Ads (Z. Ruzsics et al., unpublished data). In the absence of further experimental data, we have inspected E3 sequences for putative sorting signals. The majority of E3 transmembrane proteins contains such motifs, and many of these are likely to be functional. This indicates that either these E3 proteins may reach a specific organelle to exert their function, or they utilize these motifs to redirect their targets into specific organelles, e.g., into lysosomes for degradation.

Many viruses have evolved mechanisms to subvert key functions of the immune system. The strategies employed by the individual viruses seem to be quite different. For example, to neutralize cytokines or proapoptotic soluble factors, some viruses such as vaccinia and pox viruses secrete a number of seemingly neutralizing cytokine-binding molecules or cytokine receptor homologues, whereas many herpes viruses express cytosolic proteins that interfere with the signaling cascade induced by the apoptosis receptors (e.g., FLIPs) (Tortorella et al. 2000). Adenoviruses seem to use mostly membrane proteins to counteract host immune mechanisms. Although the virus does not require cellular membranes for replication or budding, the cellular membrane system is preserved and is required to remove critical recognition structures of the immune system from the cell surface. The individual mechanism employed for each target is different: MHC molecules are removed by inhibition of ER export, whereas apoptosis receptors are removed by re-routing them into lysosomes. Both examples demonstrate how Ads have exploited the intracellular trafficking machinery for immune evasion. In this report, we show that many other E3 proteins contain putative transport motifs. Therefore, it can be expected that E3 proteins will provide more examples of deregulation and exploitation of intracellular trafficking pathways for manipulating immunologically important key molecules. Other viruses also use this strategy. For instance, the HCMV US3 protein binds to and retains HLA molecules in the ER in a similar way as the E3/19K protein, whereas the MCMV M6 protein escorts MHC molecules to lysosomes (Jones et al. 1996; Reusch et al. 1999). Perhaps the most prominent example is the HIV Nef protein that targets several cellular proteins, e.g., CD4 and MHC molecules, but in contrast to the E3 proteins acts from the cytosol. Nef is recruited to the lipid membranes by virtue of its N-terminal myristate and connects the cytoplasmic tail of MHC class I molecules with PACS-1 (Fig. 2). Thereby, MHC class I molecules are redirected to the TGN (Tortorella et al. 2000; Piguet and Trono 2001). Down-regulation of CD4 is mediated by bridging its tail to AP-2 and subsequently to a COPI subset in endosomes, resulting in its transport to and degradation in lysosomes. The HIV Nef protein also illustrates the marked consequences of these mechanisms: Nef is a major pathogenicity factor (Greenway et al. 2000). Therefore, we hypothesize that exploitation of cellular trafficking pathways by Ad E3 proteins might also contribute to Ad pathogenesis and disease (Burgert and Blusch 2000; Burgert et al. 2002).

Studies on the life cycle of intracellular bacteria have illustrated very nicely how these pathogens disrupt the cytoskeleton, signal-transduction pathways, and vesicle transport to block phagocytic uptake or direct its uptake into a protected niche (Knodler et al. 2001). Moreover, these studies provided useful tools for cell biologists. As each pathogen has its own lifestyle, many fascinating molecules have been discovered that are useful to dissect basic host processes. It seems that a direct link between viral immune evasion and the manipulation of intracellular transport pathways, as exemplified here for E3 proteins, is only beginning to be appreciated.

Acknowledgments. We appreciate the excellent technical assistance of Andrea Osterlehner. This work was supported by a grant from the Deutsche Forschungsgemeinschaft (DFG) to H.-G. Burgert (Bu 642/1).

References

Alcami A, Koszinowski UH (2000) Viral mechanisms of immune evasion. Immunol Today 21:447–455

Aguilar RC, Boehm M, Gorshkova I, Crouch RJ, Tomita K, Saito T, Ohno H, Bonifacino JS (2001) Signal-binding specificity of the mu4 subunit of the adaptor protein complex AP-4. J Biol Chem 276:13145–13152

Anderson RG (1998) The caveolae membrane system. Annu Rev Biochem 67:199–225

Andersson AM, Melin L, Bean A, Pettersson RF (1997) A retention signal necessary and sufficient for Golgi localization maps to the cytoplasmic tail of a Bunyaviridae (Uukuniemi virus) membrane glycoprotein. J Virol 71:4717–4727

Andersson H, Kappeler F, Hauri HP (1999) Protein targeting to endoplasmic reticulum by dilysine signals involves direct retention in addition to retrieval. J Biol Chem 274:15080–15084

Andersson M, McMichael A, Peterson PA (1987) Reduced allorecognition of adenovirus-2 infected cells. J Immunol 138:3960–3966

Andersson M, Pääbo S, Nilsson T, Peterson PA (1985) Impaired intracellular transport of class I MHC antigens as a possible means for adenoviruses to evade immune surveillance. Cell 43:215–222

Arnberg N, Edlund K, Kidd AH, Wadell G (2000) Adenovirus type 37 uses sialic acid as a cellular receptor. J Virol 74:42–48

Arneson LS, Miller J (1995) Efficient endosomal localization of major histocompatibility complex class II-invariant chain complexes requires multimerization of the invariant chain targeting sequence. J Cell Biol 129:1217–1228

Banting G, Ponnambalam S (1997) TGN38 and its orthologues: roles in post-TGN vesicle formation and maintenance of TGN morphology. Biochim Biophys Acta 1355:209–217

Barclay AN, Brown MH, Law SKA, McNight AJ, Tomlinson MG, van der Merwe PA (1997) *The Leukocyte Antigen Factsbook*. Academic Press, San Diego, CA, USA

Barlowe C (2000) Traffic COPs of the early secretory pathway. Traffic 1:371–377

Barr FA (1999) A novel Rab6-interacting domain defines a family of Golgi-targeted coiled-coil proteins. Curr Biol 9:381–384

Basler CF, Droguett G, Horwitz MS (1996) Sequence of the immunoregulatory early region-3 and flanking sequences of adenovirus type-35. Gene 170:249–254

Beier DC, Cox JH, Vining DR, Cresswell P, Engelhard VH (1994) Association of human class I MHC alleles with the adenovirus E3/19K protein. J Immunol 152:3862–3872

Benedict CA, Norris PS, Prigozy TI, Bodmer JL, Mahr JA, Garnett CT, Martinon F, Tschopp J, Gooding LR, Ware CF (2001) Three adenovirus E3 proteins cooperate to evade apoptosis by tumor necrosis factor-related apoptosis-inducing ligand receptor-1 and -2. J Biol Chem 276:3270–3278

Bennett EM, Bennink JR, Yewdell JW, Brodsky FM (1999) Cutting edge: adenovirus E19 has two mechanisms for affecting class I MHC expression. J Immunol 162:5049–5052

Bergelson JM, Cunningham JA, Droguett G, Kurt Jones EA, Krithivas A, Hong JS, Horwitz MS, Crowell RL, Finberg RW (1997) Isolation of a common receptor for Coxsackie B viruses and adenoviruses 2 and 5. Science 275:1320–1323

Biron CA, Brossay L (2001) NK cells and NKT cells in innate defense against viral infections. Curr Opin Immunol 13:458–464

Black MW, Pelham HR (2001) Membrane traffic: How do GGAs fit in with the adaptors? Curr Biol 11: R460-R462

Blusch J, Deryckere F, Windheim M, Ruzsics Z, Arnberg N, Adrian T, Burgert H-G (2002) The novel early region 3 protein E3/49K is specifically expressed by adenoviruses of subgenus D: Implications for epidemic keratokonjunctivitis and adenovirus evolution. Virology 296:94–106

Boll W, Ohno H, Songyang Z, Rapoport I, Cantley LC, Bonifacino JS, Kirchhausen T (1996) Sequence requirements for the recognition of tyrosine-based endocytic signals by clathrin AP-2 complexes. EMBO J 15:5789–5795

Bonifacino JS, Dell-Angelica EC (1999) Molecular bases for the recognition of tyrosine-based sorting signals. J Cell Biol 145:923–926

Brodsky JL (1998) Translocation of proteins across the endoplasmic reticulum membrane. Int Rev Cytol 178:277–328

Bruder JT, Jie T, McVey DL, Kovesdi I (1997) Expression of gp19K increases the persistence of transgene expression from an adenovirus vector in the mouse lung and liver. J Virol 71:7623–7628

Burgert H-G (1996) Subversion of the MHC class I antigen-presentation pathway by adenoviruses and herpes simplex viruses. Trends Microbiol 4:107–112

Burgert H-G, Blusch JH (2000) Immunomodulatory functions encoded by the E3 transcription unit of adenoviruses. Virus Genes 21:13–25

Burgert H-G, Kvist S (1985) An adenovirus type 2 glycoprotein blocks cell surface expression of human histocompatibility class I antigens. Cell 41:987–997

Burgert H-G, Kvist S (1987) The E3/19K protein of adenovirus type 2 binds to the domains of histocompatibility antigens required for CTL recognition. EMBO J 6:2019–2026

Burgert H-G, Maryanski JL, Kvist S (1987) "E3/19K" protein of adenovirus type 2 inhibits lysis of cytolytic T lymphocytes by blocking cell-surface expression of histocompatibility class I antigens. Proc Natl Acad Sci USA 84:1356–1360

Burgert H-G, Ruzsics Z, Obermeier S, Hilgendorf A, Windheim M, Elsing A (2002) Subversion of host defense mechanisms by adenoviruses. Curr Top Microbiol Immunol, 269:273–318

Calvo PA, Frank DW, Bieler BM, Berson JF, Marks MS (1999) A cytoplasmic sequence in human tyrosinase defines a second class of di-leucine-based sorting signals for late endosomal and lysosomal delivery. J Biol Chem 274:12780–12789

Carlin CR, Tollefson AE, Brady HA, Hoffman BL, Wold WS (1989) Epidermal growth factor receptor is down-regulated by a 10,400 MW protein encoded by the E3 region of adenovirus. Cell 57:135–144

Carroll KS, Hanna J, Simon I, Krise J, Barbero P, Pfeffer SR (2001) Role of Rab9 GTPase in facilitating receptor recruitment by TIP47. Science 292:1373–1376

Chen HJ, Yuan J, Lobel P (1997) Systematic mutational analysis of the cation-independent mannose 6- phosphate/insulin-like growth factor II receptor cytoplasmic domain. An acidic cluster containing a key aspartate is important for function in lysosomal enzyme sorting. J Biol Chem 272:7003–7012

Chen YA, Scheller RH (2001) SNARE-mediated membrane fusion. Nat Rev Mol Cell Biol 2:98–106

Collawn JF, Stangel M, Kuhn LA, Esekogwu V, Jing SQ, Trowbridge IS, Tainer JA (1990) Transferrin receptor internalization sequence YXRF implicates a tight turn as the structural recognition motif for endocytosis. Cell 63:1061–1072

Copier J, Kleijmeer MJ, Ponnambalam S, Oorschot V, Potter P, Trowsdale J, Kelly A (1996) Targeting signal and subcellular compartments involved in the intracellular trafficking of HLA-DMB. J Immunol 157:1017–1027

Cosson P, Letourneur F (1994) Coatomer interaction with di-lysine endoplasmic reticulum retention motifs. Science 263:1629–1631

Cox JH, Bennink JR, Yewdell JW (1991) Retention of adenovirus E19 glycoprotein in the endoplasmic reticulum is essential to its ability to block antigen presentation. J Exp Med 174:1629–1637

Cox JH, Yewdell JW, Eisenlohr LC, Johnson PR, Bennink JR (1990) Antigen presentation requires transport of MHC class I molecules from the endoplasmic reticulum. Science 247:715–718

Crooks D, Kil SJ, McCaffery JM, Carlin C (2000) E3-13.7 integral membrane proteins encoded by human adenoviruses alter epidermal growth factor receptor trafficking by interacting directly with receptors in early endosomes. Mol Biol Cell 11:3559–3572

Crump CM, Xiang Y, Thomas L, Gu F, Austin C, Tooze SA, Thomas G (2001) PACS-1 binding to adaptors is required for acidic cluster motif- mediated protein traffic. EMBO J 20:2191–2201

De Jong JC, Wermenbol AG, Verweij-Uijterwaal MW, Slaterus KW, Wertheim-Van Dillen P, Van Doornum GJ, Khoo SH, Hierholzer JC (1999) Adenoviruses from human immunodeficiency virus-infected individuals, including two strains that represent new candidate serotypes Ad50 and Ad51 of species B1 and D, respectively. J Clin Microbiol 37:3940–3945

Degli-Esposti M (1999) To die or not to die–the quest of the TRAIL receptors. J Leukoc Biol 65:535–542

Dell'Angelica EC, Shotelersuk V, Aguilar RC, Gahl WA, Bonifacino JS (1999) Altered trafficking of lysosomal proteins in Hermansky-Pudlak syndrome due to mutations in the beta 3A subunit of the AP-3 adaptor. Mol Cell 3:11–21

Deryckere F, Burgert H-G (1996) Early region 3 of adenovirus type 19 (subgroup D) encodes an HLA- binding protein distinct from that of subgroups B and C. J Virol 70:2832–2841

Dittrich E, Haft CR, Muys L, Heinrich PC, Graeve L (1996) A di-leucine motif and an upstream serine in the interleukin-6 (IL-6) signal transducer gp130 mediate ligand-induced endocytosis and down- regulation of the IL-6 receptor. J Biol Chem 271:5487–5494

Doronin K, Kuppuswamy M, Toth K, Tollefson AE, Krajcsi P, Krougliak V, Wold WS (2001) Tissue-specific, tumor-selective, replication-competent adenovirus vector for cancer gene therapy. J Virol 75:3314–3324

Efrat S, Fejer G, Brownlee M, Horwitz MS (1995) Prolonged survival of pancreatic islet allografts mediated by adenovirus immunoregulatory transgenes. Proc Natl Acad Sci USA 92:6947–6951

Ehrlich M, Shmuely A, Henis YI (2001) A single internalization signal from the dileucine family is critical for constitutive endocytosis of the type II TGF-beta receptor. J Cell Sci 114:1777–1786

Ellgaard L, Molinari M, Helenius A (1999) Setting the standards: quality control in the secretory pathway. Science 286:1882–1888

Elsing A, Burgert H-G (1998) The adenovirus E3/10.4K-14.5K proteins down-modulate the apoptosis receptor Fas/Apo-1 by inducing its internalization. Proc Natl Acad Sci USA 95:10072–10077

Fabbri M, Fumagalli L, Bossi G, Bianchi E, Bender JR, Pardi R (1999) A tyrosine-based sorting signal in the beta2 integrin cytoplasmic domain mediates its recycling to the plasma membrane and is required for ligand-supported migration. EMBO J 18:4915–4925

Feuerbach D, Burgert H-G (1993) Novel proteins associated with MHC class I antigens in cells expressing the adenovirus protein E3/19K. EMBO J 12:3153–3161

Feuerbach D, Etteldorf S, Ebenau-Jehle C, Abastado JP, Madden D, Burgert H-G (1994) Identification of amino acids within the MHC molecule important for the interaction with the adenovirus protein E3/19K. J Immunol 153:1626–1636

Flomenberg P, Gutierrez E, Hogan KT (1994) Identification of class I MHC regions which bind to the adenovirus E3–19K protein. Mol Immunol 31:1277–1284
Flomenberg P, Piaskowski V, Truitt RL, Casper JT (1996) Human adenovirus-specific CD8+ T-cell responses are not inhibited by E3–19K in the presence of gamma interferon. J Virol 70:6314–6322
Francis MJ, Jones EE, Levy ER, Martin RL, Ponnambalam S, Monaco AP (1999) Identification of a di-leucine motif within the C terminus domain of the Menkes disease protein that mediates endocytosis from the plasma membrane. J Cell Sci 112:1721–1732
Gabathuler R, Kvist S (1990) The endoplasmic reticulum retention signal of the E3/19K protein of adenovirus type 2 consists of three separate amino acid segments at the carboxy terminus. J Cell Biol 111:1803–1810
Geisler C, Dietrich J, Nielsen BL, Kastrup J, Lauritsen JP, Odum N, Christensen MD (1998) Leucine-based receptor sorting motifs are dependent on the spacing relative to the plasma membrane. J Biol Chem 273:21316–21323
Gewurz BE, Gaudet R, Tortorella D, Wang EW, Ploegh HL, Wiley DC (2001) Antigen presentation subverted: Structure of the human cytomegalovirus protein US2 bound to the class I molecule HLA-A2. Proc Natl Acad Sci USA 98:6794–6799
Ginsberg HS, Lundholm-Beauchamp U, Horswood RL, Pernis B, Wold WS, Chanock RM, Prince GA (1989) Role of early region 3 (E3) in pathogenesis of adenovirus disease. Proc Natl Acad Sci USA 86:3823–3827
Glick BS, Malhotra V (1998) The curious status of the Golgi apparatus. Cell 95:883–889
Greenway AL, Holloway G, McPhee DA (2000) HIV-1 Nef: a critical factor in viral-induced pathogenesis. Adv Pharmacol 48:299–343
Gu F, Gruenberg J (1999) Biogenesis of transport intermediates in the endocytic pathway. FEBS Lett 452:61–66
Harty JT, Tvinnereim AR, White DW (2000) CD8+ T cell effector mechanisms in resistance to infection. Annu Rev Immunol 18:275–308
Hauri HP, Kappeler F, Andersson H, Appenzeller C (2000) ERGIC-53 and traffic in the secretory pathway. J Cell Sci 113:587–596
Hausmann J, Ortmann D, Witt E, Veit M, Seidel W (1998) Adenovirus death protein, a transmembrane protein encoded in the E3 region, is palmitoylated at the cytoplasmic tail. Virology 244:343–351
Hawkins LK, Wilson-Rawls J, Wold WS (1995) Region E3 of subgroup B human adenoviruses encodes a 16-kilodalton membrane protein that may be a distant analog of the E3–6.7K protein of subgroup C adenoviruses. J Virol 69:4292–4298
Hawkins LK, Wold WS (1995) The E3–20.5K membrane protein of subgroup B human adenoviruses contains O-linked and complex N-linked oligosaccharides. Virology 210:335–344
Heilker R, Spiess M, Crottet P (1999) Recognition of sorting signals by clathrin adaptors. Bioessays 21:558–567
Hermiston TW, Tripp RA, Sparer T, Gooding LR, Wold WS (1993) Deletion mutation analysis of the adenovirus type 2 E3-gp19K protein: identification of sequences within the endoplasmic reticulum lumenal domain that are required for class I antigen binding and protection from adenovirus-specific cytotoxic T lymphocytes. J Virol 67:5289–5298

Hicke L (1999) Gettin' down with ubiquitin: turning off cell-surface receptors, transporters and channels. Trends Cell Biol 9:107–112
Hinshaw JE (2000) Dynamin and its role in membrane fission. Annu Rev Cell Dev Biol 16:483–519
Hoffman P, Carlin C (1994) Adenovirus E3 protein causes constitutively internalized epidermal growth factor receptors to accumulate in a prelysosomal compartment, resulting in enhanced degradation. Mol Cell Biol 14:3695–3706
Höning S, Sandoval IV, von Figura K (1998) A di-leucine-based motif in the cytoplasmic tail of LIMP-II and tyrosinase mediates selective binding of AP-3. EMBO J 17:1304–1314
Horwitz MS (2001) Adenovirus immunoregulatory genes and their cellular targets. Virology 279:1–8
Hunziker W, Fumey C (1994) A di-leucine motif mediates endocytosis and basolateral sorting of macrophage IgG Fc receptors in MDCK cells. EMBO J 13:2963–2967
Ilan Y, Droguett G, Chowdhury NR, Li Y, Sengupta K, Thummala NR, Davidson A, Chowdhury JR, Horwitz MS (1997) Insertion of the adenoviral E3 region into a recombinant viral vector prevents antiviral humoral and cellular immune responses and permits long-term gene expression. Proc Natl Acad Sci USA 94:2587–2592
Jackson MR, Nilsson T, Peterson PA (1990) Identification of a consensus motif for retention of transmembrane proteins in the endoplasmic reticulum. EMBO J 9:3153–3162
Jackson MR, Nilsson T, Peterson PA (1993) Retrieval of transmembrane proteins to the endoplasmic reticulum. J Cell Biol 121:317–333
Johnson AO, Ghosh RN, Dunn KW, Garippa R, Park J, Mayor S, Maxfield FR, McGraw TE (1996) Transferrin receptor containing the SDYQRL motif of TGN38 causes a reorganization of the recycling compartment but is not targeted to the TGN. J Cell Biol 135:1749–1762
Jones TR, Wiertz EJ, Sun L, Fish KN, Nelson JA, Ploegh HL (1996) Human cytomegalovirus US3 impairs transport and maturation of major histocompatibility complex class I heavy chains. Proc Natl Acad Sci USA 93:11327–11333
Keller P, Toomre D, Diaz E, White J, Simons K (2001) Multicolour imaging of post-Golgi sorting and trafficking in live cells. Nat Cell Biol 3:140–149
Kil SJ, Carlin C (2000) EGF receptor residues leu(679), leu(680) mediate selective sorting of ligand-receptor complexes in early endosomal compartments. J Cell Physiol 185:47–60
Kil SJ, Hobert M, Carlin C (1999) A leucine-based determinant in the epidermal growth factor receptor juxtamembrane domain is required for the efficient transport of ligand- receptor complexes to lysosomes. J Biol Chem 274:3141–3150
Kirchhausen T (1999) Adaptors for clathrin-mediated traffic. Annu Rev Cell Dev Biol 15:705–732
Kirchhausen T (2000a) Clathrin. Annu Rev Biochem 69:699–727
Kirchhausen T (2000b) Three ways to make a vesicle. Nat Rev Mol Cell Biol 1:187–198
Kjer-Nielsen L, Teasdale RD, van Vliet C, Gleeson PA (1999) A novel Golgi-localisation domain shared by a class of coiled-coil peripheral membrane proteins. Curr Biol 9:385–388

Klumperman J (2000) Transport between ER and Golgi. Curr Opin Cell Biol 12:445–449

Knodler LA, Celli J, Finlay BB (2001) Pathogenic Trickery: Deception of host cell processes. Nat Rev Mol Cell Biol 2:578–588

Körner H, Burgert HG (1994) Down-regulation of HLA antigens by the adenovirus type 2 E3/19K protein in a T-lymphoma cell line. J Virol 68:1442–1448

Kuivinen E, Hoffman BL, Hoffman PA, Carlin CR (1993) Structurally related class I and class II receptor protein tyrosine kinases are down-regulated by the same E3 protein coded for by human group C adenoviruses. J Cell Biol 120:1271–1279

Kvist S, Östberg L, Persson H, Philipson L, Peterson PA (1978) Molecular association between transplantation antigens and a cell surface antigen in an adenovirus-transformed cell line. Proc Natl Acad Sci USA 75:5674–5678

Le Borgne R, Alconada A, Bauer U, Hoflack B (1998) The mammalian AP-3 adaptor-like complex mediates the intracellular transport of lysosomal membrane glycoproteins. J Biol Chem 273:29451–29461

Letourneur F, Klausner RD (1992) A novel di-leucine motif and a tyrosine-based motif independently mediate lysosomal targeting and endocytosis of CD3 chains. Cell 69:1143–1157

Lewis MJ, Sweet DJ, Pelham HR (1990) The ERD2 gene determines the specificity of the luminal ER protein retention system. Cell 61:1359–1363

Li Y, Wold WS (2000) Identification and characterization of a 30 K protein (Ad4E3-30 K) encoded by the E3 region of human adenovirus type 4. Virology 273:127–138

Lin S, Naim HY, Roth MG (1997) Tyrosine-dependent basolateral sorting signals are distinct from tyrosine-dependent internalization signals. J Biol Chem 272:26300–26305

Lippincott-Schwartz J, Roberts TH, Hirschberg K (2000) Secretory protein trafficking and organelle dynamics in living cells. Annu Rev Cell Dev Biol 16:557–589

Locker JK, Klumperman J, Oorschot V, Horzinek MC, Geuze HJ, Rottier PJ (1994) The cytoplasmic tail of mouse hepatitis virus M protein is essential but not sufficient for its retention in the Golgi complex. J Biol Chem 269:28263–28269

Mahr JA, Gooding LR (1999) Immune evasion by adenoviruses. Immunol Rev 168:121–130

Marks MS, Woodruff L, Ohno H, Bonifacino JS (1996) Protein targeting by tyrosine- and di-leucine-based signals: evidence for distinct saturable components. J Cell Biol 135:341–354

Martinez-Menarguez JA, Geuze HJ, Slot JW, Klumperman J (1999) Vesicular tubular clusters between the ER and Golgi mediate concentration of soluble secretory proteins by exclusion from COPI- coated vesicles. Cell 98:81–90

Mauxion F, Le Borgne R, Munier-Lehmann H, Hoflack B (1996) A casein kinase II phosphorylation site in the cytoplasmic domain of the cation-dependent mannose 6-phosphate receptor determines the high affinity interaction of the AP-1 Golgi assembly proteins with membranes. J Biol Chem 271:2171–2178

Mei YF, Wadell G (1992) The nucleotide sequence of adenovirus type 11 early 3 region: comparison of genome type Ad11p and Ad11a. Virology 191:125–133

Mellman I, Warren G (2000) The road taken: past and future foundations of membrane traffic. Cell 100:99–112

Meyer C, Zizioli D, Lausmann S, Eskelinen EL, Hamann J, Saftig P, von Figura K, Schu P (2000) mu1A-adaptin-deficient mice: lethality, loss of AP-1 binding and rerouting of mannose 6-phosphate receptors. EMBO J 19:2193–2203

Molloy SS, Anderson ED, Jean F, Thomas G (1999) Bi-cycling the furin pathway: from TGN localization to pathogen activation and embryogenesis. Trends Cell Biol 9:28–35

Muniz M, Morsomme P, Riezman H (2001) Protein sorting upon exit from the endoplasmic reticulum. Cell 104:313–320

Munro S (1998) Localization of proteins to the Golgi apparatus. Trends Cell Biol 8:11–15

Munro S, Nichols BJ (1999) The GRIP domain—a novel Golgi-targeting domain found in several coiled-coil proteins. Curr Biol 9:377–380

Nagata S (1997) Apoptosis by death factor. Cell 88:355–365

Nemerow GR (2000) Cell receptors involved in adenovirus entry. Virology 274:1-4

Nielsen H, Engelbrecht J, Brunak S, von Heijne G (1997) Identification of prokaryotic and eukaryotic signal peptides and prediction of their cleavage sites. Protein Eng 10:1-6

Nielsen MS, Madsen P, Christensen EI, Nykjaer A, Gliemann J, Kasper D, Pohlmann R, Petersen CM (2001) The sortilin cytoplasmic tail conveys Golgi-endosome transport and binds the VHS domain of the GGA2 sorting protein. EMBO J 20:2180–2190

Nilsson T, Jackson M, Peterson PA (1989) Short cytoplasmic sequences serve as retention signals for transmembrane proteins in the endoplasmic reticulum. Cell 58:707–718

Ohno H, Aguilar RC, Yeh D, Taura D, Saito T, Bonifacino JS (1998) The medium subunits of adaptor complexes recognize distinct but overlapping sets of tyrosine-based sorting signals. J Biol Chem 273:25915–25921

Ohno H, Fournier MC, Poy G, Bonifacino JS (1996) Structural determinants of interaction of tyrosine-based sorting signals with the adaptor medium chains. J Biol Chem 271:29009–29015

Owen DJ, Evans PR (1998) A structural explanation for the recognition of tyrosine-based endocytotic signals. Science 282:1327–1332

Pääbo S, Nilsson T, Peterson PA (1986a) Adenoviruses of subgenera B, C, D, and E modulate cell-surface expression of major histocompatibility complex class I antigens. Proc Natl Acad Sci USA 83:9665–9669

Pääbo S, Weber F, Kämpe O, Schaffner W, Peterson PA (1983) Association between transplantation antigens and a viral membrane protein synthesized from a mammalian expression vector. Cell 33:445–453

Pääbo S, Weber F, Nilsson T, Schaffner W, Peterson PA (1986b) Structural and functional dissection of an MHC class I antigen-binding adenovirus glycoprotein. EMBO J 5:1921–1927

Pamer E, Cresswell P (1998) Mechanisms of MHC class I-restricted antigen processing. Annu Rev Immunol 16:323–358

Pelham HR, Rothman JE (2000) The debate about transport in the Golgi–two sides of the same coin? Cell 102:713–719

Petris MJ, Camakaris J, Greenough M, LaFontaine S, Mercer JF (1998) A C-terminal di-leucine is required for localization of the Menkes protein in the trans-Golgi network. Hum Mol Genet 7:2063–2071

Pfeffer SR (1999) Transport-vesicle targeting: tethers before SNAREs. Nat Cell Biol 1: E17-E22

Piguet V, Trono D (2001) Living in oblivion: HIV immune evasion. Semin Immunol 13:51–57

Piguet V, Wan L, Borel C, Mangasarian A, Demaurex N, Thomas G, Trono D (2000) HIV-1 Nef protein binds to the cellular protein PACS-1 to downregulate class I major histocompatibility complexes. Nat Cell Biol 2:163–167

Pond L, Kuhn LA, Teyton L, Schutze MP, Tainer JA, Jackson MR, Peterson PA (1995) A role for acidic residues in di-leucine motif-based targeting to the endocytic pathway. J Biol Chem 270:19989–19997

Puertollano R, Aguilar RC, Gorshkova I, Crouch RJ, Bonifacino JS (2001) Sorting of mannose 6-phosphate receptors mediated by the GGAs. Science 292:1712–1716

Reusch U, Muranyi W, Lucin P, Burgert H-G, Hengel H, Koszinowski UH (1999) A cytomegalovirus glycoprotein re-routes MHC class I complexes to lysosomes for degradation. EMBO J 18:1081–1091

Roberts DL, Weix DJ, Dahms NM, Kim J-JP (1998) Molecular basis of lysosomal enzyme recognition: Three dimensional structure of the cation-dependent mannose 6-phosphate receptor. Cell 93:639–648

Robinson MS, Bonifacino JS (2001) Adaptor-related proteins. Curr Opin Cell Biol 13:444–453

Rodriguez-Boulan E, Gonzalez A (1999) Glycans in post-Golgi apical targeting: sorting signals or structural props? Trends Cell Biol 9:291–294

Rohn WM, Rouille Y, Waguri S, Hoflack B (2000) Bi-directional trafficking between the trans-Golgi network and the endosomal/lysosomal system. J Cell Sci 113:2093–2101

Rohrer J, Schweizer A, Russell D, Kornfeld S (1996) The targeting of Lamp1 to lysosomes is dependent on the spacing of its cytoplasmic tail tyrosine sorting motif relative to the membrane. J Cell Biol 132:565–576

Roquemore EP, Banting G (1998) Efficient trafficking of TGN38 from the endosome to the trans-Golgi network requires a free hydroxyl group at position 331 in the cytosolic domain. Mol Biol Cell 9:2125–2144

Routes J, Ryan S, Clase A, Miura T, Kuhl A, Potter T, Cook J (2000) Adenovirus E1A oncogene expression in tumor cells enhances killing by TNF-related apoptosis-inducing ligand. J Immunol 165:4522–4527

Russell WC (2000) Update on adenovirus and its vectors. J Gen Virol 81:2573–2604

Sandoval IV, Martinez-Arca S, Valdueza J, Palacios S, Holman GD (2000) Distinct reading of different structural determinants modulates the dileucine-mediated transport steps of the lysosomal membrane protein LIMPII and the insulin-sensitive glucose transporter GLUT4. J Biol Chem 275:39874–39885

Scaria A, Tollefson AE, Saha SK, Wold WS (1992) The E3–11.6K protein of adenovirus is an Asn-glycosylated integral membrane protein that localizes to the nuclear membrane. Virology 191:743–753

Schowalter DB, Tubb JC, Liu M, Wilson CB, Kay MA (1997) Heterologous expression of adenovirus E3-gp19K in an E1a-deleted adenovirus vector inhibits MHC I expression in vitro, but does not prolong transgene expression in vivo. Gene Ther 4:351–360

Schweizer A, Kornfeld S, Rohrer J (1996) Cysteine34 of the cytoplasmic tail of the cation-dependent mannose 6- phosphate receptor is reversibly palmitoylated and required for normal trafficking and lysosomal enzyme sorting. J Cell Biol 132:577–584

Sester M, Burgert H-G (1994) Conserved cysteine residues within the E3/19K protein of adenovirus type 2 are essential for binding to major histocompatibility complex antigens. J Virol 68:5423–5432

Sester M, Feuerbach D, Frank R, Preckel T, Gutermann A, Burgert H-G (2000) The amyloid precursor-like protein 2 associates with the major histocompatibility complex class I molecule K(d). J Biol Chem 275:3645–3654

Shenk T (1996) In B. N. Fields, D. M. Knipe and P. M. Howley (eds.), *Adenoviridae: The Viruses and Their Replication*, Lippincott-Raven Publishers, Philadelphia, New York, pp. 2111–2148

Shisler J, Yang C, Walter B, Ware CF, Gooding LR (1997) The adenovirus E3-10.4K/14.5K complex mediates loss of cell surface Fas (CD95) and resistance to Fas-induced apoptosis. J Virol 71:8299–8306

Signäs C, Akusjarvi G, Pettersson U (1986) Region E3 of human adenoviruses; differences between the oncogenic adenovirus-3 and the non-oncogenic adenovirus-2. Gene 50:173–184

Singer SJ (1990) The structure and insertion of integral proteins in membranes. Annu Rev Cell Biol 6:247–296

Sorkin A, Waters CM (1993) Endocytosis of growth factor receptors. Bioessays 15:375–382

Sorkina T, Bild A, Tebar F, Sorkin A (1999) Clathrin, adaptors and eps15 in endosomes containing activated epidermal growth factor receptors. J Cell Sci 112:317–327

Spiliotis ET, Osorio M, Zuniga MC, Edidin M (2000) Selective export of MHC class I molecules from the ER after their dissociation from TAP. Immunity 13:841–851

Stephens DJ, Crump CM, Clarke AR, Banting G (1997) Serine 331 and tyrosine 333 are both involved in the interaction between the cytosolic domain of TGN38 and the mu2 subunit of the AP2 clathrin adaptor complex. J Biol Chem 272:14104–14109

Stewart AR, Tollefson AE, Krajcsi P, Yei SP, Wold WS (1995) The adenovirus E3 10.4K and 14.5K proteins, which function to prevent cytolysis by tumor necrosis factor and to down-regulate the epidermal growth factor receptor, are localized in the plasma membrane. J Virol 69:172–181

Storch S, Braulke T (2001) Multiple C-terminal motifs of the 46-kDa mannose 6-phosphate receptor tail contribute to efficient binding of medium chains of AP-2 and AP-3. J Biol Chem 276:4298–4303

Suomalainen M, Nakano MY, Keller S, Boucke K, Stidwill RP, Greber UF (1999) Microtubule-dependent plus- and minus end-directed motilities are competing processes for nuclear targeting of adenovirus. J Cell Biol 144:657–672

Teasdale RD, Jackson MR (1996) Signal-mediated sorting of membrane proteins between the endoplasmic reticulum and the golgi apparatus. Annu Rev Cell Dev Biol 12:27–54

Tikkanen R, Obermuller S, Denzer K, Pungitore R, Geuze HJ, von Figura K, Höning S (2000) The dileucine motif within the tail of MPR46 is required for sorting of the receptor in endosomes. Traffic 1:631–640

Tollefson AE, Hermiston TW, Lichtenstein DL, Colle CF, Tripp RA, Dimitrov T, Toth K, Wells CE, Doherty PC, Wold WS (1998) Forced degradation of Fas inhibits apoptosis in adenovirus-infected cells. Nature 392:726–730

Tollefson AE, Ryerse JS, Scaria A, Hermiston TW, Wold WS (1996a) The E3–11.6-kDa adenovirus death protein (ADP) is required for efficient cell death: characterization of cells infected with adp mutants. Virology 220:152–162

Tollefson AE, Scaria A, Hermiston TW, Ryerse JS, Wold LJ, Wold WS (1996b) The adenovirus death protein (E3–11.6K) is required at very late stages of infection for efficient cell lysis and release of adenovirus from infected cells. J Virol 70:2296–2306

Tollefson AE, Scaria A, Saha SK, Wold WS (1992) The 11,600-MW protein encoded by region E3 of adenovirus is expressed early but is greatly amplified at late stages of infection. J Virol 66:3633–3642

Tollefson AE, Stewart AR, Yei SP, Saha SK, Wold WS (1991) The 10,400- and 14,500-dalton proteins encoded by region E3 of adenovirus form a complex and function together to down-regulate the epidermal growth factor receptor. J Virol 65:3095–3105

Tollefson AE, Toth K, Doronin K, Kuppuswamy M, Doronina OA, Lichtenstein DL, Hermiston TW, Smith CA, Wold WS (2001) Inhibition of TRAIL-induced apoptosis and forced internalization of TRAIL receptor 1 by adenovirus proteins. J Virol 75:8875–8887

Tomko RP, Xu R, Philipson L (1997) HCAR and MCAR: the human and mouse cellular receptors for subgroup C adenoviruses and group B coxsackieviruses. Proc Natl Acad Sci USA 94:3352–3356

Tortorella D, Gewurz BE, Furman MH, Schust DJ, Ploegh HL (2000) Viral subversion of the immune system. Annu Rev Immunol 18:861–926

Trapani JA, Davis J, Sutton VR, Smyth MJ (2000) Proapoptotic functions of cytotoxic lymphocyte granule constituents in vitro and in vivo. Curr Opin Immunol 12:323–329

Traub LM, Kornfeld S (1997) The trans-Golgi network: a late secretory sorting station. Curr Opin Cell Biol 9:527–533

von Herrath MG, Efrat S, Oldstone MB, Horwitz MS (1997) Expression of adenoviral E3 transgenes in beta cells prevents autoimmune diabetes. Proc Natl Acad Sci USA 94:9808–9813

Voorhees P, Deignan E, van Donselaar E, Humphrey J, Marks MS, Peters PJ, Bonifacino JS (1995) An acidic sequence within the cytoplasmic domain of furin functions as a determinant of trans-Golgi network localization and internalization from the cell surface. EMBO J 14:4961–4975

White E (1998) Regulation of apoptosis by adenovirus E1A and E1B oncogenes. Semin Virol 505–513

White SH (1994) Global statistics of protein sequences: implications for the origin, evolution, and prediction of structure. Annu Rev Biophys Biomol Struct 23:407–439

Whittaker GR, Kann M, Helenius A (2000) Viral entry into the nucleus. Annu Rev Cell Dev Biol 16:627–651

Wilson-Rawls J, Wold WS (1993) The E3-6.7K protein of adenovirus is an Asn-linked integral membrane glycoprotein localized in the endoplasmic reticulum. Virology 195:6-15

Windheim M, Burgert H-G (2002) Characterization of E3/49K, a novel highly glycosylated E3 protein of the epidemic keratokonjunctivitis causing adenovirus 19a. J Virol 76:75–76

Wold WS, Doronin K, Toth K, Kuppuswamy M, Lichtenstein DL, Tollefson AE (1999) Immune responses to adenoviruses: viral evasion mechanisms and their implications for the clinic. Curr Opin Immunol 11:380–386

Wold WS, Tollefson AE, Hermiston TW (1995) E3 transcription unit of adenovirus. Curr Top Microbiol Immunol 199 (Pt 1): 237–274

Yewdell JW, Bennink JR (1999) Mechanisms of viral interference with MHC class I antigen processing and presentation. Annu Rev Cell Dev Biol 15:579–606

Zerangue N, Malan MJ, Fried SR, Dazin PF, Jan YN, Jan LY, Schwappach B (2001) Analysis of endoplasmic reticulum trafficking signals by combinatorial screening in mammalian cells. Proc Natl Acad Sci USA 98:2431–2436

Zerial M, McBride H (2001) Rab proteins as membrane organizers. Nat Rev Mol Cell Biol 2:107–117

Zhu Y, Doray B, Poussu A, Lehto VP, Kornfeld S (2001) Binding of GGA2 to the lysosomal enzyme sorting motif of the mannose 6- phosphate receptor. Science 292:1716–1718

Zwart DE, Brewer CB, Lazarovits J, Henis YI, Roth MG (1996) Degradation of mutant influenza virus hemagglutinins is influenced by cytoplasmic sequences independent of internalization signals. J Biol Chem 271:907–917

The Coxsackie-Adenovirus Receptor—A New Receptor in the Immunoglobulin Family Involved in Cell Adhesion

L. Philipson[1] · R. F. Pettersson[2]

[1] Department of Cell and Molecular Biology, Karolinska Institute, Box 285, 17177 Stockholm, Sweden
E-mail: lennart.philipson@cmb.ki.se

[2] Ludwig Institute for Cancer Research, Stockholm Branch, Karolinska Institute, Box 240, 17177 Stockholm, Sweden

Abstract The physiological and cell biological aspects of the Coxsackie-Adenovirus Receptor (CAR) is discussed in this review. The receptor obviously recognizes the group C adenoviruses in vivo, but also fibers from other groups except group B in vitro. The latter viruses seem to utilize a different receptor. The receptor accumulates at, or close to, the tight junction in polarized epithelial cells and probably functions as a cell-cell adhesion molecule. The cytoplasmic tail of the receptor is not required for virus attachment and uptake. Although there is a correlation between CAR and uptake of adenoviruses in several human tumor cells, evidence of an absolute requirement for integrins has not been forthcoming. The implication of these findings for adenovirus gene therapy is discussed.

1 Introduction

The history of research on nonenveloped virus-receptor interaction in mammalian cells spans more than 40 years, and monoclonal antibodies, recombinant DNA techniques, and high-resolution structures were essential for progress in identifying the molecular events.

In the 1960s, it was established that picornaviruses could be divided into four receptor families based on competition studies between radioactively labeled and unlabeled virus and protease sensitivity of the receptors (for a review see Crowell and Lonberg-Holm 1986). Monoclonal antibodies against the polio- and rhinovirus receptors helped to identify them. After subsequent cDNA cloning, mutated variants hinted at the mode of interaction between the virus attachment moiety and specific regions in the receptor. The crystallographic resolution of the polio- and rhinoviruses identified the canyon in the capsid (Rossman et al. 1985) that binds to the receptor, although the physiological roles of the receptors still remain to be established. The polio- (PRV or CD 155) and rhinovirus (ICAM-1 or CD 54) receptors belong to the immunoglobulin superfamily (IgSF) and have three and five IgG-like domains, respectively, in the extracellular domain. The first N-terminal IgG domain in both cases interacts with the depression or canyon structure around each of the five-fold axes of the particle.

Around 25 years ago it was shown that group B coxsackieviruses (CBV) and subgroup C adenoviruses (Ad) seem to share the same receptor (Lonberg-Holm et al. 1976). A monoclonal antibody (Rmcb) was generated against the human receptor (Hsu et al. 1988), and a polyclonal antiserum (αp46) was made against the mouse variant (Xu et al. 1995). Concurrently, it was demonstrated that hemagglutinating variants of CBV that preferentially multiply in rhabdomyosarcoma cells can utilize the same receptor as some ECHO viruses, namely the decay accelerating factor (DAF or CD 55), which enhances the decay in the complement system (Bergelson et al. 1995; Shafren et al. 1995). The high-resolution structure of the Coxsackie B3 virus (Muckelbauer et al. 1995) revealed a similar canyon structure in CBV as in other picornaviruses.

For the Ad viruses, it was established in 1968 that the protruding fiber is the viral attachment protein (Philipson et al. 1968), and crystallographic studies revealed conclusively (Xia et al. 1994) that the C-terminal trimeric knob of the fiber interacts with the receptor. The monoclonal antibody generated to isolate the CBV receptor obviously also recognized the Ad receptor. Against this background, we will review the recent and very rapidly accumulating results characterizing the Coxsackie-adenovirus receptor, often referred to as CAR, but which for taxonomic reasons would probably be better called CVADR (Coxsackie Virus and ADenovirus Receptor), since the CAR acronym has also been used for other genes, including a clotting factor. Because the IgSF is still expanding, with new members with diverse functions being identified, it may be appropriate to clarify the nomenclature of the members within the IgSF and its subfamilies. We therefore suggest that the name of a particular receptor should also specify the number of Ig-loops in the ectodomain. The CAR receptor contains two Ig-loops, and in addition has a unique additional *DiSulfide-Link* (DSL; Fig. 1). Thus, the new name for CAR would be CAR(IG2DSL). In this review we will, however, use the traditional name CAR.

In addition to binding to a primary receptor, most viruses require the interaction with a secondary receptor for productive internalization. This appears to be true also for Ad. Thus, the fiber knob of subgroup C Ad viruses first interact with CAR, followed by an RGD-mediated interaction of the penton base with integrins. The $\alpha_v\beta_3$ or $\alpha_v\beta_5$ integrins seem to be preferred (Wickham et al. 1993), although other integrins may also be utilized (Wickham 2000). As discussed below, susceptibility to Ad infection does not directly correlate with integrin expression in cell lines. It has also been reported that Ad5 may use MHC class I as a high-affinity receptor (Hong et al. 1997), but this has not been confirmed in other reports (Davison et al. 1999; McDonald et al. 1999). Ad37 has been shown to interact with cell surface sialic acid (Arnberg et al. 2000a,b). As discussed in the chapter by Wadell, (Vol. 272 of this series) other Ad subtypes may utilize different, as yet unknown receptors. In this review we will only discuss the role of the CAR in virus entry.

Because another contribution in this volume deals with the structural aspect of Ad-CAR interactions, we will focus on the cell biological and physiological aspects of the CAR molecule.

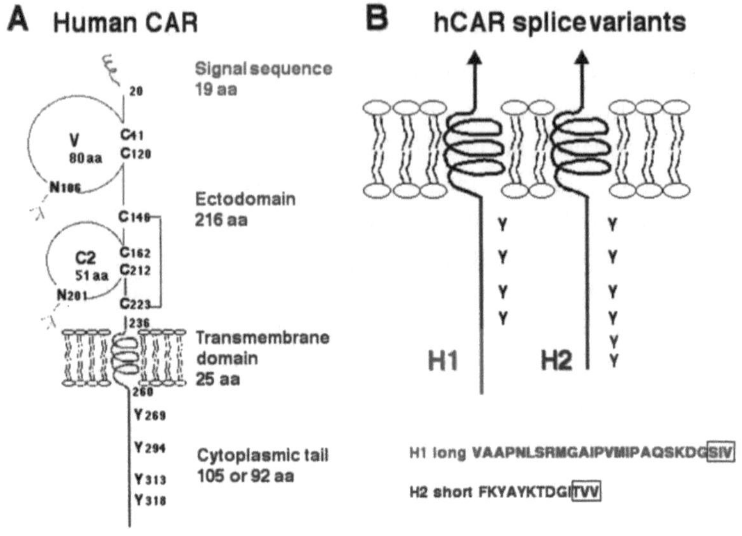

Fig. 1A, B Schematic representation of the hCAR protein and its two isoforms. Due to alternative splicing, the C-terminus of the cytoplasmic tail are different in the isoforms H1 and H2. Numbering of the residues starts with the Met at the N-terminus of the signal peptide, and is according to Tomko et al. (1997) and Bergelson et al. (1997)

2
The CAR cDNA and Gene Structure

Although cloning of the CAR cDNA was a high priority for more than two decades, the first reports describing such cloning appeared only 3 months apart in 1997 (Bergelson et al. 1997; Tomko et al. 1997). One group affinity-purified the protein from HeLa cells using a monoclonal antibody (Rmcb), followed by peptide sequencing. Based on peptide sequences, oligonucleotides were then used to isolate the cDNA (Bergelson et al. 1997). Using the αp46 polyclonal antiserum against the mouse receptor, the other group identified the cDNA in a lambda expression library made from a mouse kidney cell line expressing large amounts of the CBV receptor (Tomko et al. 1997). A 1-kb cDNA was isolated, and when its sequence was compared with human EST clones, extensive similarity was observed with two clones from the IMAGE consortium. One of these was 1.4 kb and obviously contained the entire coding region. However, expression studies in several tissues revealed 6- and 1.4-kb mRNAs in the mouse and 6- and 2.4-kb mRNAs in humans, suggesting that additional mRNAs may exist. As later revealed from the genomic structure, it appears that the longer mRNA is the predominant form, and the shorter forms may arise because of the utilization of different polyadenylation signals, through alternative splicing, or RNA degradation.

The human gene encoding CAR is located on chromosome 21q11.2, and the exon boundaries have been identified (Bowles et al. 1999). The same chromosome harbors at least two pseudogenes, and additional similar genes are distributed across both the human and mouse genomes as discussed below regarding CTX-family genes. Two independent human genome sequences have been reported (GenBank acc. no. AF 200465 and 242862–65, respectively, see Andersson et al. 2000 for a comparison). The transcribed region seems to extend over 57 kbp, about 4.55 mbp from the centromere (Hattori et al 2000), and the gene is transcribed towards the telomere. Notable differences between the sequences are six length-differences in mono- and dinucleotide repeats, as well as 17 single-base differences. Three of the latter are located in the promoter region, which may have implications for studies on transcriptional controls. In a study from our laboratory (Andersson et al. 2000), an additional exon was identified 27 kbp downstream of exon 7, which may code for an alternative C-terminus. Comparison with a BAC clone of the mouse gene showed extensive similarity in the first 5 kbp of the genomic clones (GenBank acc. no. 242861). This led to the identification of five conserved regions between human and mouse adjacent to the promoter that may be involved in transcriptional regulation. Direct identification of alternative C-terminal ends revealed three different alternative splice sites in the mouse genome (referred to as M1–3) that corresponded to very similar sites in the human genome (H1–3). The H3 sequence, however, lacked an acceptor site in the human genome, and this variant may therefore not be expressed. It is noteworthy that when the mouse and human cDNA sequences were first compared (Tomko et al. 1997; Bergelson et al. 1997), only the M1 splice variant was detected in the mouse and the H2 splice variant in humans. Genome sequencing and additional expression studies revealed that both variants are expressed in mouse and humans (Fechner et al. 1999; Andersson et al. 2000). The amino acid sequences unique to the C-termini of the H1 and H2 isoforms are shown in Fig. 1.

3
Characteristics of the CAR Protein

As deduced from the sequence of the cDNA, CAR is a type I integral transmembrane protein with sequence similarity to other members of the IgSF (Bergelson et al. 1997; Tomko et al. 1997). The full-length open reading frame (ORF) of hCAR encodes 365 amino acids representing a predicted short leader sequence (19 residues), an ectodomain of about 216 residues, a transmembrane domain of about 26 residues, and depending on alternative splicing, a cytoplasmic tail of either 105 or 92 residues (Fig. 1). The ectodomain consists of two Ig-loops (IG1 and IG2) of the variable (V, 80 residues; formed between Cys41 and Cys120) and constant (C2, 51 residues; Cys 162–Cys212) type. The C2 domain contains an extra disulfide bridge formed between Cys146 and Cys223 (numbering encompasses the whole ORF). The predicted mature protein has a molecular mass of about 40 kDa, but due to two N-linked glycans (at Asn106 and Asn 201), the M_r on SDS-gels is about 46 kDa. The C-terminal amino acid sequence of the tail, its length, and the number of tyrosines that could be involved in signal transduction depend on alternative splicing (Andersson et al. 2000). At the extreme C-terminal end of both the M1 and M2, as well as the H1 and H2 isoforms (Fig. 1), PDZ binding motifs (-TTV-COOH, or -SIV-COOH) are present that have the potential to interact with intracellular PDZ-proteins, as further discussed below in the cell biology section. The overall homology between human, mouse, rat, dog, and pig CAR is about 90% (Tomko et al. 1997; Bergelson et al. 1998; Fechner et al. 1999). The IG1 domain is more conserved (91%–94%) than the IG2 domain (83%–89%), whereas the cytoplasmic tail is about 95% identical in these species (Fechner et al. 1999). The transmembrane domain is less conserved, being 77% identical between human and mouse (Tomko et al. 1997).

4
Structure of CAR and Interaction with the Viruses

In analogy with other picornaviruses, the IG1 domain of the CAR was assumed to be the interacting partner with the two viruses sharing the same receptor. The purified IG1 domain produced in *E. coli* was demonstrated to bind to the knob of the Ad12 fiber and prevent infection of the Ad2 virus (Freimuth et al. 1999). Concurrently it was established that

CAR could bind to the fiber knob of Ad in the subgroups A, C, D, E, and F, but not in subgroup B, suggesting that CAR may function as a receptor for viruses belonging to all members of these groups (Roelvink et al. 1998). However, because members of subgroups A and C–F can cause different patterns of disease, it is unlikely that the fiber-receptor interaction is the sole determinant of viral tropism. Modeling of the CAR IG1 region and attempts to dock it to the known high-resolution structure of the Ad5 fiber knob suggested that three IG1 domains may bind to each fiber knob, and identified a region in the side valley of the fiber knob that may be recognized by the CAR IG1 domain (Tomko et al. 2000). More careful mutational analysis identified this region to the AB loop of the side valley in the fiber knob (Roelvink et al. 1999). Two mutations in the CAR IG1 domain (V70A+I72D) which restricted Ad infection were identified (Tomko et al. 2000), and the interaction between the two entities became outlined. The ultimate proof of the precise interaction was provided by the high-resolution structure of the Ad12 fiber knob in complex with the IG1 domain of CAR (Bewley et al. 1999). Figure 2 shows a ribbon diagram of this complex viewed down the viral fiber. Thus, Ad interaction with the CAR IG1 domain was established and can be used as a basis for attempts to enhance or inhibit viral infectivity.

The interaction between the CBV and the CAR is less well defined. Numerous mutants in the IG1 domain of the CAR did not interfere with the ability of CBV to infect cells, and the antibody αp46 that specifically inhibits CBV infection appears to interact with a protein containing both the IG1 and IG2 domains of the CAR (Tomko et al 2000). The original studies claiming a shared receptor for Ad and CBV (Lonberg-Holm et al. 1976) demonstrated that one virus could protect the cells against the other. This finding may be due to steric hindrance if two different but closely spaced domains (or binding sites) are involved. However, a recent report (He et al. 2001) now appears to establish, based on cryo-electron microscopic studies, that in contrast to the fiber knob the opposite side of the IG1 domain of CAR interacts with CB3 virus. This region may in fact be in juxtaposition to the IG2 domain of the CAR. The conclusion is therefore quite clear-cut: both CBV and the Ad fiber knob interact with the IG1 domain of the CAR.

Neither the transmembrane region nor the cytoplasmic tail seems to be necessary for virus interaction, or infection, because a GPI-anchored, or truncated, receptor is equally active in mediating virus uptake of both viruses (Wang and Bergelson 1999). This finding strongly suggests that

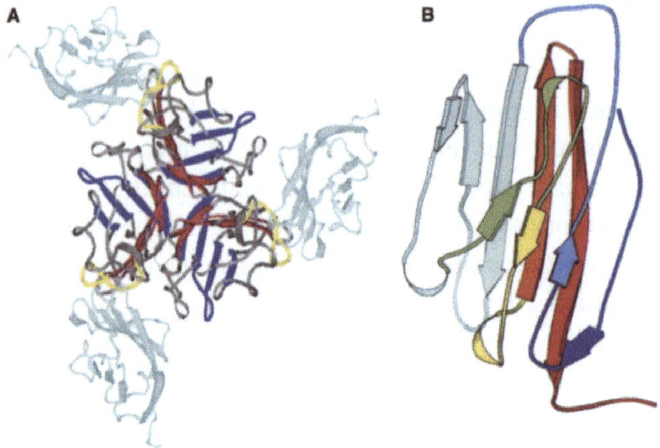

Fig. 2A, B Structure of the Ad12 fiber knob bound to the CAR D1 loop. **A** Ribbon diagram of the Ad12 knob-CAR D1 complex is viewed down the viral fiber. **B** Ribbon diagram of the CAR D1 domain from the complex colored in a rainbow from *blue* to *red*. Strands in the foreground are D, E, B, and A, and strands in the background are C", C', C, F, and G from *left to right*, respectively. For details, see Bewley et al. (1999), from which these diagrams have been reprinted with permission from the American Association for the Advancement of Science

secondary receptors, such as the α_v integrins, are important for virus uptake, or penetration.

5
CAR Is a Member of the CTX Subfamily Within the Ig Superfamily

At about the same time that the cDNAs for hCAR and mCAR were cloned (Bergelson et al. 1997; Tomko et al. 1997), several other CAR-related cDNAs were identified. These include CTX (a cortical thymocyte marker in *Xenopus laevis*; Chretien et al. 1998) and A33 (a marker of intestinal epithelial cells and colorectal tumors; Heath et al. 1997). The proteins encoded by these cDNAs all display the same basic features, i.e., they consist of two Ig-loops of the V and C2 type with an extra disulfide link (DSL) in the C2 domain, a transmembrane region, and a cytoplasmic tail of variable length. Similar proteins in this new group of IgSFs have been named the CTX-subfamily (Chrétien et al. 1998; Du Pasquier et al. 1999). Following the cloning of the frog CTX cDNA, chicken (chT1; Katevuo et al. 1999), mouse (CTM), and human (CTH; Chrétien

Table 1 Amino acid sequence comparisono between hCAR, hA33, and CTH

	hA33		CTH	
hCAR	ID (%)	SIM (%)	ID (%)	SIM (%)
Total sequence	23	40	24	38
Ectodomain	27	48	28	42
V domain	30	50	25	41
C2 domain	31	51	29	43
Intracellular domain	12	28	13	31

ID, identity; SIM, similarity (includes conserved amino acid substitutions).
The data were derived by using the Clustal W alignment program and the "similarity matrix" program BLOSUM62.

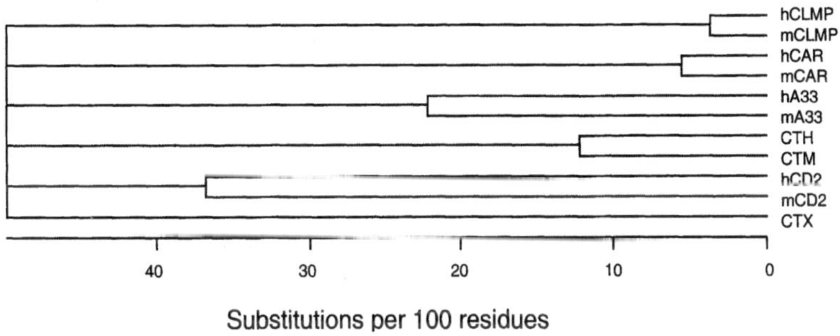

Fig. 3 Dendrogram showing the evolutionary relationship between members of the CTX subfamily within the IgSF. The dendrogram was derived using DNAStar and the Clustal W alignment programs

et al. 1998) homologs were identified. In addition to CTX and A33, a number of related sequences have been identified in database searches (Chretien et al. 1998), suggesting that there may be several additional members in this subfamily. We recently identified another CAR/CTX-like molecule, which we have named hCLMP (human) and mCLMP (mouse) *C*AR/CTX-*L*ike *M*embrane protein; E. Raschperger, U. Engstrom, R. Pettersson, J. Fuxe, manuscript in preparation). HCLMP shows 34% identity to hCAR and 31% to A33. Table 1 shows the sequence relationship between CAR, A33, and CTH, and Fig. 3 shows the evolutionary relationship of the various members. CTX-proteins all have a predicted molecular mass of about 33–40 kDa, but due to N-linked glycans (2–4 putative sites in different members), the mature proteins have an Mr of 43–63 kDa in SDS-gels. Sequence and structural similarities to other proteins, such as CD2 (T-lymphocyte cell adhesion molecule) and JAM (junctional adhesion molecule) have also been noted (Johnstone et al. 2000), but because critical signatures typical for CTX-members are missing from these molecules, they may not be considered true members of the CTX subfamily. To date, none of the CAR-related CTX-members have been found to be able to serve as a receptor for adeno-, Coxsackie B, or any other viruses.

6 Cellular Function and Cell Biology of CAR

The presence of the V and C2 Ig-loops in CAR and members of CTX-subfamily has hinted at the possibility that CAR might serve as a cell adhesion molecule mediating interaction either with other members of the IgSF, or with itself (homophilic interaction; Chrétien et al. 1998). To date, no CAR-specific ligand has been identified. However, data are rapidly accumulating supporting the notion that CAR can form dimers and thereby mediate homophilic interaction. Thus, C6 rat glioma cells transfected with mCAR aggregated in a CAR-dependent manner (Honda et al. 2000). Aggregation occurred only between CAR-positive cells, suggesting homophilic interaction. Furthermore, aggregation could be inhibited by an ectodomain-specific peptide antiserum. Similarly, hCAR was recently shown to mediate homotypic aggregation of CHO cells stably transfected with CAR (Cohen et al. 2001). Full-length CAR, and CAR in which the cytoplasmic tail was replaced by a GPI-anchor, were concentrated at sites of cell-cell contact. Immunolocalization at the light microscopy and electron microscopy levels showed that CAR is specifically confined to tight junctions in polarized epithelial cells (e.g., human tracheobronchial airway cells, a colonic cell line, or MDCK cells), where it colocalized with the PDZ-domain protein ZO-1/PSD95/dlg, a well-known component of cell adherens junctions. Interestingly, polarized cells could not be infected with Ad unless the tight junctions were disrupted with EDTA (Cohen et al. 2001). This finding has important implications for in vivo infection of epithelia with adenovirus. The model of homophilic interaction is also supported by the fact that CAR forms homodimers both in crystals and in solution (van Raaij et al. 2000).

Like cell adhesion molecules in general, it is likely that CAR also signals to the cell nucleus and participates in the regulation of cell proliferation and/or differentiation (see below). Nothing is known yet about such putative signaling pathway(s).

The inefficiency of Ad to infect lung epithelial cells became apparent during the early attempts to transduce airway epithelia with the cDNA encoding the cystic fibrosis transmembrane conductance regulator (CFTR) by using Ad as a vector (Grubb et al. 1994; Walters et al. 1999). The basis for this was revealed when it was shown that well-differentiated airway epithelial cells could not be infected with adenovirus through the apical route, due to the lack of CAR (and α_v integrins) on this cell surface (Pickles et al. 1998) and the presence of a glycocalyx layer that serves as an effective physical barrier (Pickles et al. 2000). Further analyses showed that CAR is expressed only on the basolateral surface of polarized cells in vivo and in transfected cultured cells (e.g., MDCK; Pickles et al. 2000; Walters et al. 1999). Enzymatic removal of the glycocalyx combined with the retargeting of CAR to the apical surface facilitated infection from the apical side (Pickles et al. 2000). Apical expression of CAR can be accomplished by deleting the cytoplasmic tail, or by replacing it with a GPI-anchor (Pickles et al. 2000; Walters et al. 2001). The localization of CAR to tight junctions, which makes the receptor inaccessible to virus binding, poses a further constraint (Cohen et al. 2001). In retrospect it is interesting to note that when Ad was originally identified (Hilleman et al. 1954) from adenoids, the cells were propagated for long periods in vitro before the virus could multiply. Perhaps the receptor had to be exposed on neighboring cells before gradual spreading of the virus was accomplished. Thus, understanding the cell biology of CAR is of great practical importance for targeting Ad vectors to epithelia.

Dissection of the signal for basolateral targeting in MDCK cells has underscored the importance of multiple regions within the cytoplasmic tail of CAR. More specifically, it was shown that a tyrosine at position 315 (which is part of a previously described basolateral sorting signal, YNQV) as well as the sequence LSRM (residues 345–348) are important determinants for basolateral sorting. Finally, the cytoplasmic tail of the two mouse CAR isoforms also direct the receptor to the basolateral surface (Cohen et al. 2001).

7
Expression of CAR in Cell Lines and Tissues

A common argument for the use of Ad viruses as gene therapy vectors is that they are able to infect a variety of cells and organs. Although partly true, different Ad viruses infect different tissues, and the susceptibility of different cell types and tissues to Ad2 and Ad5, the most commonly used gene therapy vectors, varies substantially (see the chapter by Wadell, Vol. 272 of this series). A large number of studies with cell lines have shown that susceptibility to infection with subgroup C Ad correlates well with the expression of CAR, whereas no correlation with the expression of $\alpha_v/\beta_{3/5}$ integrins has been demonstrated (Hemmi et al. 1998; Mori et al. 1999; Fechner et al. 2000; Asaoka et al. 2000; You et al. 2001; Fuxe et al. 2003). Transfection of the CAR cDNA into CAR-negative cells makes them susceptible to Ad infection (Tomko et al. 1997; Li et al. 1999). Information on the tissue distribution and regulation of CAR expression during development and under different physiological conditions is therefore of great importance for targeting replication-competent or -incompetent Ad to desired tissues. However, it should be emphasized that some tissues may, despite of expression of CAR, be refractory to Ad infection due to physical or anatomical barriers (Fechner et al. 1999; Hutchin et al. 2000; Pickles et al. 2000; Cohen et al. 2001). Breaking such a physical barrier by treatment with ethanol (Engler et al. 1999), or polyamides (Connor et al. 2001) has been shown to enhance the susceptibility of the bladder epithelium to Ad-mediated gene transfer. As discussed above, another important aspect is that CAR may be expressed in a domain of the cell surface (i.e., basolateral or tight junctions) that is inaccessible to virus attachment (Pickles et al. 2000; Walters et al. 1999).

CAR expression levels have been primarily determined by mRNA analyses (RT-PCR, Northern blotting, in situ hybridization) and to a much lesser extent by protein analyses (immunohistochemistry, immunoblotting). Data published thus far indicate a great variability in CAR mRNA and protein expression, among both tissues and cultured tumor cell lines. There seem to be some differences in the expression pattern of CAR between mouse and human (and other animals). This is of importance in particular when using mice as human disease models to assess the efficacy of gene therapy. Messenger RNAs of multiple sizes are expressed in human and rodent cells (Tomko et al. 1997; Bergelson et al.

1998; Fechner et al. 1999). The structure of these RNA species, which could represent differentially spliced or polyadenylated species, or even transcripts from pseudogenes, has not yet been fully established. It is therefore still unclear whether all of these mRNAs are translated into a functional protein. In the absence of this information, one must be cautious in proposing a correlation between mRNA and protein levels.

7.1 Rodents

Using a polyclonal mCAR antiserum, strong CAR-immunoreactivity (ir) was demonstrated throughout the nervous system, including the CNS, dorsal root ganglia, and peripheral nerves of the developing mouse (embryonic day 15; Tomko et al. 2000). In adult rat, CAR expression was more restricted. In the brain, CAR expression was low or absent, and restricted to the ependymal cell layer lining the ventricular system. CAR was abundantly expressed in the epithelia of trachea, bronchi, kidney, liver, and intestine, and scattered was also seen in pancreas and heart (Tomko et al. 2000). High to moderate levels of CAR mRNA expression have been demonstrated in liver, kidney, brain, lung, and heart, very low levels in testis, and no expression in spleen and skeletal muscle (Tomko et al. 1997; Fechner et al. 1999). High levels of mCAR and mRNA protein were found in total brain of developing (from embryonic day 10.5) and postnatal (to day 7) mice, as well as in newborn heart (Honda et al. 2000). Similar patterns have been reported for rats (Fechner et al. 1999). Ito et al. (2000) on the other hand showed that CAR, which was expressed to high levels in developing rat heart, was undetectable in adult heart as determined by both immunohistochemistry and RT-PCR. The fact that CBV can only give rise to encephalitis in newborn but not in adult mice (Dalldorf and Sickles, 1948) may thus be explained by the presence of the receptor in brain tissue.

7.2 Humans

Variable levels of hCAR mRNA expression have been found in human liver, kidney, lung, brain, heart, colon, small intestine, testis, prostate and pancreas, whereas no expression was found in skeletal muscle, spleen, ovary, thymus, or placenta (Tomko et al. 1997; Fechner et al.

1999). Generally, hematopoietic cells, in particular T and B cells, are notoriously difficult to infect with serotype C adenoviruses, although monocytes are partially susceptible (Huang et al. 1997). Primary human T lymphocytes express very low levels of CAR (Chen et al. 2002). Rebel et al. (2000) found that hCAR is expressed on 40% of human bone marrow cells, but not on lymphoid cells. Of the $CD34^+$ cells, 10%–15% expressed CAR, whereas more primitive hematopoietic cells seem not to display CAR. As described below, CAR expression can in some cells or tissues be induced during differentiation, or activation.

7.3
Cell Lines

CAR is expressed in many commonly used established cells lines, such as the human 293, HeLa, and mouse TCMK-1 cells, whereas others are CAR-negative [mouse L cells, CHO (hamster); Tomko et al. 1997).

7.4
Human Tumors and Tumor Cell Lines

Expression of CAR in cell lines derived from a broad range of tumors varies enormously, the levels ranging from high to undetectable. Examples include ovarian (You et al. 2001), glioma (Miller et al. 1998; Mori et al. 1999; Asaoka et al. 2000; Fuxe et al. 2003), melanoma (Hemmi et al. 1998), bladder (Li et al. 1999), prostate (Okegawa et al. 2000), rhabdomyosarcoma (Cripe et al. 2001), and gastrointestinal (Fechner et al. 2000) tumor cell lines. In all these studies, CAR expression generally correlated well with susceptibility to Ad infection.

Few reports on the expression of CAR in primary tumors have appeared to date. Based on real-time PCR analysis, we have found a great variability in hCAR expression in primary human astrocytic tumors (Fuxe et al. 2003). Grade IV tumors expressed significantly lower levels of hCAR than grade II/III tumors. Interestingly, the level of hCAR expression has been found to inversely correlate with tumorigenicity (Okegawa et al. 2000). These studies suggest that CAR may function as a tumor suppressor. Loskog et al. (2002) found that 27 out of 27 biopsies from human urinary bladder cancer were positive for hCAR, with expression levels varying between tumors. Cells from patients with chronic myeloic leukemia can be quite efficiently infected. On the other hand, chronic lymphatic leukemia cells (B-CLL) are completely resistant to Ad infection, but they can be rendered susceptible upon activation with IL-4 and CD40 ligand. Whether this correlates with an upregulation of CAR is not known (Huang et al. 1997).

From the published literature one can conclude that no comprehensive and systematic studies of CAR expression during development, in various adult tissues, and primary tumors have yet been undertaken. One would in particular like to see analyses of hCAR and mCAR protein expression by immunohistochemistry, combined with mRNA expression using in situ hybridization.

8
Regulation of CAR Expression

Limited information is available on the regulation CAR expression during development and under various inducing conditions. As mentioned above, CAR seems to be developmentally regulated in mice. Although high levels of mCAR expression are observed throughout the developing brain, expression is almost undetectable in the adult brain (Tomko et al. 2000; Honda et al. 2000). Interestingly, mCAR apparently remains expressed in the ependymal cell layer of the mouse brain ventricles, because these cells can be infected with Ad (Johansson et al. 1999). Expression of CAR is also downregulated in mouse skeletal muscle during maturation (Nalbantoglu et al. 1999). A low level of CAR expression was detected on postnatal day 3 in striated muscles, but CAR was undetectable in adult skeletal muscle. In contrast, moderate levels of CAR mRNA were found at all stages of the developing and in the adult heart. In a tissue culture model system, Hutchin et al. (2000) found that hCAR is downregulated in oropharyngeal epithelium during differentiation; the most differentiated cells expressed no detectable hCAR. In contrast, stratified head and neck squamous carcinoma cells, which did not differentiate, had similar levels of hCAR in superficial and basal layers. These studies thus indicate that CAR expression is downregulated during maturation and differentiation of some cell lineages.

CAR expression may also be modulated by other conditions. hCAR expression has been reported to increase as human umbilical vein endothelial (HUVEC) cells reach confluence (Carson et al. 1999). In an experimental rat autoimmune myocarditis, CAR was found to be strongly induced as a consequence of inflammation. Furthermore, conditioned medium from rat splenocytes activated by concanavalin A induced CAR expression (mRNA and protein) in cultivated cardiomyocytes, suggesting that cytokines might regulate CAR expression. Binding of Ad or purified fiber to human peripheral lymphocytes, which is very low on native cells, could be enhanced by stimulation with the phorbol ester PHA and IL-2, a finding probably reflecting induction of CAR expression (Mentel et al. 1997). As mentioned above, human B-CLL cells could be rendered susceptible to Ad infection following activation with the combination of IL-4 and CD40 ligand, or a CD40 antibody (Huang et al. 1997)

9
A Transgenic Model for Adenovirus Vectors

Since the CAR expression is restricted in adult animals and several studies suggested that CAR is important for virus attachment, it was tempting to establish a transgenic mouse that expressed CAR in all organs. Taking advantage of the reports that a truncated form of the receptor (Leon et al. 1998; Wang and Bergelson 1999) allowed for both attachment and uptake of Ad viruses, it was logical to use a truncated form of the CAR without the cytoplasmic tail. Any signaling via this domain, potentially causing unwanted effects, would thereby be eliminated.

Using a promoter for the abundantly expressed ubiquitin in front of the truncated hCAR cDNA followed by the β-globin 3′ end, several founder mice were identified (Tallone et al. 2001). One mouse line expressing high levels of CAR was analyzed in detail. No abnormal phenotype was observed in the transgenic animals, and they all expressed hCAR abundantly in most tissues. Using a GFP Ad vector, it was demonstrated that the virus could be targeted to several different tissues after both peritoneal and intravenous injection. This simple CAR-expressing mouse may allow several gain-of-function studies of specific genes prior to the use of Ad vectors in humans, and perhaps most important provide a negative control for any attempt using targeted Ad therapy.

10
CAR Expression in Relation to Adenovirus-Mediated Gene Therapy

As discussed above, susceptibility of cultured cells to infection with subgroup C Ad correlates very well with the expression of CAR. Although the mechanism of uptake of Ad into cultured cells is rather well understood, the situation in vivo is much more complicated. Information on CAR expression in different cell types within a given tissue is still incomplete. In addition, there is no information as to the minimum number of CAR molecules expressed per cell required for efficient infection. In vivo, the level of CAR expression varies substantially among different normal tissues, between tumors of the same kind, as well as among different types of tumors. The in vivo situation may be further complicated by anatomical or physical barriers preventing the binding of Ad to CAR on target cells. In addition, CAR-independent uptake mechanisms may exist in vivo. Such mechanisms may be responsible for the uptake of Ad into Kupffer cells in the liver (Wickham 2000). The route of administration (e.g., systemic vs. local) is also a critical parameter for the distribution of Ad in tissues. These are all factors that must be considered in assessing the efficiency of Ad-mediated gene transfer in vivo. Although the broad expression of CAR enables the infection of a variety of tissues, one would in many cases like to avoid expression in certain tissues. Instead, it would be advantageous to target Ad vectors to specific cells or tissues that do not express CAR, while at the same time prevent infection of CAR-positive tissues.

Given the above constraints and limitations, efforts to change the specificity of Ad attachment to target cells are being pursued (Wickham 2000; Krasnykh et al. 2000). One way of altering the specificity is to use non-subgroup C Ad with different tissue-tropism as vectors, or to swap just the fiber knob (Stevenson et al. 1997). The preferred strategy for altering Ad tissue-specificity is to knock out the fiber knob-CAR interaction and at the same time to redirect the virus to a new receptor. Knowledge of the 3D structure (Bewley et al. 1999) of the fiber knob-CAR interaction is of enormous help in engineering such retargeted vectors. Two main strategies are being pursued (Wickham 2000; Krasnykh et al. 2000). The first involves genetic modification of the fiber knob (e.g., the HI-loop) such that a motif (e.g., a growth factor, the integrin-binding RGD peptide, or some other homing peptide) known to be recognized by a receptor on the desired target cells is introduced with, or without, concomitant inactivation of the CAR-binding site (Dmitriev et al. 1998; Cripe et al. 2001; Wessling et al. 2001).

The second strategy involves engineering a hybrid linker molecule that bridges the fiber knob with a receptor on the target cell. Such "adaptors" may consist of a fiber knob-binding antibody, or soluble CAR ectodomain, linked to a peptide ligand that binds to a receptor mediating attachment and internalization (Douglas et al. 1999; Trepel et al. 2000; Dmitriev et al. 2000). Alternatively, a bispecific antibody binding on the one hand the fiber knob and on the other hand a cell surface molecule on the target cell has been used (Heideman et al. 2001).

Although recombinant Ad viruses may be administered in vivo at rather high doses, the local spread of virus in tissues, e.g., within tumors, is still quite inefficient. This has lead to the design and use of replication-competent oncolytic Ad vectors (Alemany et al. 2000). Such vectors can also be retargeted as briefly described above, or the transgene can be expressed under a tissue-specific, or conditionally regulated promoter (Curiel 2000). As recent developments in this area are described in the chapter by Dobbelstein in this volume, we will not discuss these strategies further here.

11 Perspectives

During the last 5 years, considerable information on the structure and expression of CAR has accumulated. It appears established that the first IgG domain of CAR recognizes both Coxsackie B viruses and subgroup C Ad viruses in vitro and in vivo. Whether other Ad subgroups use the CAR as a receptor under in vivo conditions remains to be established, but obviously all subgroups except group B can bind to CAR in vitro. The high-resolution structure of the first IgG domain in complex with the fiber knob defines the interaction epitopes and provides the basis for designing ligands that may enhance or prevent Ad infection.

On the other hand, the physiology of the CAR in cell-to-cell interaction definitely requires more penetrating studies. Although it appears that CAR expression is developmentally regulated with high expression in early stages of the embryo and a gradual suppression during differentiation, the correlation between mRNA and protein expression must be clarified in more detail. The multiple species of mRNAs must definitely be characterized and it must be established whether they arise by differential splicing, alternative polyadenylation, or are derived from transcribed pseudogenes. The possibility of stable processing intermediates should not be ruled out.

It may also be rewarding to initiate more careful studies of the induction or suppression of CAR transcription. If the ligands, the signal transduction pathway as well as the transcription factors involved were identified for both CAR and other CTX genes, common events may be revealed. Furthermore if induction could be achieved from the outside or through small molecules intracellularly, more efficient viral vectors may be designed.

The possible contribution of CAR to forming or maintaining tight junctions in polarized cells should be analyzed in depth. By eliminating some of the signals for basolateral localization of the receptors, it may be possible to interfere with the polarization events if CAR is directly involved in this process. The first attempt to make a transgenic mouse with a CAR truncated in the cytoplasmic region suggests that apical expression can be achieved in vivo. The same results were also reported from in vitro experiments.

Last but not least, it may be rewarding to study the developmental control of CAR expression. The abundant expression in neuroepithelial cells in the developing mouse compared to the restricted expression in the ependymal layer in the ventricles of the brain in the adult mouse suggests that CAR expression may be confined to neuronal stem (or progenitor) cells, and may perhaps serve as a good marker for this cell population, at least in the central nervous system. In fact, the cells showing CAR expression in the ependymal layer can readily be infected with Ad viruses and show stem cell properties with other techniques (Johansson et al. 1999).

Lastly, the conclusion is that viral receptors in addition to their specific role in viral attachment and uptake have profound roles in cell biology and immunology. Both functions are of fundamental importance, and the virus systems may again and again help to identify and characterize basic mechanisms in biology.

Acknowledgments. We wish to thank Jonas Fuxe and Elisabeth Raschperger for help in preparing the figures and the table.

References

Alemany R. Balague C, Curiel DT (2000) Replicative adenoviruses for cancer therapy. Nat Biotechnol 18:723–7

Andersson B, Tomko RP, Edwards K, Mirza M, Darban H, Oncu D, Sonhammer E, Sollerbrant K, Philipson L (2000) Putative regulatory domains in the human and mouse CVADR genes. Gene Funct Dis 2:11–15

Arnberg N, Edlund K, Kidd H, Wadell, G (2000a) Adenovirus type 37 uses sialic acid as a cellular receptor. J Virol 74:42–8

Arnberg N, Kidd AH, Edlund K, Olfat F, Wadell G (2000b) Initial interactions of subgenus D adenoviruses with A549 cellular receptors: sialic acid versus alpha(v) integrins. J Virol 74:7691–3

Asaoka K, Tada M, Sawamura Y, Ikeda J, Abe H (2000) Dependence of efficient adenoviral gene delivery in malignant glioma cells on the expression levels of the Coxsackievirus and adenovirus receptor. J Neurosurg 92:1002–8

Bergelson JM, Cunningham JA, Droguett G, Kurt-Jones EA, Krithivas A, Hong JS, Horwitz MS, Crowell RL, Finberg RW (1997) Isolation of a common receptor for Coxsackie B viruses and adenoviruses 2 and 5. Science 275:1320–3

Bergelson JM, Krithivas A, Celi L, Droguett G, Horwitz MS, Wickham T, Crowell RL, Finberg RW (1998) The murine CAR homolog is a receptor for coxsackie B viruses and adenoviruses. J Virol 72:415–9

Bergelson JM, Modlin JF, Wieland-Alter W, Cunningham JA, Crowell RL, Finberg RW (1997) Clinical coxsackievirus B isolates differ from laboratory strains in their interaction with two cell surface receptors. J Infect Dis 175:697–700

Bergelson JM, Mohanty JG, Crowell RL, St John N. F, Lublin DM, Finberg RW (1995) Coxsackievirus B3 adapted to growth in RD cells binds to decay- accelerating factor (CD55). J Virol 69:1903–6

Bewley MC, Springer K, Zhang YB, Freimuth P, Flanagan JM(1999) Structural analysis of the mechanism of adenovirus binding to its human cellular receptor, CAR. Science 286:1579–83

Bowles KR, Gibson J, Wu J, Shaffer LG, Towbin JA, Bowles NE (1999) Genomic organization and chromosomal localization of the human Coxsackievirus B-adenovirus receptor gene. Hum Genet 105:354–9

Carson SD, Hobbs JT, Tracy SM, Chapman NM (1999) Expression of the coxsackievirus and adenovirus receptor in cultured human umbilical vein endothelial cells: regulation in response to cell density. J Virol 73:7077–9

Chen Z, Ahonen M, Hamalainen H, Bergelson JM, Kahari VM, Lahesma, R (2002) High-efficiency gene transfer to primary T lymphocytes by recombinant adenovirus vectors. J Immunol Methods 260:79–89

Chretien I, Marcuz A, Courtet M, Katevuo K, Vainio O, Heath JK, White SJ, Du Pasquier L (1998) CTX, a Xenopus thymocyte receptor, defines a molecular family conserved throughout vertebrates. Eur J Immunol 28:4094–104

Cohen CJ, Gaetz J, Ohman T, Bergelson JM (2001) Multiple regions within the coxsackievirus and adenovirus receptor cytoplasmic domain are required for basolateral sorting. J Biol Chem 276:25392–8

Cohen CJ, Shieh JT, Pickles RJ, Okegawa T, Hsieh JT, Bergelson JM (2001) The coxsackievirus and adenovirus receptor is a transmembrane component of the tight junction. Proc Natl Acad Sci U S A 98:15191–6

Connor RJ, Engler H, Machemer T, Philopena JM, Horn MT, Sutjipto S, Maneval DC, Youngster S, Chan TM, Bausch J, McAuliffe JP, Hindsgaul O,Nagabhushan TL (2001) Identification of polyamides that enhance adenovirus-mediated gene expression in the urothelium. Gene Ther 8:41–8

Cripe TP, Dunphy EJ, Holub AD, Saini A, Vasi NH, Mahller YY, Collins MH, Snyder JD, Krasnykh V, Curiel DT, Wickham TJ, DeGregori J, Bergelson JM, Currier MA

(2001) Fiber knob modifications overcome low, heterogeneous expression of the coxsackievirus-adenovirus receptor that limits adenovirus gene transfer and oncolysis for human rhabdomyosarcoma cells. Cancer Res 61:2953–60

Crowell RL, Lonberg.-Holm. K (1986) Virus attachment and entry into cells. 216 pages, American Society for Microbiology, Washington DC

Curiel DT (2000) The development of conditionally replicative adenoviruses for cancer therapy. Clin Cancer Res 6:3395–9

Dalldorf G, Sickles GM (1948) An unidentified, filtrable agent isolated from the faeces of children with paralysis. Science 108:61–62

Davison E, Kirby I, Elliott T, Santis G (1999) The human HLA-A*0201 allele, expressed in hamster cells, is not a high- affinity receptor for adenovirus type 5 fiber. J Virol 73:4513–7

Dmitriev I, Kashentseva E, Rogers BE, Krasnykh V, Curiel DT (2000) Ectodomain of coxsackievirus and adenovirus receptor genetically fused to epidermal growth factor mediates adenovirus targeting to epidermal growth factor receptor-positive cells. J Virol 74:6875–84

Dmitriev I, Krasnykh V, Miller CR, Wang M, Kashentseva E, Mikheeva G, Belousova N, Curiel DT (1998) An adenovirus vector with genetically modified fibers demonstrates expanded tropism via utilization of a coxsackievirus and adenovirus receptor-independent cell entry mechanism. J Virol 72:9706–13

Douglas J. T, Miller CR, Kim M, Dmitriev I, Mikheeva G, Krasnykh V, Curiel DT (1999) A system for the propagation of adenoviral vectors with genetically modified receptor specificities. Nat Biotechnol 17:470–5

Du Pasquier L, Courtet M, Chretien I (1999) Duplication and MHC linkage of the CTX family of genes in Xenopus and in mammals. Eur J Immunol 29:1729–39

Engler H, Anderson SC, Machemer TR, Philopena JM, Connor RJ, Wen SF, Maneval DC (1999) Ethanol improves adenovirus-mediated gene transfer and expression to the bladder epithelium of rodents. Urology 53:1049–53

Fechner H, Haack A, Wang H, Wang X, Eizema K, Pauschinger M, Schoemaker R, Veghel R, Houtsmuller A, Schultheiss HP, Lamers J, Poller W (1999) Expression of coxsackie adenovirus receptor and alphav-integrin does not correlate with adenovector targeting in vivo indicating anatomical vector barriers. Gene Ther 6:1520–35

Fechner H, Wang X, Wang H, Jansen A, Pauschinger M, Scherubl H, Bergelson JM, Schultheiss HP, Poller W (2000) Trans-complementation of vector replication versus Coxsackie-adenovirus- receptor overexpression to improve transgene expression in poorly permissive cancer cells. Gene Ther 7:1954–68

Freimuth P, Springer K, Berar, C, Hainfeld J, Bewley M, Flanagan J (1999) Coxsackievirus and adenovirus receptor amino-terminal immunoglobulin V- related domain binds adenovirus type 2 and fiber knob from adenovirus type 12. J Virol 73:1392–8

Fuxe J, Lui L, Malin S, Philipson L, Collins P, Pettersson RF (2003) Expression of the Coxsackie and adenovirus receptor in human astrocytic tumors and xenografts. Int J Cancer 103:729–729

Grubb BR, Pickles RJ, Ye H, Yankaskas JR, Vick RN, Engelhardt JF, Wilson JM, Johnson LG, Boucher RC (1994) Inefficient gene transfer by adenovirus vector to cystic fibrosis airway epithelia of mice and humans. Nature 371:802–6

Hattori M, Fujiyama A, Taylor TD, Watanabe H, Yada T, Park HS, Toyoda A, Ishii K, et al (2000) The DNA sequence of human chromosome 21. Nature 405:311–9

He Y, Chipman PR, Howitt J, Bator CM, Whitt MA, Baker TS, Kuhn RJ, Anderson CW, Freimuth P, Rossmann MG (2001) Interaction of coxsackievirus B3 with the full length coxsackievirus- adenovirus receptor. Nat Struct Biol 8:874–8

Heath JK, White SJ, Johnstone CN, Catimel B, Simpson RJ, Moritz RL, Tu GF, Ji H, Whitehead RH, Groenen LC, Scott AM, Ritter G, Cohen L, Welt S, Old LJ, Nice EC, Burgess AW (1997) The human A33 antigen is a transmembrane glycoprotein and a novel member of the immunoglobulin superfamily. Proc Natl Acad Sci U S A 94,:469–74

Heideman DA, Snijders PJ, Craanen ME, Bloemena E, Meijer CJ, Meuwissen SG, van Beusechem VW, Pinedo HM, Curiel DT, Haisma HJ, Gerritsen WR (2001) Selective gene delivery toward gastric and esophageal adenocarcinoma cells via Ep-CAM-targeted adenoviral vectors. Cancer Gene Ther 8:342–51

Hemmi S, Geertsen R, Mezzacasa A, Peter I,Dummer R (1998) The presence of human coxsackievirus and adenovirus receptor is associated with efficient adenovirus-mediated transgene expression in human melanoma cell cultures. Hum Gene Ther 9:2363–73

Hilleman MR Werner JH (1954) Recovery of a new agent from patients with acute respiratory illness. Proc Soc Exp Biol Med 85:183–188

Honda T, Saitoh H, Masuko M, Katagiri-Abe T, Tominaga K, Kozakai I, Kobayashi K, Kumanishi T, Watanabe YG, Odani S, Kuwano R (2000) The coxsackievirus-adenovirus receptor protein as a cell adhesion molecule in the developing mouse brain. Brain Res Mol Brain Res 77:19–28

Hong SS, Karayan L, Tournier J, Curiel DT, Boulanger PA (1997) Adenovirus type 5 fiber knob binds to MHC class I alpha2 domain at the surface of human epithelial and B lymphoblastoid cells. EMBO J 16:2294–306

Hsu KH, Lonberg-Holm K, Alstein B, Crowell RL (1988) A monoclonal antibody specific for the cellular receptor for the group B coxsackieviruses. J Virol 62:1647–52

Huang MR, Olsson M, Kallin A, Pettersson U, Totterman TH (1997) Efficient adenovirus-mediated gene transduction of normal and leukemic hematopoietic cells. Gene Ther 4:1093–9

Hutchin ME, Pickles RJ, Yarbrough WG (2000) Efficiency of adenovirus-mediated gene transfer to oropharyngeal epithelial cells correlates with cellular differentiation and human coxsackie and adenovirus receptor expression. Hum Gene Ther 11:2365–75

Ito M, Kodama M, Masuko M, Yamaura M, Fuse K, Uesugi Y, Hirono S, Okura Y, Kato K, Hotta Y, Honda T, Kuwano R, Aizawa Y(2000) Expression of coxsackievirus and adenovirus receptor in hearts of rats with experimental autoimmune myocarditis. Circ Res 86:275–80

Johansson CB, Momma S, Clarke DL, Risling M, Lendahl U, Frisen J (1999) Identification of a neural stem cell in the adult mammalian central nervous system. Cell 96:25–34

Johnstone CN, Tebbutt NC, Abud HE, White SJ, Stenvers KL, Hall NE, Cody SH, Whitehead RH, Catimel B, Nice EC, Burgess AW, Heath JK (2000) Characterization of mouse A33 antigen, a definitive marker for basolateral surfaces of intestinal epithelial cells. Am J Physiol Gastrointest Liver Physiol 27: G500–10

Katevuo K, Imhof BA, Boyd R, Chidgey A, Bean A, Dunon D, Gobel TW, Vainio O (1999) ChT1, an Ig superfamily molecule required for T cell differentiation. J Immunol 162:5685–94

Krasnykh V, Dmitriev I, Mikheeva G, Miller CR, Belousova N, Curiel DT (1998 Characterization of an adenovirus vector containing a heterologous peptide epitope in the HI loop of the fiber knob. J Virol 72:1844–52

Krasnykh V, Dmitriev I, Navarro JG, Belousova N, Kashentseva E, Xiang J, Douglas JT, Curiel DT (2000) Advanced generation adenoviral vectors possess augmented gene transfer efficiency based upon coxsackie adenovirus receptor-independent cellular entry capacity. Cancer Res 60:6784–7

Leon, R. P., Hedlund, T., Meech, S. J., Li, S., Schaack, J., Hunger, S. P., Duke, R. C., and DeGregori, J. (1998). Adenoviral-mediated gene transfer in lymphocytes. Proc Natl Acad Sci U S A 95, 13159–64

Li E, Brown SL, von Seggern DJ, Brown GB, Nemerow GR (2000) Signaling antibodies complexed with adenovirus circumvent CAR and integrin interactions and improve gene delivery. Gene Ther 7:1593–9

Li Y, Pong RC, Bergelson JM, Hall MC, Sagalowsky AI, Tseng CP, Wang Z, Hsieh JT (1999) Loss of adenoviral receptor expression in human bladder cancer cells: a potential impact on the efficacy of gene therapy. Cancer Res 59:325–30

Lonberg-Holm K, Crowell R, Philipson L (1976) Unrelated animal viruses share receptors. Nature (London) 259:679–681

Loskog A, HT, Wester K, de la Torre M, Philipson L, Malmstrom P-U, Totterman TH (2002) Human urinary bladder carcinomas express adenovirus attachment and internalization receptors. Gene Ther, in press.

McDonald D, Stockwin L, Matzow T, Blair Zajdel ME, Blair GE (1999) Coxsackie and adenovirus receptor (CAR)-dependent and major histocompatibility complex (MHC) class I-independent uptake of recombinant adenoviruses into human tumour cells. Gene Ther 6:512–9

Mentel R, Dopping G, Wegner U, Seidel W, Liebermann H, Dohner L (1997) Adenovirus-receptor interaction with human lymphocytes. J Med Virol 51:252–7

Miller CR, Buchsbaum DJ, Reynolds PN, Douglas JT, Gillespie GY, Mayo MS, Raben D, Curiel DT (1998) Differential susceptibility of primary and established human glioma cells to adenovirus infection: targeting via the epidermal growth factor receptor achieves fiber receptor-independent gene transfer. Cancer Res 58:5738–48

Mori T, Arakaw, H, Tokin, T, Mineura K, Nakamura Y (1999) Significant increase of adenovirus infectivity in glioma cell lines by extracellular domain of hCAR. Oncol Res 11:513–21

Muckelbauer JK, Kremer M, Minor I, Diana G, Dutko FJ, Groarke J, Pevear DC, Rossmann MG (1995) The structure of coxsackievirus B3 at 3.5 A resolution. Structure 3:653–67

Nalbantoglu J, Pari G, Karpati G, Holland PC (1999) Expression of the primary coxsackie and adenovirus receptor is downregulated during skeletal muscle maturation and limits the efficacy of adenovirus-mediated gene delivery to muscle cells. Hum Gene Ther 10:1009-19

Okegawa T, Li Y, Pong RC, Bergelson JM, Zhou J, Hsieh JT (2000) The dual impact of coxsackie and adenovirus receptor expression on human prostate cancer gene therapy. Cancer Res 60:5031–6

Okegawa T, Pong RC, Li Y, Bergelson JM, Sagalowsky AI, Hsieh JT (2001) The mechanism of the growth-inhibitory effect of coxsackie and adenovirus receptor (CAR) on human bladder cancer: a functional analysis of car protein structure. Cancer Res 61:6592–600

Philipson L, Lonberg-Holm K, Pettersson U (1968) Virus-receptor interaction in an adenovirus system. J Virol 2:1064–75

Pickles RJ, Fahrner JA, Petrella JM, Boucher RC, Bergelson JM (2000) Retargeting the coxsackievirus and adenovirus receptor to the apical surface of polarized epithelial cells reveals the glycocalyx as a barrier to adenovirus-mediated gene transfer. J Virol 74:6050–7

Pickles RJ, McCarty D, Matsui H, Hart PJ, Randell SH, Boucher RC (1998) Limited entry of adenovirus vectors into well-differentiated airway epithelium is responsible for inefficient gene transfer. J Virol 72:6014–23

Rebel VI, Hartnett S, Denham J, Chan M, Finberg R, Sieff CA (2000) Maturation and lineage-specific expression of the coxsackie and adenovirus receptor in hematopoietic cells. Stem Cells 18:176–82

Roelvink PW, Lizonova A, Lee JG, Li Y, Bergelson JM, Finberg RW, Brough DE, Kovesdi I, Wickham TJ (1998) The coxsackievirus-adenovirus receptor protein can function as a cellular attachment protein for adenovirus serotypes from subgroups A, C, D, E, and F. J Virol 72:7909–15

Roelvink PW, Lee GM, Einfeld DA, Kovesdi I, Wickham TJ (1999) Identification of a conserved receptor-binding site on the fiber proteins of CAR-recognizing adenoviridae. Science 286:1568–71

Rossmann MG, Arnold E, Erickson JW, Frankenberger EA, Griffith JP, Hecht HJ, Johnson JE, Kamer G, Luo M, Mosser AG, et al. (1985) Structure of a human common cold virus and functional relationship to other picornaviruses. Nature 317:145–53

Shafren DR, Bates RC, Agrez MV, Herd RL, Burns GF, Barry RD (1995) Coxsackieviruses B1, B3, and B5 use decay accelerating factor as a receptor for cell attachment. J Virol 69:3873–7

Stevenson SC, Rollence M, Marshall-Neff J, McClelland A (1997) Selective targeting of human cells by a chimeric adenovirus vector containing a modified fiber protein. J Virol 71:4782–90

Tallone T, Malin S, Samuelsson A, Wilbertz J, Miyahara M, Okamoto K, Poellinger L, Philipson L, and Pettersson S (2001) A mouse model for adenovirus gene delivery. Proc Natl Acad Sci U S A 98:7910–5

Tomko RP, Johansson CB, Totrov M, Abagyan R, Frisen J, Philipson L (2000) Expression of the adenovirus receptor and its interaction with the fiber knob. Exp Cell Res 255:47–55

Tomko RP, Xu R, Philipson L (1997) HCAR and MCAR: the human and mouse cellular receptors for subgroup C adenoviruses and group B coxsackieviruses. Proc Natl Acad Sci USA 94:3352–6

Trepel M, Grifman M, Weitzman MD, Pasqualini R (2000) Molecular adaptors for vascular-targeted adenoviral gene delivery. Hum Gene Ther 11:1971–81

van Raaij MJ, Chouin E, van der Zandt H, Bergelson JM, Cusack S (2000) Dimeric structure of the coxsackievirus and adenovirus receptor D1 domain at 1.7 A resolution. Structure Fold Des 8:1147–55

Walters RW, Grunst T, Bergelson JM, Finberg RW, Welsh MJ, Zabner J (1999) Basolateral localization of fiber receptors limits adenovirus infection from the apical surface of airway epithelia. J Biol Chem 274:10219–26

Walters RW, van't Hof W, Yi SM, Schroth MK, Zabner J, Crystal RG, Welsh MJ (2001) Apical localization of the coxsackie-adenovirus receptor by glycosyl- phosphatidylinositol modification is sufficient for adenovirus-mediated gene transfer through the apical surface of human airway epithelia. J Virol 75:7703–11

Wang X, Bergelson JM(1999) Coxsackievirus and adenovirus receptor cytoplasmic and transmembrane domains are not essential for coxsackievirus and adenovirus infection. J Virol 73:2559–62

Wesseling JG, Bosma PJ, Krasnykh V, Kashentseva EA, Blackwell J., Reynold, PN, Li H, Parameshwar M, Vickers SM, Jaffee EM, Huibregtse K, Curiel DT, Dmitriev I (2001) Improved gene transfer efficiency to primary and established human pancreatic carcinoma target cells via epidermal growth factor receptor and integrin-targeted adenoviral vectors. Gene Ther 8:969–76

Wickham TJ (2000) Targeting adenovirus. Gene Ther 7:110–4

Wickham TJ, Mathias P, Cheresh DA, Nemerow GR (1993) Integrins alpha v beta 3 and alpha v beta 5 promote adenovirus internalization but not virus attachment. Cell 73:309–19

Xia D, Henry L J, Gerard RD, Deisenhofer J (1994) Crystal structure of the receptor-binding domain of adenovirus type 5 fiber protein at 1.7 A resolution. Structure 2:1259–70

Xu R, Mohanty JG, Crowell RL (1995) Receptor proteins on newborn Balb/c mouse brain cells for coxsackievirus B3 are immunologically distinct from those on HeLa cells. Virus Res 35:323–40

You Z, Fischer DC, Tong X, Hasenburg A, Aguilar-Cordova E, Kieback D G (2001) Coxsackievirus-adenovirus receptor expression in ovarian cancer cell lines is associated with increased adenovirus transduction efficiency and transgene expression. Cancer Gene Ther 8:168–75

[illegible]

Mechanisms of E3 Modulation of Immune and Inflammatory Responses

S. P. Fessler · F. Delgado-Lopez · M. S. Horwitz

Department of Microbiology and Immunology, Albert Einstein College of Medicine, 1300 Morris Park Ave., Bronx, New York, NY 10461, USA

E-mail: horwitz@aecom.yu.edu

Abstract Adenoviruses contain genes that have evolved to control the host immune and inflammatory responses; however, it is not clear whether these genes function primarily to facilitate survival of the virus during acute infection or during its persistent phase. These issues have assumed greater importance as the use of adenoviruses as vectors for gene therapy has been expanded. This review will focus on the mechanism of immune evasion mediated by the proteins encoded within the early region 3 (E3) transcription region, which affect the functions of a number of cell surface receptors including Fas, intracellular cell signaling events involving NF-κB, and the secretion of pro-inflammatory molecules such as chemokines. The successful use of E3 genes in facilitating allogeneic transplantation and in preventing autoimmune diabetes in several transgenic mouse models will also be described.

1 Introduction

Fifty-one different adenovirus (Ad) types that infect humans have been isolated to date. Although many of these Ads do not cause disease, infections from a number of them are pathogenic and sometimes fatal (Horwitz 2001). Studies of the morbidity and mortality resulting from this virus are certainly of interest, and infections of immunosuppressed patients are especially problematic (Kojaoghlanian et al. 2003; Flomenberg et al. 1994). In addition to their clinical relevance, the potential use of Ads as gene therapy vectors to achieve both short- and long-term therapeutic effects has been extensively studied by many groups (Roy-Chowdhury and Horwitz 2002; Hitt and Graham 2000). Relevant to both issues is the contribution of a number of viral genes responsible for blocking the antiviral arms of the host immune system.

Several Ad early genes such as E1, E3, E4, the late protein L4–100K, and the Pol III transcript VA RNA are involved in counteracting the host immune response (reviewed in Horwitz 2001; Wold et al. 1999; Andrade et al. 2001). The functions of AdE1 and E4 genes are discussed in other chapters of this book. Of interest to our laboratory is the E3 region, which contains a number of critical immunoregulatory genes. Deletion of E3 genes in the context of infection with either a virus or viral vector has effects on acute viral pathogenesis and possibly on vector persistence. In the cotton rat (*Sigmodon hispidus*) model of viral infection, deletion of E3 genes resulted in an increase of inflammation and cellular infiltrate in the lung (Ginsberg et al. 1989). A more recent paper by Wen et al. (2001) demonstrates a similar consequence of E3 deletion in vectors used in arterial gene transfer in a mouse model.

The E3 region of adenovirus encodes a number of different proteins that regulate apoptosis (Fig. 1). Only one E3 protein, 11.6K, also known as the adenovirus death protein (ADP) is known to promote cell death and release of progeny virus from the nucleus late in infection (Tollefson et al. 1996a,b). The known functions of the other genes within the E3 region act to protect infected cells from cytolysis induced by a number of immune and inflammatory mechanisms. In this review, the known im-

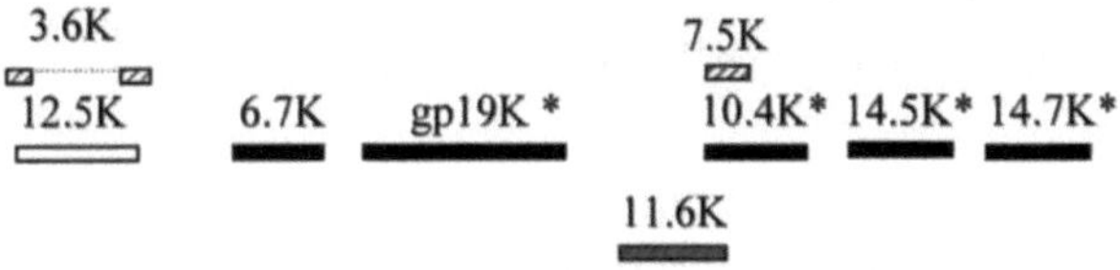

- gp19K downregulates cell-surface expression of MHC I by binding and retaining it in ER; also inhibits tapasin.
- 14.7K inhibits TNF and Fas cytolysis, arachidonic acid release by TNF
- 10.4K/14.5K (RID α/β = Receptor Internalization and Degradation). Targets Fas and TRAIL receptors for endosomal internalization and lysosomal degradation. Blocks cytolysis and arachidonic acid release by TNF. Blocks chemokine synthesis and NF-κB activation by TNF.
- 6.7K may cooperate with RID to internalize and degrade TRAIL-R1
- 11.6K (ADP = Adenovirus Death Protein) promotes cell death and viral release late in infection.

Unknown Function
Putative ORF
Proapoptotic Protein
Immunosuppressive Protein
* Conserved Among Multiple Groups

Fig. 1 E3 open reading frames. Shown is a map of E3 ORFs and the functions of the proteins encoded. This map is based on the E3 region of Ad2 or Ad5

munosuppressive mechanisms of action of the members of the E3 region will be discussed, including recent data on a number of new functions of some of these proteins as well as data on the successful use of these proteins to control the autoimmune response in type one diabetes (TID). Functions of these proteins will be discussed and classified by their sites of action. For example, at the cell surface, different cellular receptors are altered; within the cell, various intracellular proteins and processes involved in signal transduction, apoptosis, and inflammation are affected; and there are also profound effects on a number of secreted molecules such as chemokines or other pro-inflammatory molecules which are inhibited by proteins coded in the E3 region. However, none of the AdE3 proteins themselves are known to be secreted.

2
Regulation of Cellular Receptors by Ad E3 Genes

Regulation of the structure and function of cell receptors involves three proteins: gp19K, a 19-kDa glycoprotein localized to the endoplasmic reticulum (ER), and 10.4K and 14.5K, which act together as a complex at the plasma membrane. The function of gp19K as an inhibitor of class I MHC has been known for a number of years and seems to be limited to this receptor. The AdE3–10.4K and 14.5K proteins (also known as RIDα and RIDβ for their effects on *r*eceptor *i*nternalization and *d*egradation) act together in a complex on the plasma membrane. RIDα/β affects a number of surface receptors and probably involves other inhibitory mechanisms which depend on signal transduction.

2.1
AdE3–gp19K

The 19K glycoprotein contained in all Ad groups except for the highly oncogenic (in rodents) Group A agents (Deryckere and Burgert 1996; reviewed in Burgert and Blusch 2000) has been shown to act on the MHC class I receptor, retaining it in the ER and preventing its transport to the cell surface (Burgert and Kvist 1985; Burgert et al. 1987; Andersson et al. 1987). This severely limits the recognition and targeted destruction of these cells by cytotoxic T lymphocytes (CTLs) (Rawle et al. 1991; Burgert et al. 1987). The affinity of gp19K for murine MHC class I haplotypes varies but it does bind to all human MHC class I haplotypes, albeit with different affinities, and promotes their cell surface downregulation (Beier et al. 1994; Korner and Burgert 1994), resulting in their protection from CTL-mediated cytolysis (Flomenberg et al. 1996). An additional function of gp19K in the inhibition of MHC class I processes was demonstrated by Bennett et al. (1999), who showed that gp19K inhibited tapasin processing of polypeptides for presentation to MHC. This results in the blocking of antigen presentation, which augments the physical downregulation of class I MHC at the cell surface by the ER retention mechanism.

2.2
AdE3-10.4K/14.5K (RIDα/β)

The heteromultimeric complex of 10.4K/14.5K proteins has been the subject of study for a number of years, since it was shown to have a role in downregulation of surface levels of the EGF receptor (Carlin et al. 1989). The retention of RIDα/β across subgroups and serotypes suggests the importance of its role in adenovirus infection despite the vast differences in the biology of the different Ads. Results of further studies of RID have suggested more pathologically relevant roles for this complex, as inhibitory effects on both Fas and TRAIL signaling have been demonstrated. Analysis of viral mutants has shown a RID-mediated protection from FasL-mediated cytolysis (Shisler et al. 1997; Tollefson et al. 1998; Elsing and Burgert 1998) accompanied by a decrease in cell surface levels of Fas. This receptor was shown to be internalized through endosomes and degraded in lysosomes. Surface levels of CD40 and TNF-R1 were unaffected. Surface expression of TNF-R1 was shown to be only minimally affected at times postinfection when RID was able to inhibit TNF-induced activation of NF-κB, suggesting a different mechanism for anticytolytic effects exhibited against TNF signaling (Friedman and Horwitz 2002). RID's effects on Fas were shown to occur in some but not all B and T lymphocytes, and protection from cytolysis was shown to correlate with receptor internalization and degradation (McNees et al. 2002).

The involvement of RID in protection against TRAIL cytolysis was demonstrated by two different groups (Benedict et al. 2001; Tollefson et al. 2001). Benedict et al. showed a requirement for E3-6.7K in internalization and degradation of TRAIL-R1; however, Tollefson and colleagues demonstrated that RID alone can target both TRAIL-R1 and R2 for destruction and demonstrated a role of the cytoplasmic domain of the target receptor in achieving this effect.

Both RIDα and RIDβ are transmembrane proteins (Stewart et al. 1995) and both are posttranslationally modified. The complex is comprised of two molecules of 10.4K which are covalently joined by a disulfide bond at Cys-31 on the extracellular side of the membrane (Hoffman et al. 1992) and one molecule of 14.5K which is not covalently linked to the complex but easily copurified (Gooding et al. 1991). RIDβ is modified by glycosylation on the extracellular side of the plasma membrane (Krajcsi et al. 1992) and phosphorylated on the cytoplasmic tail domain (Krajcsi and Wold 1992). Proteolytic processing of both molecules occurs: 14.5K is processed between residues 17 and 18 (Krajcsi et al. 1992), and one of the two molecules of the 10.4K complex is cleaved between Ala-22 and Ala-23 (Krajcsi et al. 1992). The lack of processing of the second 10.4K molecule results in its spanning the membrane a second time within the uncleaved portion of the protein. The proper location sites of proteolytic processing and other modification of these proteins are dependent upon the presence of both of these proteins (Krajcsi and Wold 1992; Krajcsi et al. 1992).

Mutagenic analysis of RIDβ demonstrated the necessity of the Tyr-124 residue in this mechanism of RID action. Stability and localization of RID were unaffected, but mutating this residue to alanine abolished downregulation of TRAIL, Fas and EGF receptors and protection from cytolytic effects of TRAIL and Fas. Tyr-124 is located within a broadly used sorting motif (Yppϕ) in clathrin-mediated trafficking, allowing for the binding of proteins to adaptor proteins (reviewed in Kirchhausen 1999). Interestingly, mutating the phosphorylation site at Ser-116 had no effect on receptor trafficking (Lichtenstein et al. 2002).

3 Regulation of Intracellular Processes

3.1 AdE3–14.7K

This protein has been the subject of study of a number of groups for its anti-TNF cytolytic activity, although the mechanism of its protection has not been completely solved. This protein was shown to be present in a number of different serotypes (Horton et al. 1990), and deletion mutants of Ad4 that included 14.7K were shown to be sensitive to TNF cytolysis (Horton et al. 1990). Our studies of the E3 region of Ad35, a member of group B, demonstrated the presence of a homologue of 14.7K, designated 15.3K by its predicted molecular weight (Basler et al. 1996; Basler and Horwitz 1996). Expression of this protein is detectable by an antibody raised in our laboratory, and this protein presumably contributes to the resistance of Ad35 to TNF cytolysis (Buntzman et al. 2003).

Protection against TNF cytolysis by E3–14.7 was described in murine cells (Gooding et al. 1990) and was shown to occur independent of other proteins (Horton et al. 1991). The TNF-induced release of arachidonic acid (AA) from membranes by cytoplasmic phospholipase A2 (cPLA2) was shown to be inhibited by AdE3–14.7K (Zilli et al. 1992); however, the translocation of cPLA2 to membranes was not inhibited (Dimitrov et al. 1997). The inhibition of AA release by the RID complex does involve a block in translocation and subsequent activation of cPLA2 (Dimitrov et al. 1997). 14.7K was shown to block Fas-induced cytolysis as well as cytolysis from transfected FADD and FLICE (caspase 8) (Chen et al. 1998). This protein was shown to interact with FLICE (Chen et al. 1998; Kim and Foster 2002), probably accounting for some of its anti-apoptotic effects. Structural analysis of this protein demonstrated that its ability to protect against TNF cytolysis was compromised as a result of most short (7–16 a.a.) deletion mutants made throughout the whole protein. Ablation of both its stability and function was achieved by mutation of a number of cysteine residues (Ranheim et al. 1993), consistent with its ability and putative functional requirement to bind zinc (Kim and Foster 2002). Interestingly, this protein was shown to exist preferentially as a nonamer in solution (Kim and Foster 2002).

Another method of studying the function of 14.7K was to identify its interacting partners. This was done in our laboratory by the yeast two-hybrid technique and resulted in the identification of a number of proteins that were novel at the time of their isolation. These proteins, designated FIPs (for *14.7K-Interacting Proteins*), have been shown by us and by others to have interesting functions and localizations (Fig. 2).

3.2
FIP-1

FIP-1 (Li et al. 1997) is a small GTPase which was also identified as Rag A (Schurmann et al. 1995) with a yeast homologue known as GTR1 (Hirose et al. 1998). Through experiments in yeast, this protein was shown to be involved in the Ran/RCC1/RanGAP pathway, which regulates the cell-cycle nuclear envelope formation and nucleocytoplasmic transport (Nakashima et al. 1996; Hirose et al. 1998). The effects of this protein on TNF signaling have not been fully elucidated; however, FIP-1 was shown to interact with a number of proteins that undergo phosphorylation in response to TNF (Li et al. 1997). Identification of interacting proteins of FIP-1 yielded two GIPs (*GTPase-Interacting Proteins*). GIP-1 is a light chain component (TCTEL) of dynein involved in transport along microtubules (Lukashok et al. 2000). GIP-2, also identified as Rag

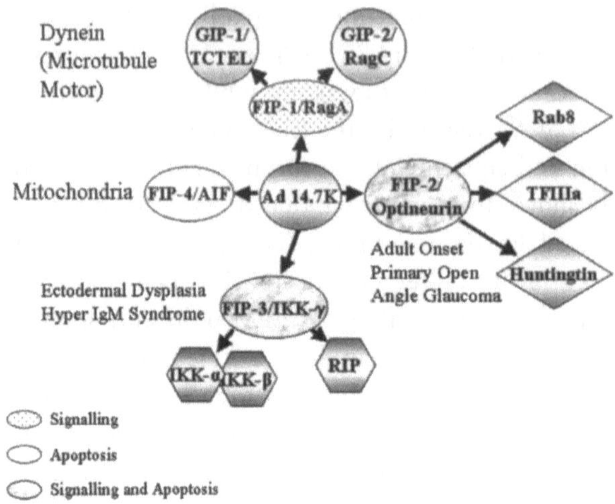

Fig. 2 E3-interacting proteins. Interactions between 14.7 and the four FIPs are depicted, along with their interacting proteins as determined by us and others. Roles of the FIPs in signal transduction (*FIPs 1–3*) and apoptosis (*FIPs 2–4*) are indicated. Subcellular localization of different proteins as well as diseases associated with mutations in FIP-2 and FIP-3 are indicated *adjacent* to the proteins

C (Sekiguchi et al. 2001), is another small GTPase whose yeast homologue, GTR2, was shown to interact with GTR1 and participate in the same pathway (Nakashima et al. 1999).

3.3
FIP-2

The involvement of FIP-2 in the TNF signaling pathway was first demonstrated by the accumulation of its RNA in response to TNF (Fig. 3) and by its ability to reverse the protective effects of 14.7K on TNF-induced cytolysis (Li et al. 1998). Subsequent work demonstrated induction of FIP-2 protein levels by interferons and TNF (Schwamborn et al. 2000) and an increase in FIP-2 phosphorylation in response to PMA, resulting in a decrease in its half-life (Schwamborn et al. 2000). This work also showed FIP-2's association with two novel kinase activities but not with IKK. FIP-2 also interacts with Huntingtin protein and Rab8 (Hattula and Peranen 2000) as well as TFIIIa (Moreland et al. 2000), and mutations within FIP-2 are associated with adult-onset primary open angle glauco-

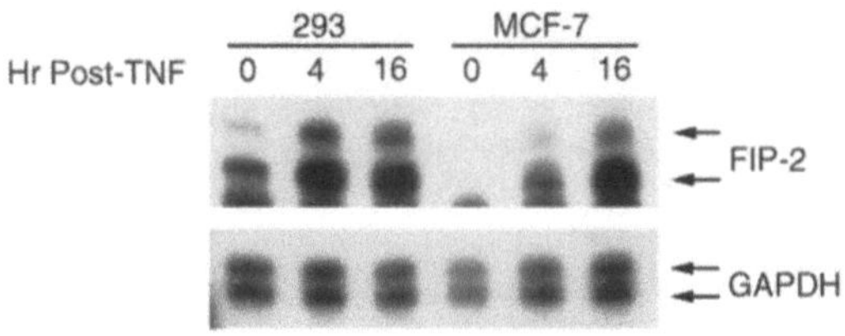

Fig. 3 Induction of FIP-2 expression by TNF. 293 cells and adenocarcinoma MCF-7 cells were treated with TNF for the indicated times, and total cellular RNA was used in a ribonuclease protection assay with a FIP-2-specific probe. The *upper band* is of predicted molecular weight, and the *lower band* probably represents an alternatively spliced form of FIP-2 message. Human glyceraldehyde-3-phosphate dehydrogenase (*GAPDH*) was used as an internal control. (Reprinted with permission from Li et al. 1998)

ma (Rezaieet al. 2002). Further examination of the potential role of FIP-2 in signaling has been undertaken in our laboratory.

3.4
FIP-3

The effects of FIP-3 on the activity of the NF-κB pathway were shown by us (Li et al. 1999) and other groups (Yamaoka et al. 1998; Rothwarf et al. 1998; Mercurio et al. 1999), who also isolated the protein, reported it as NEMO, IKK-γ and IKKAP-1 and characterized the protein as an integral subunit of the IKK complex. Although NEMO/IKKγ is required for NF-κB activity, we have shown that the overexpression of FIP-3 resulted in the repression of basal and TNF-stimulated NF-κB activity. Its nuclear localization upon overexpression, induction of apoptosis, and formation of inclusion bodies of co-transfected GFP were counteracted by co-transfection with 14.7K plasmid (Li et al. 1999) and that of its Ad35 homologue 15.3K (data not shown). Apoptotic induction by FIP-3 is not dependent upon its signaling activity (Ye et al. 2000), but is sensitive to mutagenesis in multiple regions of the protein (Fig. 4A). The stimulation by FIP-3 overexpression of AP-1 activation by c-jun phosphorylation has also been shown by us and others. FIP-3 is also phosphorylated in response to TNF on several serine residues (Tarassishin and Horwitz 2001) (Fig. 4B). FIP-3 has been shown to be mutated in several ectodermal dysplasias (Zonana et al. 2000) including a fatal form of incontinentia pigmenti in male infants (Aradhya et al. 2001; Dufke et al. 2001). The

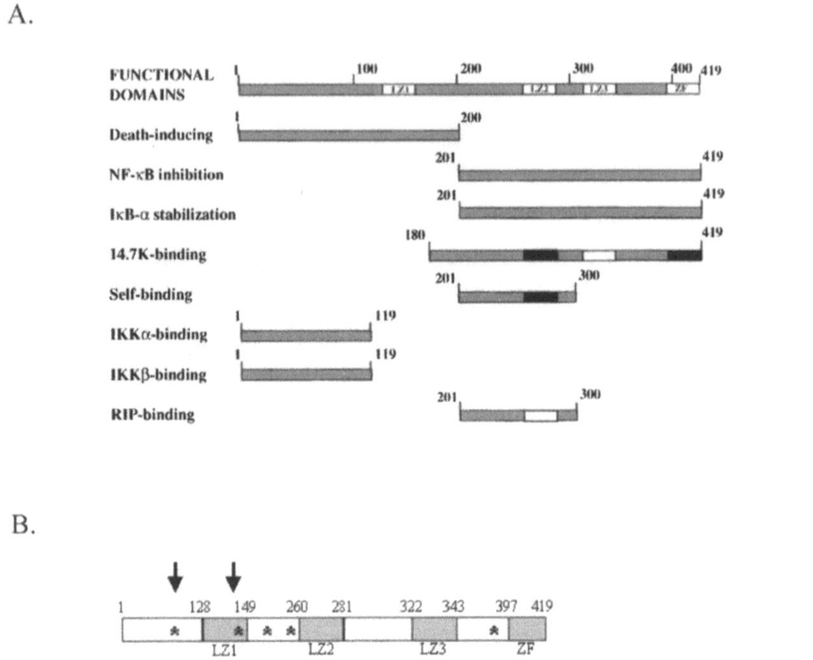

Fig. 4 Functional domains of FIP-3. **A** Schematic diagram showing the domains of FIP-3 responsible for its functions. *Shaded areas* indicate regions important for a specified function. LZ or ZF subdomains that are important for a specified interaction are further *darkened*, and those that do not appear to be important for a particular function are left *blank*. Reprinted with permission from Ye et al. 2000. **B** Map of FIP-3 indicating potential phosphorylation sites (*) mutated in experiments by Tarassishin and Horwitz (2001). *Arrows* indicate residues necessary for IKK-β activity. *LZ*, conserved leucine zipper motif; *ZF*, conserved zinc finger motif. (Reprinted from Tarassishin and Horwitz 2001 (copyright 2001) with permission from Elsevier)

gene for FIP-3 is located at the end of the X-chromosome adjacent to the G6PD locus. FIP-3 is also mutated in one form of the hyper-IgM deficiency syndrome, and such patients do not class-switch the products of their B cells from IgM to IgG to make higher-affinity antibodies (Jain et al. 2001). Patients with ectodermal dysplasia resulting from FIP-3 mutation also have a deficiency in natural killer (NK) cell cytotoxicity (Orange et al. 2002).

3.5 FIP-4

The interaction of FIP-4 and 14.7K was recognized in our laboratory (Li 1996) before the identification of FIP-4 by Susin et al. (1999) as apoptosis-inducing factor (AIF), a mitochondrial protein whose translocation into the nucleus in response to apoptotic stimuli is a critical step in apoptosis. The significance of the interaction of 14.7K with FIP-4/AIF with respect to the anti-apoptotic effects of 14.7K has not been determined. However, FIP-4/AIF is released from mitochondria and appears in the nucleus very late in the viral infectious cycle (Fessler and Horwitz 2003).

3.6 NF-κB

The adenovirus RID complex was shown to foster protection of a number of murine cell lines from TNF-induced cytolysis independent of the presence of AdE3–14.7K (Gooding et al. 1991). Other work describes RID's ability to block AA release from the plasma membrane in response to TNF stimulation by blocking the activation and translocation of cPLA2 early in infection. However, at this time postinfection NF-κB activation was not blocked (Dimitrov et al. 1997). Recent results from our laboratory demonstrated that RID can inhibit TNF-induced activation of NF-κB and the resultant induction of chemokines. The interest in TNF signaling in our laboratory began with the discovery of FIP-3's activity on NF-κB and AP-1 signaling and the inhibition of the pro-apoptotic effects of transfected FIP-3 by cotransfection of 14.7K. It was hypothesized that Ads may regulate TNF activation of NF-κB and that the E3 region, specifically 14.7K, may be involved in this regulation. In work by Friedman and Horwitz (2002) it was shown that expression of the E3 region in a variety of human cells types blocked the TNF-dependent activation of the NF-κB pathway; specifically the activity of the IKK complex, the resultant degradation of IκB-α, and the binding of NF-κB to its cognate site (Fig. 5). However, further work in our laboratory using viral mutants has shown that this effect is dependent upon RID and no other E3 proteins. The mechanism of action of RID on the TNF signaling pathway is under further study in our laboratory.

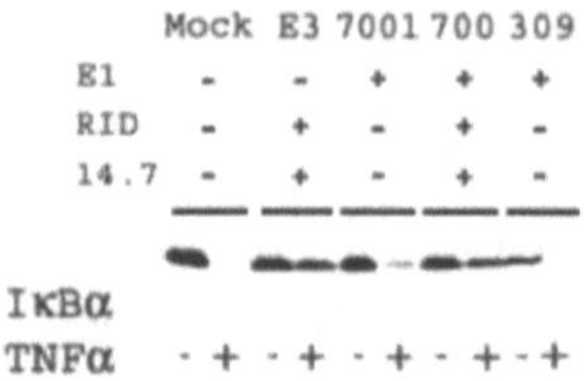

Fig. 5 The Ad E3 but not the Ad E1 region prevents the degradation of IkB-α after TNF-treatment. HeLa cells were uninfected (Mock) or infected with Ad CMV-E3 *dl*7001 (deletion of entire E3 region) rec700 (wild type) or *dl*309 (gp19K$^+$ ADP$^+$ 10.4K$^-$ 14.5K$^-$ 14.7K$^-$) at an MOI of 2,000 particles/cell for 16 h. Relevant E1 or E3 mutations are as indicated. At 30 min prior to harvest, cells were treated with TNF (20 ng/ml). The lysates were analyzed by SDS-PAGE, and Western blotting was done with the IκB-α antibody. Other Ad mutants mapped the inhibitory effect to RIDα/β and not to 14.7K. Equal loading of samples was verified by hybridization with a β-tubulin antibody (data not shown). (Reprinted with permission from Friedman and Horwitz 2002)

4 Regulation of Secreted Factors

4.1 Chemokines

The inhibition of AA release by 14.7K and the RID complex is important in Ad's blocking of inflammation, as AA is converted into a number of pro-inflammatory and leukotactic prostaglandins and leukotrienes (reviewed in Funk 2001). Work in our laboratory has demonstrated a role of the E3 region in blocking the release of other pro-inflammatory secreted factors.

Mononuclear cells accumulate at the site of inflammation in response to the early events triggered by pro-inflammatory molecules such as TNF and IL-1. Most cells express the receptors for these cytokines, and many react to these stimuli by producing a number of chemokines that bind to specific receptors and attract macrophages, neutrophils, and T lymphocytes. The inflammatory response is a strong host defense in the clearance of pathogens, including adenoviruses, and there exists a host of information about the induction of this response by adenovirus and adenovirus-derived vectors, including the induction of chemokine synthesis. For example, in vitro experiments by Leland and Metcalf (1999) showed an induction of IL-8 by Ad7 (group B) but not Ad5 soon after infection. However, it has also been demonstrated that infection of another cell type with an Ad5-derived viral vector lacking E1 and E3 also resulted in early induction of IL-8 (Bruder and Kovesdi 1997). In vivo experiments done by Muruve et al. (1999) also demonstrate an early induction [1–6 h postinfection (hpi)] of chemokines including IP-10 and MCP-1 in mouse liver, resulting from administration of high doses of E1/E3-deleted Ad vectors resulting in a dose-dependent neutrophil-mediated toxicity.

Other studies demonstrated similar results after a much longer period of infection and relatively lower doses of virus. IL-8 was shown to be elevated in vivo in the bronchial alveolar lavage fluid (BALF) of macaque monkeys 3–28 days after infection with an E1/E3-deleted Ad vector used to deliver the cystic fibrosis (CF) gene (Wilmott et al. 1996). Similarly, administration of an Ad vector into the CSF of rhesus monkeys led to elevated levels of IL-8 and IL-6 (Driesse et al. 2000). The lack of immunomodulatory effects normally provided by the E1 and E3 regions may contribute to inflammation after prolonged infection, as shown in these experiments, concordant with a number of studies showing a large reduction in inflammation dependent upon inclusion of the E3 region (Wen et al. 2001; Ginsberg et al. 1989). Work done in our laboratory has focused on the E3 inhibition of TNF-induced chemokines and is summarized below.

Several studies done in our laboratory suggest that Ad E3 proteins expressed from transgenes can abrogate inflammatory cell infiltration in immune-mediated disease processes such as allogeneic islet-cell rejection and autoimmune (Type I) diabetes, as summarized in the next section (Efrat et al. 1995, 2001; von Herrath et al. 1997). Based on this knowledge, we initiated studies to determine whether Ads containing E3 genes could affect the synthesis of a number of pro-inflammatory chemokines, and which of the adenovirus E3 genes might be responsible for this effect independent of the previously described inhibitory effect of the E1 region. We have shown that when the astrocytoma line U373 was activated by the pro-inflammatory molecule TNF, the increase in the chemokines MCP-1, IL-8, and IP-10 transcripts is blocked by a recombinant Ad expressing the E3 genes under CMV promoter control (Lesokhin et al. 2002) (Fig. 6). Further experiments using similar vectors expressing only some of the E3 genes in various combinations to map the genes that are necessary and sufficient for this inhibitory effect are currently underway. Although data exist on the induction of chemokine

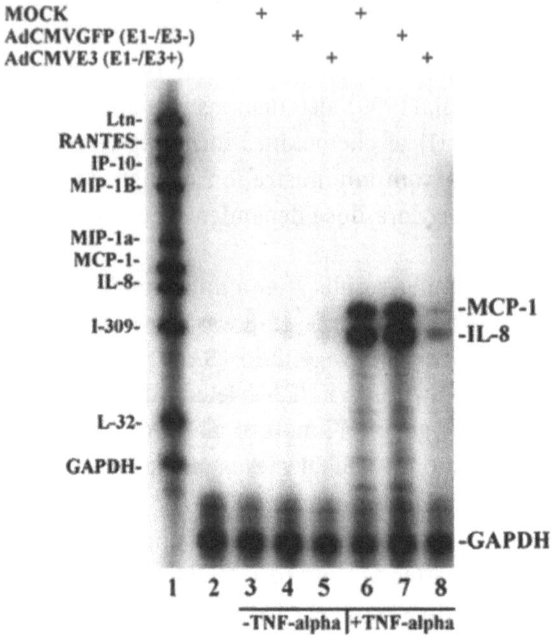

Fig. 6 Ad E3 inhibition of TNF-induced chemokine synthesis in astrocytoma cells. U373 cells were infected with 4,000 particles of virus per cell. TNF was added at 12 hpi, and the samples were harvested at 16 hpi. RNase protection was performed with ^{32}P-labeled probes synthesized from the hck-5 probe set (Pharmingen) using 5 µg of sample RNA. *Lane 1*, undigested hck-5 probe; *lane 2*, probes from lane 1 digested with RNase without RNA sample; *lanes 3–5*, hybridization to RNA from cells that were mock-infected or infected with AdCMVGFP and AdCMVE3 viruses without TNF treatment; *lanes 6–8*, hybridization to RNA from cells that were mock-infected or infected with AdCMVGFP and AdCMVE3 viruses with the subsequent addition of TNF. The protected bands indicated by the labels on the *right* (MCP-1, IL-8, and GAPDH) migrate faster than undigested probes, as expected. (Reprinted from Lesokhin, Delgado-Lopez et al. 2002 with permission)

→

Fig. 7 Expression of Ad E3 genes in β cells inhibits autoimmune diabetes. Fast-onset (**A**) or slow-onset (**B**) LCMV-induced insulin-dependent diabetes mellitus (*IDDM*) is inhibited in mice that express Ad E3 under the control of the rat insulin promoter (*RIP-E3*). Each of four groups (RIP-LCMV, RIP-E3, RIP-LCMV × RIP-E3, and non-transgenic littermates) consisted of ten mice, five of which were infected with LCMV between 6 and 8 weeks of age. Blood glucose was measured at indicated intervals after induction of diabetes by LCMV infection, and blood glucose values exceeding 350 mg/dl were considered to be diabetic. **A** Fast-onset IDDM occurred in RIP-GP

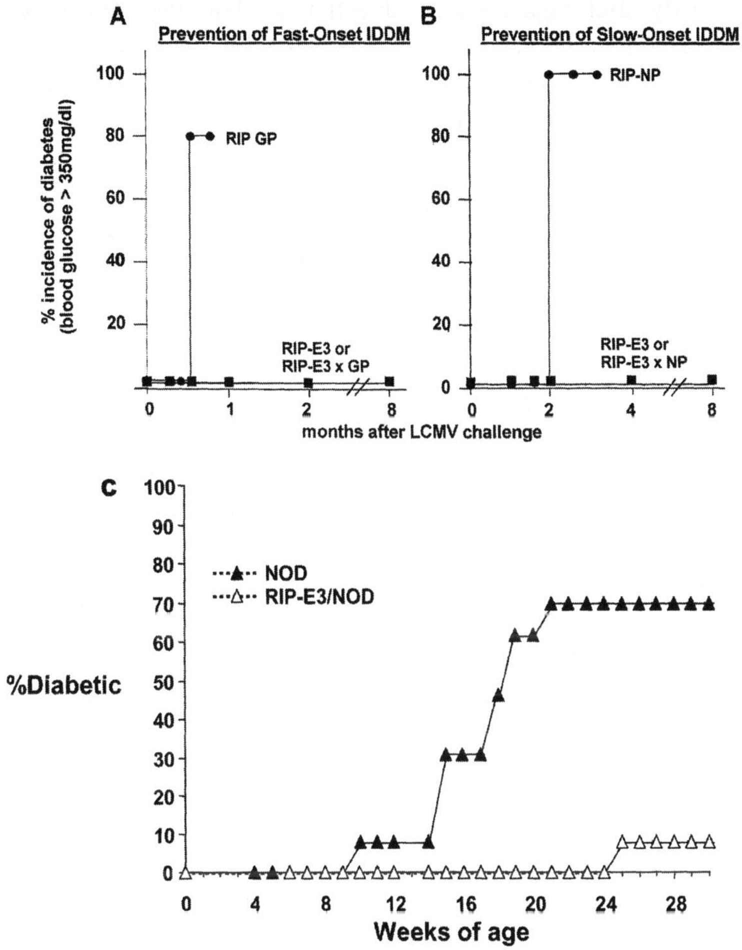

(●) transgenic mice but not in RIP-E3×GP (■) double transgenic mice or single transgenic RIP-E3 controls. **B** Slow-onset IDDM occurred in RIP-NP (●) mice but not in RIP-E3 × NP (■) double transgenic mice or single transgenic RIP-E3 controls. Nontransgenic controls did not show any evidence of diabetes in either the LCMV-infected or uninfected group, and higher LCMV dosages given to double transgenic RIP-E3 × GP mice also did not result in IDDM (data not shown). (**A**, **B**, Reprinted from von Herrath et al. 1997 with permission from the American Diabetes copyright © 2001. American Diabetes Foundation). **C** Decreased incidence of diabetes in the RIP-E3/NOD (nonobese diabetes) transgenic mice. RIP-E3/NOD (△) and nontransgenic NOD (▲) littermate female mice were monitored weekly for plasma glucose levels or glycosuria. Values over 400 mg/dl on two consecutive weekly determinations were considered to be diabetic. (Modified from Efrat et al. 2001, and reprinted with permission copyright 2001, National Academy of Sciences, USA)

synthesis early after high MOI Ad infection as described above, our work also demonstrates that accumulation of the RNA of chemokines to robust levels is only observed in response to cytokines and other external pro-inflammatory molecules and is not seen as a result of viral infection later than 6 hpi.

5 Targeted Immunosuppression by E3 Proteins, Independent of Adenovirus Infection

5.1 Allogeneic Transplantation and Autoimmune Diabetes

Experiments from our laboratory have been directed to utilize the immunoregulatory functions of the AdE3 region to control the immune response during allogeneic transplantation and autoimmune diabetes. For example, we have taken the AdE3 genes out of the context of virus infections and expressed them as transgenes in pancreatic islet β cells under islet-specific promoters. Islets containing AdE3 transgenes were protected from allogeneic rejection following their transplantation under the renal capsule of recipient mice. In two models of murine diabetes, the nonobese diabetic (NOD) mouse and the lymphocytic choriomeningitis virus (LCMV)-induced model, diabetes could be prevented entirely (Fig. 7A, B) or very significantly reduced (Fig. 7C) in the presence of the AdE3 transgenes (von Herrath et al. 1997; Efrat et al. 2001). In the NOD model, in which partial deletions of the AdE3 genes have also been used to construct new transgenic animals, it has been shown that deletions of the E3B proteins (RIDα/β and 14.7K) reduced the protective effect more than deletions of gp19K (Pierce et al. 2003); however, it appears that all of the AdE3 genes were necessary for the maximal protective effects. In addition to the local effect of tolerizing islets by adding E3 genes, splenocytes isolated from animals that possess the Ad E3 transgenes in β cells were partially tolerized. When such splenocytes are transfused into NOD-*scid* animals, they reject islets much more slowly than splenocytes which matured in animal in the absence of the AdE3 (Efrat et al. 2001).

6
Conclusions

The evolutionary conservation of immunomodulatory proteins, which are present in most or all adenovirus subgroups, is remarkable considering the difference in biology and pathogenesis of adenoviruses. Although the effects of E3 proteins can be quite rapid and protect cells undergoing productive infection from cytokine-induced cytolysis (Tollefson et al. 2001), other properties of these proteins and their expression highlight the complexity of the life cycle of this virus. For example, the presence of NF-κB sites in the E3 promoter and the resulting inducibility of the synthesis of these proteins in response to TNF (Deryckere and Burgert 1996) as well as T-cell activation (Mahr et al. 2003) suggest a role of the E3 region in persistent or latent infections, during which periodic inflammatory challenges would be a challenge to viral survival. Exploring the mechanism of action of the E3 proteins continues to reveal novel inflammatory pathways controlled during adenovirus infection. The ability that cloning offers to move AdE3 genes into novel biological situations, such as facilitation of foreign tissue transplantation and prevention of autoimmunity, promises new approaches to the therapy of nonviral diseases in which control of the immune system is important.

References

Andersson, M., McMichael, A., and Peterson, P. A. (1987) Reduced allorecognition of adenovirus 2 infected cells. Journal of Immunology 138, 3960–3966

Andrade, F., Bull, H. G., Thornberry, N. A., Ketner, G. W., Casciola-Rosen, L. A., and Rosen, A. (2001) Adenovirus L4-100 K assembly protein is a granzyme B substrate that potently inhibits granzyme B-mediated cell death. Immunity. 14, 751–761

Aradhya, S., Courtois, G., Rajkovic, A., Lewis, R., Levy, M., Israel, A., and Nelson, D. (2001) Atypical forms of incontinentia pigmenti in male individuals result from mutations of a cytosine tract in exon 10 of NEMO (IKK-gamma) Am.J Hum.Genet. 68, 765–771

Basler, C. and Horwitz, M. S. (1996) Subgroup B adenovirus type 35 early region 3 mRNAs differ from those of the subgroup C adenoviruses. Virol. 216, 165–177

Basler, C. F., Droguett, G., and Horwitz, M. S. (1996) Sequence of the immunoregulatory early region 3 and flanking sequences of adenovirus type 35. Gene 170, 249–254

Beier, D. C., Cox, J. H., Vining, D. R., Cresswell, P., and Engelhard, V. H. (1994) Association of human class I MHC alleles with the adenovirus E3/19K protein. J Immunol. 152, 3862–3872

Benedict, C. A., Norris, P. S., Prigozy, T. I., Bodmer, J. L., Mahr, J. A., Garnett, C. T., Martinon, F., Tschopp, J., Gooding, L. R., and Ware, C. F. (2001) Three adenovirus E3 proteins cooperate to evade apoptosis by tumor necrosis factor-related apoptosis-inducing ligand receptor-1 and –2. J Biol.Chem 276, 3270–3278

Bennett, E. M., Bennink, J. R., Yewdell, J. W., and Brodsky, F. M. (1999) Cutting edge: adenovirus E19 has two mechanisms for affecting class I MHC expression. Journal of Immunology 162, 5049–5052

Bruder, J. T. and Kovesdi, I. (1997) Adenovirus infection stimulates the Raf/MAPK signaling pathway and induces interleukin-8 expression. J Virol. 71, 398–404

Buntzman, A, Friedman, J., Fessler, S., and Horwitz, M. S. 2003. Ref Type: Unpublished Work

Burgert, H. G. and Blusch, J. H. (2000) Immunomodulatory functions encoded by the E3 transcription unit of adenoviruses. Virus Genes 21, 13–25

Burgert, H. G. and Kvist, S. (1985) An adenovirus type 2 glycoprotein blocks cell surface expression of human histocompatibility class I antigens. Cell 41, 987–997

Burgert, H. G., Maryanski, J. L., and Kvist, S. (1987) "E3/19K" protein of adenovirus type 2 inhibits lysis of cytolytic T lymphocytes by blocking cell-surface expression of histocompatibility class I antigens. Proceedings of the National Academy of Sciences of the United States of America 84, 1356–1360

Carlin, C. R., Tollefson, A. E., Brady, H. A., Hoffman, B. L., and Wold, W. S. M. (1989) Epidermal growth factor receptor is down-regulated by a 10,400 mw protein encoded by the E3 region of adenovirus. Cell 57, 135–144

Chen, P., Tian, J., Kovesdi, I., and Bruder, J. T. (1998) Interaction of the adenovirus 14.7-kDa protein with FLICE inhibits Fas ligand-induced apoptosis. Journal of Biological Chemistry 273, 5815–5820

Deryckere, F. and Burgert, H. G. (1996) Early region 3 of adenovirus type 19 (subgroup D) encodes an HLA-binding protein distinct from that of subgroups B and C. Journal of Virology 70, 2832–2841

Dimitrov T, Krajcsi P, Hermiston TW, Tollefson AE, Hannink M, a., and Wold W.S.M. (1997) Adenovirus E3-10.4/14.5K protein complex inhibits tumor necrosis factor-induced translocation of cytosolic phospholipase A2 to membranes. Journal of Virology 71, 2830–2837

Driesse, M. J., Esandi, M. C., Kros, J. M., Avezaat, C. J., Vecht, C., Zurcher, C., van, d. V., I, Valerio, D., Bout, A., and Sillevis Smitt, P. A. (2000) Intra-CSF administered recombinant adenovirus causes an immune response- mediated toxicity. Gene Ther. 7, 1401–1409

Dufke, A., Vollmer, B., Kendziorra, H., Mackensen-Haen, S., Orth, U., Orlikowsky, T., and Gal, A. (2001) Hydrops fetalis in three male fetuses of a female with incontinentia pigmenti. Prenat.Diagn. 21, 1019–1021

Efrat, S., Fejer, G., Brownlee, M., and Horwitz, M. S. (1995) Prolonged survival of pancreatic islet allografts mediated by adenovirus immunoregulatory transgenes. Proceedings of the National Academy of Sciences of the United States of America 92, 6947–6951

Efrat, S., Serreze, D. V., Svetlanov, A., Post, C. M., Johnson, E. A., Herold, K., and Horwitz M.S. (2001) Adenovirus early region 3 (E3) immunomodulatory genes decrease the incidence of autoimmune diabetes in nonobese diabetic (NOD) mice. Diabetes 50, 980–984

Elsing, A. and Burgert, H. G. (1998) The adenovirus E3/10.4K-14.5K proteins down-modulate the apoptosis receptor Fas/Apo-1 by inducing its internalization. Proc.-Natl.Acad.Sci.U.S.A 95, 10072–10077

Fessler, S. and Horwitz M.S. 2003 Ref Type: Unpublished Work

Flomenberg, P., Babbitt, J., Drobyski, W. R., Ash, R. C., Carrigan, D. R., Sedmak, G. V., McAuliffe, T., Camitta, B., Horowitz, M. M., Bunin, N., and Casper, J. T. (1994) Increasing incidence of adenovirus disease in bone marrow transplant recipients. Journal of Infectious Diseases 169, 775–781

Flomenberg, P., Piaskowski, V., Truitt, R. L., and Casper, J. T. (1996) Human adenovirus-specific CD8+ T-cell responses are not inhibited by E3–19K in the presence of gamma interferon. J Virol. 70, 6314–6322

Friedman, J. M. and Horwitz, M. S. (2002) Inhibition of tumor necrosis factor alpha-induced NF-kappa B activation by the adenovirus E3–10.4/14.5K complex. J Virol. 76, 5515–5521

Funk, C. D. (2001) Prostaglandins and leukotrienes: advances in eicosanoid biology. Science 294, 1871–1875

Ginsberg, H. S., Lundholm-Beauchamp, U., Horswood, R. L., Pernis, B., Wold, W. S. M., Chanock, R. M., and Prince, G. A. (1989) Role of early region 3 (E3) in pathogenesis of adenovirus disease. Proceedings of the National Academy of Sciences of the United States of America 86, 3823–3827

Gooding, L. R., Ranheim, T. S., Tollefson, A. E., Aquino, L., Duerksen-Hughes, P. J., Horton, T. M., and Wold, W. S. M. (1991) The 10,400- and 14,500-dalton proteins encoded by region E3 of adenovirus function together to protect many but not all mouse cell lines against lysis by tumor necrosis factor. Journal of Virology 65(8), 4114–4123

Gooding, L. R., Sofola, I. O., Tollefson, A. E., Duerksen-Hughes, P. J., and Wold, W. S. M. (1990) The adenovirus E3–14.7K protein is a general inhibitor of tumor necrosis factor-mediated cytolysis. Journal of Immunology 145, 3080–3086

Hattula, K. and Peranen, J. (2000) FIP-2, a coiled-coil protein, links Huntingtin to Rab8 and modulates cellular morphogenesis. Curr.Biol. 10, 1603–1606

Hirose, E., Nakashima, N., Sekiguchi, T., and Nishimoto, T. (1998) RagA is a functional homologue of S. cerevisiae Gtr1p involved in the Ran/Gsp1-GTPase pathway. J.Cell Sci. 111, 11–21

Hitt, M. M. and Graham, F. L. (2000) Adenovirus vectors for human gene therapy. Adv.Virus Res. 55, 479–505

Hoffman, P., Yaffe, M. B., Hoffman, B. L., Yei, S., Wold, W. S. M., and Carlin, C. R. (1992) Characterization of the adenovirus E3 protein that down-regulates the epidermal growth factor receptor. Journal of Biological Chemistry 267(19), 13480–13487

Horton, T. M., Ranheim, T. S., Aquino, L., Kusher, D. I., Saha, S. K., Ware, C. F., Wold, W. S. M., and Gooding, L. R. (1991) Adenovirus E3 14.7K protein functions in the absence of other adenovirus proteins to protect transfected cells from tumor necrosis factor cytolysis. Journal of Virology 65(5), 2629–2639

Horton, T. M., Tollefson, A. E., Wold, W. S. M., and Gooding, L. R. (1990) A protein serologically and functionally related to the group C E3 14,700-kilodalton protein is found in multiple adenovirus serotypes. Journal of Virology 64(3), 1250–1255

Horwitz M.S. (2001) Adenoviruses. In "Fields Virology" (D. M. Knipe and P. M. Howley, Eds.), pp. 2301–2326. Lippincott-Williams and Wilkins

Horwitz, M. S. (2001) Adenovirus immunoregulatory genes and their cellular targets. Virology 279, 1–8

Jain, A., Ma, C. A., Liu, S., Brown, M., Cohen, J., and Strober, W. (2001) Specific missense mutations in NEMO result in hyper-IgM syndrome with hypohydrotic ectodermal dysplasia. Nat.Immunol. 2, 223–228

Kim, H. J. and Foster, M. P. (2002) Characterization of Ad5 E3-14.7K, an adenoviral inhibitor of apoptosis: structure, oligomeric state, and metal binding. Protein Sci. 11, 1117–1128

Kirchhausen, T. (1999) Adaptors for clathrin-mediated traffic. Annu.Rev.Cell Dev.-Biol. 15, 705–732

Kojaoghlanian, T., Flomenberg, P., and Horwitz M.S. (2003) The Impact of Adenovirus Infection on the Immunocompromised Host. Reviews in Medical Virology 76, 8236–8243

Korner, H. and Burgert, H. G. (1994) Down-regulation of HLA antigens by the adenovirus type 2 E3/19K protein in a T-lymphoma cell line. Journal of Virology 68, 1442–1448

Krajcsi, P., Tollefson, A. E., Anderson, C. W., Stewart, R., Carlin, C. R., and Wold, W. S. M. (1992a) The E3–10.4K protein of adenovirus is an integral membrane protein that is partially cleaved between Ala_{22} and Ala_{23} and has a C_{cyt} orientation. Virology 187, 131–144

Krajcsi, P., Tollefson, A. E., Anderson, C. W., and Wold, W. S. M. (1992b) The adenovirus E3 14.5-kilodalton protein, which is required for down-regulation of the epidermal growth factor receptor and prevention of tumor necrosis factor cytolysis, is an integral membrane protein oriented with its C terminus in the cytoplasm. Journal of Virology 66(3), 1665–1673

Krajcsi, P., Tollefson, A. E., and Wold, W. S. M. (1992c) The E3–14.5K integral membrane protein of adenovirus that is required for down-regulation of the EGF receptor and for prevention of TNF cytolysis is O-glycosylated but not N-glycosylated. Virology 188, 570–579

Krajcsi, P. and Wold, W. S. M. (1992) The adenovirus E3–14.5K protein which is required for prevention of TNF cytolysis and for down-regulation of the EGF receptor contains phosphoserine. Virology 187, 492–498

Leland, B. J. and Metcalf, J. P. (1999) Type-specific induction of interleukin-8 by adenovirus. Am.J Respir.Cell Mol.Biol. 21, 521–527

Lesokhin, A. M., Delgado-Lopez, F., and Horwitz, M. S. (2002) Inhibition of chemokine expression by adenovirus early region three (e3) genes. J Virol. 76, 8236–8243

Li, Y. Identification and characterization of cellular proteins which interact with adenovirus E3–14.7kDA protein, an antagonist of TNF-alpha. 1996. Albert Einstein College of Medicine (thesis for Ph.D. degree)

Li, Y., Kang, J., Friedman, J., Tarassishin, L., Ye, J., Kovalenko, A., Wallach, D., and Horwitz, M. S. (1999) Identification of a cell protein (FIP-3) as a modulator of

NF-kappaB activity and as a target of an adenovirus inhibitor of tumor necrosis factor alpha-induced apoptosis. Proc.Natl.Acad.Sci.U.S.A 96, 1042–1047

Li, Y., Kang, J., and Horwitz M.S. (1998) Interaction of an adenovirus E3-14.7 kDa protein with a novel TNF-alpha inducible cellular protein containing leucine zipper domains. Mol.Cell.Biol. 18, 1601–1610

Li, Y., Kang, J., and Horwitz, M. S. (1997) Interaction of an adenovirus 14.7 kDa protein inhibitor of TNF-alpha cytolysis with a new member of the GTPase superfamily of signal transducers. Journal of Virology 71, 1576–1582

Lichtenstein, D. L., Krajcsi, P., Esteban, D. J., Tollefson, A. E., and Wold, W. S. (2002) Adenovirus RIDbeta subunit contains a tyrosine residue that is critical for RID-mediated receptor internalization and inhibition of Fas- and TRAIL-induced apoptosis. J Virol. 76, 11329–11342

Lukashok, S. A., Tarassishin, L., Li, Y., and Horwitz, M. S. (2000) An adenovirus inhibitor of tumor necrosis factor alpha-induced apoptosis complexes with dynein and a small GTPase. Journal of Virology 74, 4705–4709

Mahr, J. A., Boss, J. M., and Gooding, L. R. (2003) The adenovirus e3 promoter is sensitive to activation signals in human T cells. J Virol. 77, 1112–1119

McNees, A. L., Garnett, C. T., and Gooding, L. R. (2002) The adenovirus E3 RID complex protects some cultured human T and B lymphocytes from Fas-induced apoptosis. J Virol. 76, 9716–9723

Mercurio, F., Murray, B. W., Shevchenko, A., Bennett, B. L., Young, D. B., Li, J. W., Pascual, G., Motiwala, A., Zhu, H., Mann, M., and Manning, A. M. (1999) IkappaB kinase (IKK)-associated protein 1, a common component of the heterogeneous IKK complex. Mol.Cell Biol. 19, 1526–1538

Moreland, R. J., Dresser, M. E., Rodgers, J. S., Roe, B. A., Conaway, J. W., Conaway, R. C., and Hanas, J. S. (2000) Identification of a transcription factor IIIA-interacting protein. Nucleic Acids Res. 28, 1986–1993

Muruve, D. A., Barnes, M. J., Stillman, I. E., and Libermann, T. A. (1999) Adenoviral gene therapy leads to rapid induction of multiple chemokines and acute neutrophil-dependent hepatic injury in vivo. Hum.Gene Ther. 10, 965–976

Nakashima, N., Hayashi, N., Noguchi, E., and Nishimoto, T. (1996) Putative GTPase Gtr1p genetically interacts with the RanGTPase cycle in Saccharomyces cerevisiae. J Cell Sci. 109 (Pt 9), 2311–2318

Nakashima, N., Noguchi, E., and Nishimoto, T. (1999) Saccharomyces cerevisiae putative G protein, Gtr1p, which forms complexes with itself and a novel protein designated as Gtr2p, negatively regulates the Ran/Gsp1p G protein cycle through Gtr2p. Genetics 152, 853–867

Orange, J. S., Brodeur, S. R., Jain, A., Bonilla, F. A., Schneider, L. C., Kretschmer, R., Nurko, S., Rasmussen, W. L., Kohler, J. R., Gellis, S. E., Ferguson, B. M., Strominger, J. L., Zonana, J., Ramesh, N., Ballas, Z. K., and Geha, R. S. (2002) Deficient natural killer cell cytotoxicity in patients with IKK- gamma/NEMO mutations. J Clin.Invest 109, 1501–1509

Pierce, M. A., Chapman, H. D., Post, C. M., Svetlanov, A., Efrat, S., Horwitz M.S., and Serreze, D. V. (2003) Adenovirus early region 3 antiapoptotic 10.4K, 14.5K, and 14.7K genes decrease the incidence of autoimmune diabetes in NOD mice. Diabetes 52(5) 1119–1127

Ranheim, T. S., Shisler, J., Horton, T. M., Wold, L. J., Gooding, L. R., and Wold, W. S. M. (1993) Characterization of mutants within the gene for the adenovirus E3 14.7-kilodalton protein which prevents cytolysis by tumor necrosis factor. Journal of Virology 67(4), 2159–2167

Rawle, F. C., Knowles, B. B., Ricciardi, R. P., Brahmacheri, V., Duerksen-Hughes, P., Wold, W. S. M., and Gooding, L. R. (1991) Specificity of the mouse cytotoxic T lymphocyte response to adenovirus 5. Journal of Immunology 146, 3977–3984

Rezaie, T., Child, A., Hitchings, R., Brice, G., Miller, L., Coca-Prados, M., Heon, E., Krupin, T., Ritch, R., Kreutzer, D., Crick, R. P., and Sarfarazi, M. (2002) Adult-onset primary open-angle glaucoma caused by mutations in optineurin. Science 295, 1077–1079

Rothwarf, D. M., Zandi, E., Natoli, G., and Karin, M. (1998) IKK-gamma is an essential regulatory subunit of the IkappaB kinase complex. Nature 395, 297–300

Roy-Chowdhury, J. and Horwitz M.S. (2002) Evolution of Adenoviruses as Gene Therapy Vectors. Molecular Therapy 5, 344

Schurmann, A., Brauers, A., Mabmann, S., Becker, W., and Joost, H. G. (1995) Cloning of a novel family of mammalian GTP-binding proteins (RagA, RagBs, RagB1) with remote similarity to the Ras-related GTPases. Journal of Biological Chemistry 270, 28982–28988

Schwamborn, K., Weil, R., Courtois, G., Whiteside, S. T., and Israel, A. (2000) Phorbol esters and cytokines regulate the expression of the NEMO- related protein, a molecule involved in a NF-kappa B-independent pathway. Journal of Biological Chemistry 275, 22780–22789

Sekiguchi, T., Hirose, E., Nakashima, N., Ii, M., and Nishimoto, T. (2001) Novel g proteins, rag c and rag d, interact with gtp-binding proteins, rag a and rag b. J Biol.Chem 276, 7246–7257

Shisler, J., Yang, C., Walter, B., Ware, C. F., and Gooding, L. R. (1997) The adenovirus E3–10.4K/14.5K complex mediates loss of cell surface Fas (CD95) and resistance to Fas-induced apoptosis. Journal of Virology 71, 8299–8306

Stewart, A. R., Tollefson, A. E., Krajcsi, P., Yei, S. P., and Wold, W. S. M. (1995) The adenovirus E3 10.4K and 14.5K proteins, which function to prevent cytolysis by tumor necrosis factor and to down-regulate the epidermal growth factor receptor, are localized in the plasma membrane. Journal of Virology 69, 172–181

Susin, S. A., Lorenzo, H. K., Zamzami, N., Marzo, I., Snow, B. E., Brothers, G. M., Mangion, J., Jacotot, E., Costantini, P., Loeffler, M., Larochette, N., Goodlett, D. R., Aebersold, R., Siderovski, D. P., Penninger, J. M., and Kroemer, G. (1999) Molecular characterization of mitochondrial apoptosis-inducing factor. Nature 397, 441–446

Tarassishin, L. and Horwitz, M. S. (2001) Sites on FIP-3 (NEMO/IKKgamma) essential for its phosphorylation and NF- kappaB modulating activity. Biochem.Biophys.Res.Commun. 285, 555–560

Tollefson, A. E., Hermiston, T. W., Lichtenstein, D. L., Colle, C. F., Tripp, R. A., Dimitrov, T., Toth, K., Wells, C. E., Doherty, P. C., and Wold, W. S. (1998) Forced degradation of Fas inhibits apoptosis in adenovirus- infected cells. Nature 392, 726–730

Tollefson, A. E., Ryerse, J. S., Scaria, A., Hermiston, T. W., and Wold, W. S. (1996) The E3-11.6-kDa adenovirus death protein (ADP) is required for efficient cell death: characterization of cells infected with adp mutants. Virology 220, 152–162

Tollefson, A. E., Scaria, A., Hermiston, T. W., Ryerse, J. S., Wold, L. J., and Wold, W. S. M. (1996) The adenovirus death protein (E3-11.6K) is required at very late stages of infection for efficient cell lysis and release of adenovirus from infected cells. Journal of Virology 70, 2296–2306

Tollefson, A. E., Toth, K., Doronin, K., Kuppuswamy, M., Doronina, O. A., Lichtenstein, D. L., Hermiston, T. W., Smith, C. A., and Wold, W. S. (2001) Inhibition of TRAIL-induced apoptosis and forced internalization of TRAIL receptor 1 by adenovirus proteins. J Virol. 75, 8875–8887

von Herrath, M., Efrat, S., Oldstone, M. B. A., and Horwitz, M. S. (1997) Expression of adenoviral E3 transgenes in β cells prevents autoimmune diabetes. Proceedings of the National Academy of Sciences of the United States of America 94, 9808–9813

Wen, S., Driscoll, R. M., Schneider, D. B., and Dichek, D. A. (2001) Inclusion of the E3 region in an adenoviral vector decreases inflammation and neointima formation after arterial gene transfer. Arterioscler.Thromb.Vasc.Biol. 21, 1777–1782

Wilmott, R. W., Amin, R. S., Perez, C. R., Wert, S. E., Keller, G., Boivin, G. P., Hirsch, R., De Inocencio, J., Lu, P., Reising, S. F., Yei, S., Whitsett, J. A., and Trapnell, B. C. (1996) Safety of adenovirus-mediated transfer of the human cystic fibrosis transmembrane conductance regulator cDNA to the lungs of nonhuman primates. Hum.Gene Ther. 7, 301–318

Wold, W. S., Doronin, K., Toth, K., Kuppuswamy, M., Lichtenstein, D. L., and Tollefson, A. E. (1999) Immune responses to adenoviruses: viral evasion mechanisms and their implications for the clinic. Curr.Opin.Immunol. 11, 380–386

Yamaoka, S., Courtois, G., Bessia, C., Whiteside, S. T., Weil, R., Agou, F., Kirk, H. E., Kay, R. J., and Israel, A. (1998) Complementation Cloning of Nemo, a Component of the lkB Kinase Complex Essential for NF-kB Activation. Cell 93, 1231–1240

Ye, J., Xie, X., Tarassishin, L., and Horwitz, M. S. (2000) Regulation of the NF-kappaB activation pathway by isolated domains of FIP3/IKKgamma, a component of the IkappaB-alpha kinase complex. Journal of Biological Chemistry 275, 9882–9889

Zilli, D., Voelkel-Johnson, C., Skinner, T., and Laster, S. M. (1992) The adenovirus E3 region 14.7 kDa protein, heat and sodium arsinate inhibit the TNF-induced release of arachidonic acid. Biochemical and Biophysical Research Communications 188, 177–183

Zonana, J., Elder, M. E., Schneider, L. C., Orlow, S. J., Moss, C., Golabi, M., Shapira, S. K., Farndon, P. A., Wara, D. W., Emmal, S. A., and Ferguson, B. M. (2000) A novel X-linked disorder of immune deficiency and hypohidrotic ectodermal dysplasia is allelic to incontinentia pigmenti and due to mutations in IKK-gamma (NEMO) Am.J Hum.Genet. 67, 1555–1562

2
Oncogenesis

Modulation of Oncogenic Transformation by the Human Adenovirus E1A C-Terminal Region

G. Chinnadurai

Institute for Molecular Virology, Saint Louis University School of Medicine, 3681 Park Ave., St. Louis, MO 63110, USA

E-mail: Chinnag@slu.edu

Abstract The *E1A* oncogene of human adenoviruses cooperates with other viral and cellular oncogenes in oncogenic transformation of primary and established cells. The N-terminal half of E1A proteins that form specific protein complexes with pRb family and p300/CBP transcriptional regulators is essential for the transforming activities of E1A. Although the C-terminal half of E1A is dispensable for the transforming activities, it negatively modulates the oncogenic activities of the N-terminal region. Mutants of E1A lacking the C-terminal half or a short C-terminal region exhibit a hyper-transforming phenotype in cooperative transformation assays with the activated *ras* oncogene. The E1A C-terminal region implicated in the oncogenesis-restraining activity interacts with a 48-kDa cellular phosphoprotein, CtBP, that functions as a transcriptional corepressor. It appears that the C-terminal region of E1A may suppress E1A-mediated oncogenic transformation by a dual mechanism of relieving repression cellular genes by CtBP, and also by antagonizing the oncogenic activities of the N-terminal half of E1A.

1 Introduction

The *E1A* oncogene of human adenoviruses is one of the most extensively studied viral oncogenes, and it serves as a model oncogene for small DNA tumor viruses. The studies on *E1A* have been instrumental in illuminating how the viral oncogenes subvert host cell-cycle control by binding to specific cellular growth regulatory proteins and have facilitated dissection of the cellular growth control machinery in finer detail. The *E1A* region of human adenovirus 2 and 5 encodes two major proteins of 289 and 243 amino acids (aa; 289R and 243R). Both proteins contain two exons and are identical except for the presence of an internal 46-aa region unique to the 289R protein. Whereas the 289R protein is required for productive viral infection, the 243R protein encodes all the functions necessary for immortalization of primary cells and for transformation of these cells in cooperation with other viral or cellular oncogenes (Zerler et al. 1986). Exon 1 of the E1A proteins is essential for these transforming activities, and controls cell proliferation and transformation by modulating gene expression through interaction with cellular growth-regulatory proteins such as the retinoblastoma gene product (pRb) and the p300/CBP transcriptional co-activator (reviewed by Moran 1993 and Nevins et al. 1997).

Studies on E1A have facilitated the dissection of the functions of cell growth-regulating proteins such as pRb and have contributed immensely to our understanding of the role of these cellular proteins in cell-cycle regulation and oncogenesis. Specifically, the work on E1A has led to the recognition of the pivotal role of transcription factor E2F in cell-cycle control (Nevins et al. 1997; Dyson 1998). Elegant studies have revealed that E2F is the target of the tumor suppressor protein pRb and related proteins p107 and p130 (Nevins et al. 1997; Dyson 1998). These studies have shown that the E1A/pRb interaction results in functional inactivation of the negative cell growth-regulatory activity of pRb by releasing the transcription factor E2F, and that this interaction mimics loss of pRb in certain human tumors (Moran 1993; Nevins et al. 1997). Similarly, studies on E1A have also led to the discovery of p300, a CBP-related transcription factor that plays an important role in cell proliferation and oncogenic transformation. p300/CBP, a transcriptional co-activator that possesses an intrinsic histone acetyl transferase (HAT) activity (Bannister and Kouzarides 1996), binds to many different transcription factors specific to various promoters. Interaction of E1A with p300 also releases another co-activator, P/CAF (acetyl transferase) from the p300-P/CAF complex (Yang et al. 1996). E1A also independently binds to P/CAF (Reid et al. 1998). Association of E1A has been shown to modulate the intrinsic HAT activity of p300 (Chakravarti et al. 1999; Hamamori et al. 1999). Recent studies suggest that interaction of both pRb and p300/CBP with E1A results in acetylation of pRb by p300/CBP, which may contribute to modulation of pRb activities during cell proliferation (Chan et al. 2001). Thus, E1A exon 1 appears to exert its effect on cell proliferation by protein complex formation with pRb family proteins and p300/CBP proteins. In contrast, the functions of exon 2 have been less intensively studied. However, studies from our laboratory and those of others indicate that E1A exon 2 has important oncogenesis-modulating activities. Additionally, exon 2 has been implicated in certain positive and negative transcriptional regulatory activities (Linder et al. 1992; Bondesson et al. 1992; Mymryk and Bayley 1993). Exon 2 contains regulatory elements that are required for efficient trans-activation of the E4 promoter by the CR3 region of the 289R protein (Bondesson et al. 1992). Exon 2 sequences have also been implicated in repression of the promoter for the metalloprotease stromelysin (Linder et al. 1992). This activity of exon 2 may contribute to the inhibitory effect of exon 2 on tumor invasion. It appears that exon 2 may have an autonomous trans-activation activity, because an Ad5 mutant that expresses only exon 2 induces expression of other early viral genes and supports limited viral replication in a cell type-dependent manner (Mymryk and Bayley 1993). Exon 2 (along with exon 1) is required for immortalization of primary cells and cooperative transformation with E1B (Subramanian et al. 1989, 1991; Quinlan and Douglas 1992), and for induction of Ad2/5-specific cytotoxic T lymphocytes (Urbanelli et al. 1989).

2 Modulation of Oncogenesis by E1A Exon 2

Studies from our laboratory have revealed that exon 2 negatively modulates in vitro transformation, tumorigenesis, and metastasis (Subramanian et al. 1989). These conclusions were based on our observation that deletions within the C-terminal 67 amino acids of the E1A 243R protein of Ad2 enhance E1A/T24 *ras* cooperative transformation in vitro (Subramanian et al. 1989; Boyd et al. 1993) and tumorigenesis of transformed cells in syngeneic and athymic rodent models (Subramanian et al. 1989). In these studies, two E1A mutants lacking the coding sequences for the C-terminal 61 or 67 residues were found to be defective in induction of foci (of proliferated cells) formation on transfected primary baby rat kidney (BRK) cells. However, cotransfection of these mutants with the activated *ras* oncogene, T24 *ras*, resulted in efficient cooperative transformation. The transformed foci were qualitatively and quantitatively distinguishable from those foci observed on cells cotransfected with *wt* 243R and T24 ras. The foci of cells expressing the E1A mutants and the T24 ras oncogene were much larger and were more numerous compared to cells expressing *wt* E1A 243R and T24 *ras*. Importantly, transformed BRK cells expressing *wt* E1A and the ras oncogene were not tumorigenic in the syngeneic Fisher rat model, whereas the cells expressing the E1A mutant and the *ras* oncogene were highly tumorigenic. These transformed cells were also more tumorigenic in the athymic mouse model. The transformed cells expressing the E1A C-terminal deletion mutants and T24 *ras* were also highly metastatic in athymic mice, whereas cells expressing *wt* E1A (243R) and T24 *ras* were not metastatic (Subramanian et al. 1989). Thus, the E1A C-terminal mutants exhibit a hyper-transforming phenotype. These results were consistent with a model in which E1A exon 2 might negatively modulate oncogenesis. This model was strengthened by analyses of other smaller deletion mutants within exon 2 (Boyd et al. 1993). In these studies, the hyper-transforming phenotype observed with larger exon 2 deletion mutants was also observed with a smaller deletion mutant (*dl*1135) that deletes only a 14-aa region near the C-terminus (residues 225–238). Quinlan and coworkers also reported that exon 2 sequences negatively modulate the extent of transformation in cooperation with the *ras* oncogene (Douglas et al. 1991). However, the sequences (aa 210–227 of 243R) identified by Quinlan's group as being responsible for this activity map upstream of the sequences that we have identified. The possibility that the sequences identified by the Quinlan group may influence the activity of the sequence that we have identified has not been formally investigated. Other laboratories have also provided evidence that exon 2 of E1A negatively regulates oncogenic transformation (Linder et al. 1992). Linder et al. observed that transformed cells that express E1A C-terminal mutants exhibit increased invasiveness in reconstituted membrane invasion assays (Linder et al. 1992). Put together, the data from various groups suggest that the E1A exon 2 negatively modulates oncogenesis.

3 Identification and Cloning of E1A C-Terminal Binding Protein

Since a short 14-aa region of the C-terminal region of E1A (residues 225–238 of the 243R protein) was sufficient for the oncogenesis-restraining activity of E1A, we undertook a search for cellular proteins that interact with this region. These studies led to the identification of a 48-kDa cellular phosphoprotein termed C-terminal binding protein (CtBP; Boyd et al. 1993). CtBP was originally identified using a GST-E1A C-terminal (aa 174–243) fusion protein affinity matrix and by immunoprecipitation studies. The cDNA for CtBP was cloned by yeast two-hybrid interaction cloning using the E1A C-terminal region as the bait. The cDNA encodes a 440-aa (48-kDa) protein. The protein encoded by the cDNA specifically interacted with E1A exon 2 in in vitro protein binding studies and in in vivo coimmunoprecipitation analysis. The predicted protein sequence of the cDNA also contained amino acid sequences obtained from two peptides prepared from biochemically purified CtBP. Polyclonal antibodies prepared against the protein encoded by the cDNA recognized biochemically purified CtBP. The CtBP antibody also recognized endogenous CtBP from primate and rodent (rat and mouse) cells, indicating that CtBP is conserved. Data bank searches indicated that CtBP is also highly conserved among various vertebrates and invertebrates. CtBP is a phosphoprotein. It is phosphorylated in a cell cycle-dependent manner (Boyd et al. 1993), suggesting that it may play some important role during the cell cycle. The CtBP that we first identified is now termed CtBP1. Subsequently, a highly homologous human protein termed CtBP2 was also identified by analysis of EST data bank sequences (Katsanis et al. 1998). The human *Ctbp1* gene maps to chromosome 4p16, and the human *Ctbp2* gene maps to chromosome 21q21.3 (Katsanis et al. 1998). More recently, an N-terminally truncated version of CtBP1 termed CtBP3/BARS 50 was also identified (Weigert et al. 1999).

Ad2	243R	222	CIEDLLHEPGQ----	PLDLS	CK	RPRP	243
Ad12	235R	209	SILDLIQEEEREQTV	PVDLS	VK	RPRCN	235
Ad7	261R	239	KLEDLLEGGDG----	PLDLS	TR	KLPRQ	261
Ad9	251R	228	KIEDLLQDMGGDE—	PLDLS	LK	RPRN	251
Ad9	251R	228	KIEDLLQDMGGDE—	PLDLS	LK	RPRN	251
Ad4	257R	235	CLDDLLQGGDE----	PLDLC	TR	KRPRH	257
Ad40	249R	225	CIEDLL—EEDPTDE	PLNLS	LK	RPKCS	249
SA7	231R	207	SLHDLI—EEVEQTV	PLDLS	LK	RSRSN	231

Fig. 1 CtBP binding motifs of E1A proteins of human and primate adenoviruses. The amino acid sequences of the C-terminal ends of various E1A proteins coded by the 12S E1A mRNA are shown. The core CtBP-binding motif and the adjoining K/R (see text) residues are *highlighted*

4
CtBP-Binding Motif

Binding studies have indicated that CtBP binds efficiently to a 9-aa region located between residues 229 and 238 (Boyd et al. 1993). Comparison of the 9-aa region of various E1A proteins of human and primate adenoviruses indicated that a 6-aa motif, PLDLSC, is relatively well conserved among the E1A proteins of human and primate adenoviruses (Fig. 1). Mutational dissection of the 6-aa region indicated that a mutation of the PL or DL residues abolished interaction of CtBP with E1A C-terminal region, and a mutation of SC residues only partially impaired interaction. Because the C residue was not conserved in Ad12 E1A (that also binds CtBP), it was concluded that PLDLS is the core CtBP-binding motif (Schaeper et al. 1995). A 5-aa motif, PLDLS of Ad2 E1A, which is conserved among the E1A proteins of various human and animal adenoviruses was identified as the core CtBP-binding motif. NMR-based physical interaction studies have confirmed the role of PLDLS and also extended the role of flanking sequences of E1A in CtBP binding (Molloy et al. 1998). Other factors such as the secondary structure, in addition to the presence of the PLDLS-like motifs, also appear to influence CtBP binding to various target proteins (Molloy et al. 2001).

5
Functions of CtBP

CtBP is present in the cytoplasm as well as in the nucleus. A possible role of CtBP in transcriptional repression was first suggested by a tethering transcriptional assay (Sollerbrant et al. 1996). In this assay, the CR1 region of E1A (aa 28–90 or 1–90) fused to a heterologous DNA-binding domain (Gal4) strongly activated a synthetic promoter containing a Gal4-binding site. Inclusion of the C-terminal 48-aa region of E1A in the chimeric Gal4-E1A construct abrogated CR1-mediated trans-activation activity. Deletion of a 14-aa region within the C-terminal region (aa 225–238) that encompasses the CtBP-binding motif relieved the repressive activity of the C-terminal region. These results suggested that interaction of CtBP with the C-terminal region antagonized the activity of CR1 in *cis*.

A number of elegant genetic and biochemical studies with the *Drosophila* homolog of CtBP (dCtBP) strongly suggested that dCtBP functions as a transcriptional corepressor for a number of DNA-binding repressors that function during *Drosophila* embryo development (Nibu et al. 1998a,b; Poortinga et al. 1998; Zhang and Levine 1999). Among the various transcriptional repressors that function during *Drosophila* embryo development, six different repressors, *Krüppel, Knips, Snail, Hairy, zfh-1,* and *Hairless* have been reported to recruit dCtBP (Nibu et al. 1998a,b; Poortinga et al. 1998; Zhang and Levine 1999; Postigo and Dean 1999; Keller et al. 2000; Morel et al. 2001). Mutations in the CtBP-binding motifs (PLDLS-related motifs) of the various repressors abolish their activity. It has been suggested that dCtBP mediates one of the two major transcriptional repression pathways operating during *Drosophila* embryo development (Nibu et al. 1998a; Mannervik and Levine 1999; Mannervik et al. 1999; Zhang and Levine 1999).

In vertebrates, mouse CtBP2 (mCtBP2) has been reported to interact with the Krüppel-like repressor BKLF (Turner and Crossley 1998). A *Xenopus* CtBP homolog (XCtBP) has been shown to complex with the *Xenopus* polycomb repressor, XPc (Sewalt et al. 1999). XCtBP has also been shown to be a corepressor of XTcf-3, a transcription factor that regulates Wnt signaling (Brannon et al. 1999). In the absence of Wnt signals, certain isoforms of mammalian Tcf appear to repress the target genes by recruiting CtBP. The zinc finger homeodomain transcription factor δEF1/ZEB (homologous to Drosophila *zfh-1*) that represses transcription of

E2 box promoters binds with both CtBP1 and CtBP2 (Furusawa et al. 1999; Postigo and Dean 1999). Mutation of the CtBP-binding motif of δEF1 attenuates its trans-repression activity. Human CtBP2 has also been shown to co-immunoprecipitate with the human polycomb protein hPc2 (Sewalt et al. 1999). A Ras-regulated transcription factor, Net (which belongs to the Ets family of oncogenes) has also been shown to mediate its repressor activity through CtBP (Criqui-Filipe et al. 1999). More recently, several other vertebrate CtBP-dependent repressors including Friend of GATA-2 (FOG-2; Svensson et al. 2000; Deconinck et al. 2000) and Ikaros (Koipally and Georgopoulos 2000) have also been described. The homeodomain protein TGIF, which suppresses TGF-β signaling, also appears to mediate transcriptional repression, in part, by recruiting CtBP (Melhuish and Wotton 2000). The loss of TGIF function is associated with holoprosencephaly (HPE) in humans. Interestingly, some of the HPE mutations obliterate CtBP binding (Melhuish and Wotton 2000). Another zinc-finger-containing repressor, Evi-1, that promotes oncogenesis by repressing TGF-β signaling also binds to CtBP (Izutsu et al. 2001; Palmer et al. 2001). In all these interactions, PLDLS-related motifs have been shown to be essential for CtBP interaction. The corepressor activity of CtBP has been directly demonstrated with dCtBP, hCtBP1, and mCtBP2 by targeting CtBP through heterologous DNA-binding domains such as Gal4 or LexA to the promoters (Nibu et al. 1998; Phippen et al. 2000; Turner and Crossley 1998; Sewalt et al. 1999; Furusawa et al. 1999).

The mechanism of CtBP-mediated transcriptional repression is unclear. Human CtBP1 has been shown to interact with the histone deacetylase HDAC-1 in transient overexpression assays (Sundqvist et al. 1998), raising the possibility that one of the mechanisms of CtBP-mediated repression involves chromatin remodeling through histone deacetylation. Repression of certain promoters has been reported to be sensitive to the HDAC inhibitor TSA (Criqui-Filipe et al. 1999; Izutsu et al. 2001), whereas repression of certain other promoters by CtBP promoters appears to be independent of HDAC (Meloni et al. 1999; Koipally and Georgopoulos 2000). The activity of dCtBP also does not appear to be fully dependent on *Drosophila* deacetylase Rpd3 (Mannervik and Levine 1999). dCtBP expressed in mammalian cells also does not appear to complex with HDAC (Phippen et al. 2000). Type II HDACs contain putative CtBP-binding motifs (PXDLR) near the N-terminus, and HDAC-4, 5 and HDAC-7 (Bertos et al. 2001) have been shown to interact with CtBP in yeast two-hybrid studies and in biochemical binding studies (Zhang et al. 2001a). Although class II HDACs contain putative CtBP-binding motifs, their potential role in CtBP-mediated transcriptional repression remains to be established by rigorous experimentation. Thus, the involvement of class I and class II HDACs in CtBP-mediated transcriptional repression is not yet fully resolved. The possible involvement of HDACs in CtBP-mediated repression may depend on the context of the promoter. Because CtBP associates with the *Xenopus* and human Polycomb proteins and co-localizes in distinct nuclear regions, it has been suggested that CtBP might repress transcription by inducing areas of heterochromatin (Sewalt et al. 1999). The observation that the lymphoid transcription factor Irkos, which represses transcription in B cells, is also localized in heterochromatin lends some support to this view.

An attractive hypothesis would be that the transcriptional regulatory activities of CtBP is related to the potential enzymatic activities of CtBP. CtBP has compelling sequence homology to a number of NAD-dependent D-isomer-specific acid dehydrogenases (Schaeper et al. 1995). With highly purified hCtBP1, we and others have observed significant NAD/H-binding activity (Zhang et al. 2000; Kumar et al. 2002; Balasubramani an et al. 2003). However, it appears that the dehydrogenase activity may not be linked to the repressor activity, because mutants in the conserved His residue (at the active site) of mCtBP and dCtBP do not appear to affect transcriptional repression in transcriptional tethering assays as fusion proteins with heterologous DNA-binding domains (Turner and Crossley 1998; Pippen et al. 2000). It is of interest that the yeast transcriptional silencer protein Sir2 binds to NAD and hydrolyzes NAD (reviewed by Moazed 2001). Although CtBP1 does not appear to hydrolyze NAD (P. Balasubramanian and G. Chinnadurai, unpublished observations), nucleotide binding influences the interaction of CtBP and its target transcription factors. It is possible that the NAD-binding activity of CtBP links the transcriptional repression with energy homeostasis. Other potential functions of CtBP1 are also possible. For example, a third mammalian homolog of CtBP designated CtBP3/BARS50, which is identical to CtBP1 except for the absence of N-terminal 20 residues, has been reported to have an acyltransferase activity (Weigert et al. 1999). It remains to be determined whether CtBP1 and CtBP2 possess a similar enzymatic activity. Thus, the mechanism by which CtBP mediates transcriptional repression remains enigmatic at present.

6 CtBP-CtIP Complex

During a search to identify cellular proteins that interact with CtBP, we identified and cloned a 125-kDa (897-aa) cellular protein, CtIP (*Ct*BP-*I*nteracting *P*rotein) that specifically interacts with CtBP (Schaeper et al. 1998). An analysis of CtIP sequences indicated that CtIP contains the 5-aa E1A CtBP-binding motif (PLDLS). Interaction studies have revealed that CtIP interacts with CtBP via the PLDLS motif, indicating that E1A and CtIP interact with CtBP through the same PLDLS motif. An in vitro competitive binding assay revealed that E1A readily competes with CtIP for CtBP binding, suggesting that one of the biochemical activities of exon 2 may include disruption protein complexes such as the CtBP/CtIP complex.

Three different groups have reported that the breast-associated tumor suppressor (BRCA-1) specifically interacts with CtIP (Wong et al. 1998; Yu et al. 1998; Li et al. 1999). Interaction between BRCA-1 and CtIP is abolished by tumor-specific mutations located within the C-terminal BRCT repeat of BRCA-1 (Wong et al, 1998; Yu et al. 1998; Li et al. 1999). This is important because mutations in the *BRCA-1* gene are responsible for nearly all of the hereditary ovarian and breast cancers (Easton et al. 1993; 1995; Gayther et al. 1995). The interaction between CtIP and BRCA-1 is also abrogated in cells treated with DNA-damaging agents such as UV, γ-irradiation, and Adriamycin (Li et al. 1999). Co-expression of CtIP and CtBP also inhibited BRCA-1-dependent trans-activation of the p21 promoter. It has been suggested that binding of the BCRT repeat of BRCA-1 to CtIP/CtBP is critical in mediating transcriptional regulation of cell-cycle regulatory genes such as *p21* in response to DNA damage (Li et al. 1999). CtIP is a target for the cell-cycle regulatory kinase ATM (Li et al. 2000). In cells treated with DNA-damaging agents, CtIP is hyperphosphorylated by ATM, resulting in dissociation of CtIP from BRCA-1, leading to relief of repression of DNA damage-response genes such as GADD45 mediated by BRCA-1 (Li et al. 2000). These results shed light on the link between ATM deficiency and breast cancer via the CtIP/BRCA-1 complex.

It is also of much interest that CtIP interacts with pRb family proteins. In a modified two-hybrid analysis where CtIP was tested for interaction with any of a large panel of approximately 1,600 known proteins (Brent and Finley 1997), CtIP was found to interact specifically with various

Rb-family proteins, pRb, p107, and p130 (Fusco et al. 1998). CtIP contains a sequence motif, LECEE, that is similar to the Rb-binding motif (LXCXE). Deletion of the LECEE motif abolished interaction of CtIP with Rb-family proteins. CtIP was also identified as an Rb-family interacting protein by conventional two-hybrid library screening (Meloni et al. 1999). It was shown that at least some events of transcriptional repression mediated by the Rb-family proteins might involve recruitment of CtBP by pRb and p130 via CtIP (Meloni et al. 1999). Because E1A interacts with the Rb-family proteins through the LXCXE motif, we hypothesize that the E1A CR2 region may also disrupt a complex of CtIP with Rb-family proteins in a fashion analogous to disruption of a complex between E2F and pRb. The observation that CtIP interacts with two different types of tumor suppressors is intriguing and suggests that CtIP may be an important regulator of oncogenesis.

7
Regulation of CtBP Interaction with Exon 2 by Exon 1

The interaction of the transcriptional coactivator p300/CBP with the N-terminal region of exon 1 is well characterized and is important for the transforming activity of E1A (reviewed by Moran 1993). It is generally believed that E1A interaction with p300/CBP contributes to the transcriptional repression activity associated with the N-terminal region of E1A. In support of this view, interaction of E1A with p300 was reported to inhibit the HAT activity of p300/CBP and disrupt the interaction of p300 with the associated HAT, P/CAF (Yang et al. 1996). Additionally, E1A was reported to directly inhibit the HAT activity of p300 and P/CAF (Chakravarthi et al. 1999). E1A was recently shown to stimulate acetylation of pRb by recruiting both pRb and p300/CBP (Chan et al. 2001), suggesting that E1A targets the acetylation activity of p300/CBP to pRb by recruiting these proteins in a multimeric protein complex. These results appear to integrate the effects of protein interactions with the N-terminus (with p300/CBP) and CR2 region (with pRb) of exon 1 in the manifestation of cell proliferation and differentiation.

It appears that interaction of nuclear acetylases (p300/CBP and P/CAF) with the N-terminal region of E1A may also regulate CtBP interaction with the C-terminus. A lysine residue (Lys 239) adjacent to the PLDLS motif of E1A (243R) was shown to be specifically acetylated by P/CAF and p300/CBP (Zhang et al. 2000). The conserved Lys residue is essential for CtBP interaction. Mutants of E1A that mimic the effect of acetylation (e.g. LysGln or Ala) were defective in CtBP binding. The specific role of acetylation in CtBP binding was addressed by peptide competition studies where acetylated E1A C-terminal peptides did not compete, whereas the unacetylated peptides efficiently competed. These results suggest that acetylation of the Lys-239 residue decreases CtBP binding. The E1A mutants that mimic the effect of acetylation were defective in repressing CREB-stimulated (CBP-dependent) transcriptional activation under certain conditions (low levels of E1A expression). It therefore appears that interaction of nuclear acetylases with exon 1 might decrease CtBP interaction with exon 2 by acetylation of the E1A C-terminus. It should be noted that the E1A proteins of certain adenovirus serotypes contain an Arg residue instead of the Lys (Fig. 1). The Arg residue is not subject acetylation regulation, and mutants of E1A which contain the LysArg mutation bind enhanced levels of CtBP in vitro (Zhang et al. 2001). Studies with the nuclear hormone receptor-interacting protein RIP140 suggest that the acetylation control of CtBP binding may be a general mechanism of regulation of a number of CtBP-dependent repressors (Vo et al. 2001).

It is possible that CtBP also modulates the activities of exon 1. In a transcriptional tethering assay, the N-terminal region (aa 28–90 or aa 1–90) of E1A (implicated in interaction with nuclear acetylases) fused to the Gal4 DNA binding domain strongly activated transcription from a synthetic promoter containing a Gal4 binding site. Inclusion of the E1A C-terminal region in the same construct inhibited the trans-activation function. Deletion of the CtBP binding region of the C-terminus relieved the repression (Sollerbrant et al. 1996). These results suggest that interaction of CtBP with the C-terminus might antagonize the activity of protein complexes with the E1A N-terminal region *in cis*. The possibility that CtBP may modulate the acetylation activity of the nuclear acetylases complex at the N-terminal region is an attractive hypothesis and remains to be investigated. Association of deacetylases with CtBP may play a role in this regulation. Interestingly, the functional consequence of protein interaction with exon 1 and exon 2 appear to be opposing. Interaction of Rb family proteins and p300/CBP contribute to the cell proliferation and transforming activities of E1A, whereas interaction of CtBP with the C-terminus appears to contribute to a tumor suppressor activity. The functions of exon 1 and exon 2 may therefore regulate the activity of each other.

8
Role of E1A-CtBP Interaction on Expression of Cellular Genes

The effect of E1A-CtBP interaction has not been examined in detail. In interesting studies, Frisch (Frisch 1994; Grooeclaes and Frisch 2000) has observed that transduction of E1A into certain human tumor cell lines conferred epithelial morphology to these cells. Consistent with the morphological change, expression of certain epithelial genes such as desmoglein-2 and plakoglobin was activated in these cells. An E1A mutant defective in interaction with CtBP failed to enhance expression of these genes. In transient transfection studies (Grooeclaes and Frisch 2000), the promoter of a third epithelial gene (E-cadherin) was activated by E1A, whereas an E1A mutant defective in CtBP interaction failed to activate. These results suggest that the expression of these epithelial genes is under negative regulation by CtBP, and that interaction with E1A relieves this negative regulation of epithelial gene expression. In cotransfection studies, expression of the C-terminal CtBP binding region of E1A has been reported to induce expression of certain viral (adenovirus E4 and major late) and cellular (PCNA and Hsp70) promoters (Sundqvist et al. 2001). Additionally, these investigators showed by gene array analysis that in cell lines that express the C-terminal region of E1A in an inducible fashion, expression of a number of endogenous genes involved in cell-cycle regulation, apoptosis regulation, and various intracellular signaling pathways was observed upon induction of expression of the C-terminal region of E1A. These results suggest that expression of the C-terminal region of E1A may activate cellular gene expression by derepression.

9
Role of CtBP in Oncogenesis by Viral and Cellular Genes

In addition to its role in E1A-mediated cooperative transformation (as discussed in Sect. 2), it appears that CtBP may play important roles in oncogenesis mediated by cellular and viral genes. However, the implicated roles may depend on the context of the oncogene. It was shown that the EBV nuclear antigen EBNA3C interacts with CtBP via a PLDLS motif. EBNA3C appears to be important for immortalization of human B cells (reviewed by Kieff, 1996). It also exerts a transcriptional repression activity in a Gal4 DNA-binding domain tethering transcription assay. The nonchimeric EBNA3C can also repress EBV Cp latency-associated promoter. Like E1A, EBNA3C can cooperate with the activated Ras oncogene in transformation of primary rat embryo fibroblasts (REFs; Parker et al. 1996). Deletion of the CtBP-binding motif resulted in modest reduction in the repressor activity in the tethering transcription assay and did not have significant effect under the context of full-length EBNA3C. However, these mutants were significantly defective in the EBNA3C-Ras cooperative transformation assay (Touitou et al. 2001). Apparently, the effect of the mutation of EBNA3C on the Ras cooperative assay (observed on REFs) is different from that observed with E1A and Ras on epithelial (kidney) cells.

The transforming activity of the cellular oncogene Evi-1 (implicated in human myeloid leukemia and myelodysplastic syndromes) appears to be linked to its ability to interact with CtBP (Izutsu et al. 2001; Palmer et al. 2001). Evi-1 is a zinc finger-containing nuclear protein that antagonizes TGF-β signaling by repressing Smad-induced transcription of TGF-β-responsive genes (Kurokawa et al. 1987). The repressor domain of Evi-1 contains two CtBP binding motifs. Mutants with deletion of the sequences containing these motifs are defective in CtBP binding and in transcriptional repression, suggesting that Evi-1 represses transcription by recruiting CtBP (Izutsu et al. 2001; Palmer et al. 2001). Mutants of Evi-1 that are defective in binding to CtBP are also defective in transformation of Rat1 fibroblasts in vitro (Palmer et al. 2001). Put together, these observations indicate that CtBP may play an important role in Evi-1-mediated leukemogenesis.

CtBP appears to modulate the activity of the Net transcription factor, whose activity is subject to regulation by the Ras oncogene (Criqui-Filipe et al. 1999). Net is a member of the Ets family of transcriptional repressors. Net represses the c-fos promoter by interaction with the serum response element (SRE). The negative regulatory effect of Net is switched to a positive regulatory effect by Ras signaling. In the absence of Ras signaling, Net mediates its repressor activity by recruiting CtBP. The mechanism by which the Ras signal relieves the repressor activity of Net remains to be elucidated.

Recent results suggest that CtBP may play a crucial role in the Wnt signaling pathway during development and oncogenesis. The canonical Wnt signaling pathway has been elucidated by genetic studies in *Drosophila* and *C. elegans* and by ectopic expression in *Xenopus* embryos (reviewed by Polakis 2000). Wnt proteins bind to the Frizzled family receptors on the plasma membrane and transduce signals that result in increased levels of β-catenin. The level of β-catenin is downregulated by the tumor suppressor gene adenomatous polyposis coli (APC). APC promotes degradation of β-catenin. Activation of Wnt signaling results in functional inactivation of glycogen synthase kinase 3β (GSK3β), thereby preventing phosphorylation of its targets that include APC and β-catenin. As a result of Wnt signaling, the level of β-catenin is increased. β-Catenin then binds to the transcription factor Tcf-4 and activates Wnt-responsive genes such as c-myc and cyclin D1. In vertebrates, β-catenin recruits the coactivator CBP to activate Wnt-responsive genes via Tcf-4 (Hecht and Kemler 2000). Tcf-4 is constitutively activated in colon cancers by mutations in APC and β-catenin genes and is constitutively repressed in normal cells through interaction of Groucho/TLE and CtBP. Human Tcf-4 contains two CtBP-binding motifs at the C-terminus. These sites also contain Lys residues adjoining the CtBP-binding motifs, and these residues may be subject to acetylation regulation by CBP. Acetylation of the CtBP-binding motifs by CBP may reduce the affinity of Tcf-4 for CtBP and therefore the ability of Tcf-4 to repress Wnt-responsive genes. An analysis of a number of colorectal cancer cell lines indicates that several of these cell lines contain nonsense or frameshift mutations that obliterate the CtBP-binding domains (Duval et al., 2000). These results suggest that mutations which disrupt interaction of Tcf-4 and CtBP may be important for colorectal carcinogenesis, and highlight the potential role of CtBP in modulating Wnt signaling during carcinogenesis.

10
Interaction of Other Cellular Proteins with Exon 2

In addition to CtBP, exon 2 also interacts with certain serine/threonine kinases related to the yeast dual-specificity kinase, Yak1p (Zhang et al. 2001b). This interaction was discovered during investigations on the functions of exon 2 in yeast. Expression of exon 2 in yeast induced pseudohyphal differentiation, which is independent of the MAPK or cAMP/PKA signaling pathways. The yeast two-hybrid interaction studies have identified the interaction of Yak1p with exon 2. Two extended regions within exon 2 (aa 141–175 and 193–238) are required for interaction of Yak1p. E1A mutants defective in interaction with Yak1p appear to be defective in pseudohyphal differentiation. In vitro binding studies have revealed that mammalian homologs of yeast Yak1p, Dyrk1A and Dyrk1B, also interact with E1A, and this interaction augments the kinase activity of Dyrk. It is possible that the oncogenesis-modulating activity of exon 2 may reflect a combinatorial effect of the interactions with CtBP as well as Dyrk.

11
Discussion and Perspective

Whereas E1A exon 1 which complexes with nuclear acetyl transferases and pRb family proteins mediates the cell-proliferation and oncogenic activities of E1A, exon 2 negatively modulates the oncogenic activities. The oncogenesis-restraining activity of exon 2 appears to be predominantly linked to interaction with CtBP. We envision that interaction of CtBP with exon 2 might modulate oncogenesis by either one of the two mechanisms. In the first model (model A in Fig. 2), we propose that interaction of CtBP with the C-terminus of E1A negatively regulates the activities of cellular proteins that interact with exon 1. This model has precedent, because a transcriptional activation function encoded by the CR1 region (aa 28–90 that predominantly binds with P/CAF) has been shown to be suppressed by the C-terminal region *in cis* and is relieved by deleting the CtBP-binding region (Sollerbrant et al. 1996). This could be achieved by regulating the extent of acetylation reactions mediated by the nuclear acetylases (p300/CBP and P/CAF) by CtBP. CtBP may negatively modulate the activities of the nuclear acetylases by recruiting factors such as class I and class II HDACs. A putative acid dehydrogenase activity of CtBP may also regulate the rate of acetylation (Berg et al. 1998) of nuclear acetylases associated with exon 1. One of the functions of P/CAF and p300/CBP would be to decrease interaction of CtBP with E1A, thereby relieving the negative regulatory effects of CtBP. In the second model (model B in Fig. 2), we envision that E1A relieves repression of certain cellular genes by disrupting CtBP-containing repressor complexes. This model also has precedent, because expression of E1A activates expression of several endogenous genes (Grooteclaes and Frisch 2000; Sundqvist et al. 2001; Schuierer et al. 2001). Further studies on the biochemical activities of CtBP and the effect of E1A interaction will serve as models for studies on cellular proteins that recruit CtBP during differentiation and oncogenesis.

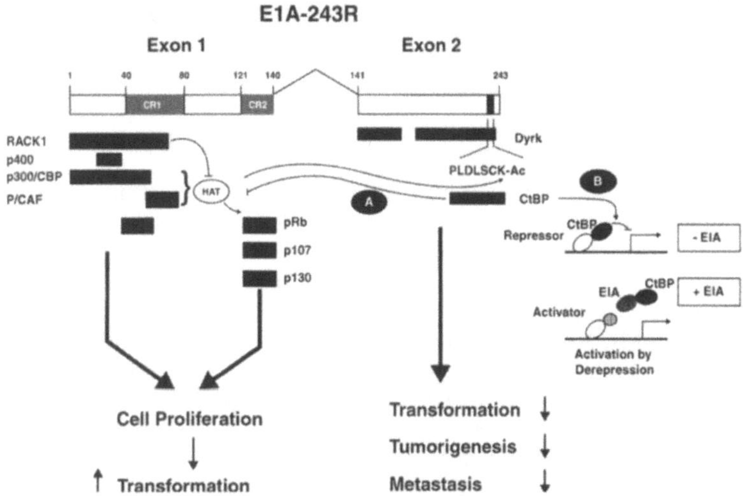

Fig. 2 Interaction of cellular proteins with E1A and modulation of oncogenesis. E1A protein region coded by exon 1 interacts with nuclear acetylases p300/CBP and P/CAF and pRb family proteins. This complex was shown to acetylate pRb in the same complex enhancing pRb interaction with Mdm2 (Chan et al. 2001), thereby reducing the growth-suppressive effect of pRb. Exon 1 region also interacts with the p400 complex that is involved in chromatin remodeling (Fuchs et al. 2001). It is believed that the combined activity of the nuclear acetylases and the pRb family proteins induces cell proliferation and contributes to the transforming activity of E1A. The activities of p400 also appear to contribute to the transforming activity of E1A under certain conditions. Based on the analysis of E1A activities in yeast, it was suggested that interaction of the receptor for activated PKC, RACK1, prevents association of the nuclear acetylases with E1A by competitive binding to E1A (Sang et al. 2001). The protein region coded by exon 2 interacts with CtBP and the mammalian protein kinases (*Dyrk*) related to the yeast kinase Yak1p (Zhang et al. 2001). This interaction appear to enhance the kinase activity of Dyrk. Deletion of short regions near the C-terminus of E1A that encompass the CtBP-binding motif (*PLDLS*) results in enhanced oncogenic transformation (Subramanian et al. 1989; Boyd et al. 1993). Interaction of CtBP appears to be regulated by the nuclear acetylases that interact with the N-terminal region (Zhang et al. 2000). Acetylation of a conserved Lys residue adjacent to the PLDLS motif decreases CtBP binding, thereby enhancing the proliferative activities of exon 1. We postulate two different models (*A*, *B*) for modulation of oncogenic transformation by CtBP as described in the text

Acknowledgement. Research in the laboratory of the author was supported by a grant from the National Cancer Institute (CA-84941).

References

Balasubramiam P, Zhao L-J, Chinnadurai G (2003) Nicotinamide adenine dinucleotide stimulates oligomerization, interaction with adenovirus and an intrinsic dehydrogenase activity of CtBP. FEBS Lett 537:157–160

Bannister AJ, Kouzarides T (1996) The CBP co-activator is a histone acetyltransferase. Nature 384: 641–643

Berg A, Westphal AH, Bosma HJ, de Kok A (1998) Kinetics and specificity of reductive acylation of wild-type and mutated lipoyl domains of 2-oxo-acid dehydrogenase complexes from *Azotobacter vinelandii.* Eur J Biochem 252:45–50

Bertos NR, Wang AH, Yang XJ (2001) Class II histone deacetylases: structure, function, and regulation. Biochem Cell Biol 79:243–252

Boyd JM, Subramanian T, Schaeper U, La Regina M, Bayley S, Chinnadurai G (1993) A region in the C-terminus of adenovirus 2/5 E1a protein is required for association with a cellular phosphoprotein and important for the negative modulation of T24 ras mediated transformation, tumorigenesis and metastasis. EMBO J 12:469–478

Bondesson M, Svensson C, Linder S, Akusjarvi G.(1992) The carboxy-terminal exon of the adenovirus E1A protein is required for E4F-dependent transcription activation. EMBO J 11:3347–3354

Brannon M, Brown JD, Bates R, Kimelman D, Moon RT (1999) XCtBP is a XTcf-3 co-repressor with roles throughout Xenopus development. Development 126:3159–3170

Brent R, Finley RL Jr (1997) Understanding gene and allele function with two-hybrid methods. Annu Rev Genet 31: 663–704

Chan HM, Krstic-Demonacos M, Smith L, Demonacos C, La Thangue NB (2001) Acetylation control of the retinoblastoma tumor-suppressor protein. Nature Cell Biol 3:667–674

Chakravarti D, Ogryzko V, Kao, HY, Nash A, Chen H, Nakatani Y, Evans RM (1999) A viral mechanism for inhibition of p300 and PCAF acetyltransferase activity. Cell 96:393–403

Criqui-Filipe P, Ducret C, Maira SM, Wasylyk B (1999) Net, a negative Ras-switchable TCF, contains a second inhibition domain, the CID, that mediates repression through interactions with CtBP and de-acetylation. EMBO J 18:3392–3403

Deconinck AE, Mead PE, Tevosian SG, Crispino JD, Katz SG, Zon LI, Orkin SH (2000) FOG acts as a repressor of red blood cell development in xenopus. Development 127:2031–2040

Douglas JL, Gopalakrishnan S, Quinlan MP (1991) Modulation of transformation of primary epithelial cells by the second exon of the E1A12S gene. Oncogene 6:2093–2103

Duval A, Rolland S, Tubacher E, Bui H, Thomas G, Hamelin R (2000) The human T-cell transcription factor-4 gene: structure, extensive characterization of alternative splicings, and mutational analysis in colorectal cancer cell lines. Cancer Res 60:3872–3879

Dyson N (1998) The regulation of E2F by pRB-family proteins. Genes Dev 12:2245–2262

Easton DF, Bishop DT, Ford D, Crockford GP (1993) Genetic linkage analysis in familial breast and ovarian cancer: results from 214 families. The Breast Cancer Linkage Consortium. Amer J Hum Genet 52: 678–701

Easton DF, Ford D, Bishop DT (1995) Breast and ovarian cancer incidence in BRCA1-mutation carriers. Breast Cancer Linkage Consortium. Amer J Hum Genet 56: 265–271

Frisch SM (1994) Ela induces the expression of epithelial characteristics. J Cell Biol 127:1085–1096

Fuchs M, Gerber J, Drapkin R, Sif S, Ikura T, Ogryzko V, Lane WS, Nakatani Y, Livingston DM (2001) The p400 complex is an essential E1A transformation target. Cell 106:297–307

Furusawa T, Moribe H, Kondoh H, Higashi Y (1999) Identification of CtBP1 and CtBP2 as corepressors of zinc finger-homeodomain factor deltaEF1. Mol Cell Biol 19:8581–8590

Fusco C, Reymond A, Zervos AS (1998) Molecular cloning and characterization of a novel retinoblastoma-binding protein. Genomics 51:351–358.

Gayther SA, Warren W, Mazoyer S, Russell PA, Harrington, PA, Chiano M, Seal S, Hamoudi R, van Rensburg EJ, Dunning AM, et al. (1995) Germline mutations of the BRCA1 gene in breast and ovarian cancer families provide evidence for a genotype-phenotype correlation. Nature Genetics 11: 428–433.

Grooteclaes ML, Frisch SM (2000) Evidence for a function of CtBP in epithelial gene regulation and anoikis. Oncogene 19:3823–3828

Hamamori Y, Sartorelli V, Ogryzko V, Puri PL, Wu HY, Wang JY, Nakatani Y, Kedes L (1999) Regulation of histone acetyltransferases p300 and PCAF by the bHLH protein twist and adenoviral oncoprotein E1A. Cell 96:405–413

Hecht A, and Kemler R (2000) Curbing the nuclear activities of beta-catenin. Control over Wnt target gene expression. EMBO Rep 1, 24–28

Izutsu K, Kurokawa M, Imai Y, Maki K, Mitani K, Hirai H (2001) The corepressor CtBP interacts with Evi-1 to repress transforming growth factor beta signaling. Blood 97:2815–2822

Katsanis N, Fisher EM, (1998) A novel C-terminal binding protein (CTBP2) is closely related to CTBP1, an adenovirus E1A-binding protein, and maps to human chromosome 21q21.3. Genomics 47:294–299

Kieff E (1996) Epstein-Barr virus and its replication. In BN Fields, DM Knipe and PM How (ed) Fields Virology. Lippincott-Raven, Philadelphia, PA

Keller SA, Mao Y, Struffi P, Margulies C, Yurk CE, Anderson AR, Amey RL, Moore S, Ebels JM, Foley K, Corado M, Arnosti DN (2000) dCtBP-dependent and -independent repression activities of the Drosophila Knirps protein. Mol Cell Biol 20:7247–7258

Koipally J, Georgopoulos K (2000) Ikaros interactions with CtBP reveal a repression mechanism that is independent of histone deacetylase activity. J Biol Chem 275:19594–19602

Kumar V, Carlson JE, Ohgi KA, Edwards TA, Rose DW, Escalante CR, Rosenfeld MG Aggarwal AK (2002) Transcription corepressor CtBP is an NAD(+)-regulated dehydrogenase. Mol Cell 10:857–869

Kurokawa M, Mitani K, Irie K, Matsuyama T, Takahashi T, Chiba S, Yazaki Y, Matsumoto K, Hirai H (1998) The oncoprotein Evi-1 represses TGF-beta signalling by inhibiting Smad3. Nature 394:92–96

Li S, Chen P-L, Subramanian T, Chinnadurai G, Tomlinson G, Osborne CK, Sharp ZD, Lee W-H (1999) Binding of CtIP to the BRCT repeats of BRCA1 involved in the transcription regulation of p21 is disrupted upon DNA damage. J Biol Chem 274:11334–11338

Li S, Ting NS, Zheng L, Chen PL, Ziv Y, Shiloh Y, Lee EY, Lee WH (2000) Functional link of BRCA1 and ataxia telangiectasia gene product in DNA damage response. Nature 406:210–215

Linder S, Popowicz P, Svensson C, Marshall H, Bondesson M, Akusjarvi G (1992) Involvement of the carboxyl-terminal exon of the adenovirus-2 E1A gene in repression of metalloprotease gene expression. Oncogene 7: 439–443

Mannervik M, Levine M (1999) The Rpd3 histone deacetylase is required for segmentation of the Drosophila embryo. Proc Natl Acad Sci USA 96:6797–6801

Mannervik M, Nibu Y, Zhang H, Levine M (1999) Transcriptional coregulators in development. Science 284:606–609

Melhuish TA, Wotton D. (2000) The interaction of the carboxyl terminus-binding protein with the Smad corepressor TGIF is disrupted by a holoprosencephaly mutation in TGIF. J Biol Chem 275:39762–39766

Meloni AR, Smith EJ, Nevins JR (1999) A mechanism for Rb/p130-mediated transcription repression involving recruitment of the CtBP corepressor. Proc Natl Acad Sci USA 96:9574–9579

Moran E (1993) DNA tumor virus transforming proteins and the cell cycle. Curr Op Gen Dev 3:63–70

Morel V, Lecourtois M, Massiani O, Maier D, Preiss A, Schweisguth F (2001) Transcriptional repression by suppressor of hairless involves the binding of a hairless-dCtBP complex in Drosophila. Curr Biol 11:789–792

Molloy DP, Milner AE, Yakub IK, Chinnadurai G, Gallimore PH, Grand RJ (1998) Structural determinants present in the C-terminal binding protein binding site of adenovirus early region 1A proteins. J Biol Chem 273:20867–20876

Molloy DP, Barral PM, Bremner KH, Gallimore PH, Grand RJ (2001) Structural determinants outside the PXDLS sequence affect the interaction of adenovirus E1A, C-terminal interacting protein and Drosophila repressors with C-terminal binding protein. Biochim Biophys Acta 1546:55–70

Moazed D (2001) Enzymatic activities of Sir2 and chromatin silencing. Curr Opin Cell Biol 13:232–238

Mymryk JS, Bayley ST (1993) Induction of gene expression by exon 2 of the major E1A proteins of adenovirus type 5. J Virol 67:6922–6928

Nevins JR, Leone G, DeGregori J, Jakoi L (1997) Role of the Rb/E2F pathway in cell growth control. J Cell Physiol 173: 233–236

Nibu Y, Zhang H, Levine M (1998) Interaction of short-range repressors with Drosophila CtBP in the embryo. Science 280:101–104

Nibu Y, Zhang H, Bajor E, Barolo S, Small S, Levine M (1998b) dCtBP mediates transcriptional repression by Knirps, Krüppel and Snail in the Drosophila embryo. EMBO J 17:7009–7020

Parker GA, Crook T, Bain M, Sara EA, Farrell PJ, Allday MJ (1996) Epstein-Barr virus nuclear antigen (EBNA)3C is an immortalizing oncoprotein with similar properties to adenovirus E1A and papillomavirus E7. Oncogene 13:2541–2549

Palmer S, Brouillet JP, Kilbey A, Fulton R, Walker M, Crossley M, Bartholomew C (2001) Evi-1 transforming and repressor activities are mediated by CtBP co-repressor proteins. J Biol Chem 276:25834–25840

Phippen TM, Sweigart AL, Moniwa M, Krumm A, Davie JR, Parkhurst SM (2000) Drosophila C-terminal binding protein functions as a context-dependent transcriptional co-factor and interferes with both mad and groucho transcriptional repression. J Biol Chem 275:37628–37637

Polakis P (2000) Wnt signaling and cancer. Genes Dev 14:1837–1851

Poortinga G, Watanabe M, Parkhurst SM (1998) Drosophila CtBP: a Hairy-interacting protein required for embryonic segmentation and hairy-mediated transcriptional repression. EMBO J 17:2067–2078

Postigo AA, Dean DC (1999) ZEB/zfh-1 represses transcription through interaction with the co-repressor CtBP. Proc Natl Acad Sci USA 96:6683–668

Quinlan MP, Douglas JL (1992) Immortalization of primary epithelial cells requires first- and second-exon functions of adenovirus type 5 12S. J Virol 66:2020–2030

Reid JL, Bannister AJ, Zegerman P, Martinez-Balbas MA, Kouzarides T (1998) E1A directly binds and regulates the P/CAF acetyltransferase. EMBO J 17:4469–4477

Sang NL, Severino A, Russo P, Baldi A, Giordano A, Mileo AM, Paggi MG, De Luca A (2001) RACK1 interacts with E1A and rescues E1A-induced yeast growth inhibition and mammalian cell apoptosis. J Biol Chem 276:27026–27033

Schaeper U, Boyd JM, Verma S, Uhlmann E, Subramanian T, Chinnadurai G (1995) Molecular cloning and characterization of a cellular phosphoprotein that interacts with a conserved C-terminal domain of adenovirus E1A involved in negative modulation of oncogenic transformation. Proc Natl Acad Sci USA 92:10667–10671

Schaeper U, Subramanian T, Lim L, Boyd JM, Chinnadurai G (1998) Interaction between a cellular protein that binds to the C-terminal region of adenovirus E1A (CtBP) and a novel cellular protein is disrupted by E1A through a conserved PLDLS motif. J Biol Chem 273:8549–8552

Schuierer M, Hilger-Eversheim K, Dobner T, Bosserhoff AK, Moser M, Turner J, Crossley, M., and Buettner, R (2001) Induction of AP-2alpha expression by adenoviral infection involves inactivation of the AP-2rep transcriptional corepressor CtBP1. J Biol Chem 276: 27944–27949.

Sewalt RG, Gunster MJ, van der Vlag J, Satijn DP, Otte AP (1999) C-Terminal binding protein is a transcriptional repressor that interacts with a specific class of vertebrate Polycomb proteins. Mol Cell Biol 19:777–787

Sollerbrant K, Chinnadurai G, Svensson C (1996) The CtBP binding domain in the adenovirus E1A protein controls CR1-dependent transactivation. Nucleic Acids Res 24:2578–2584

Sundqvist A, Sollerbrant K, Svensson C (1998) The carboxy-terminal region of adenovirus E1A activates transcription through targeting of a C-terminal binding protein-histone deacetylase complex. FEBS Lett 429:183–188

Sundqvist A, Bajak E, Kurup SD, Sollerbrant K, Svensson C (2001) Functional knockout of the corepressor CtBP by the second exon of adenovirus E1a relieves repression of transcription. Exp Cell Res 268:284–293

Svensson EC, Huggins GS, Dardik FB, Polk CE, Leiden JM (2000) A Functionally Conserved N-terminal Domain of the Friend of GATA 2 (FOG-2) Protein Represses GATA4-Dependent Transcription. J Biol Chem 275:13721–13726

Subramanian T, Malstrom SE, Chinnadurai G (1991) Requirement of the C-terminal region of adenovirus E1a for cell transformation in cooperation with E1b. Oncogene 6:1171–1173

Subramanian T, LaRegina M, Chinnadurai G (1989) Enhanced ras oncogene mediated cell transformation and tumorigenesis by adenovirus 2 mutants lacking the C-terminal region of E1a protein. Oncogene 4:415–520

Touitou R, Hickabottom M, Parker G, Crook T, Allday MJ (2001) Physical and Functional Interactions between the Corepressor CtBP and the Epstein-Barr Virus Nuclear Antigen EBNA3C. J Virol 75:7749–7755

Turner J, Crossley M (1998) Cloning and characterization of mCtBP2, a co-repressor that associates with basic Kruppel-like factor and other mammalian transcriptional regulators. EMBO J 17:5129–5140

Urbanelli D, Sawada Y, Raskova J, Jones NC, Shenk T, Raska K (1989) C-terminal domain of the adenovirus E1A oncogene product is required for induction of cytotoxic T lymphocytes and tumor-specific transplantation immunity. Virology 173:607–614

Vo N, Fjeld C, Goodman RH (2001) Acetylation of nuclear hormone receptor-interacting protein RIP140 regulates binding of the transcriptional corepressor CtBP. Mol Cell Biol 21:6181–6188

Wen Y, Nguyen D, Li Y, Lai ZC (2000) The N-terminal BTB/POZ domain and C-terminal sequences are essential for Tramtrack69 to specify cell fate in the developing Drosophila eye. Genetics 156:195–203

Weigert R, Silletta MG, Spano S, Turacchio G, Cericola C, Colanzi A, Senatore S, Mancini R, Polishchuk EV, Salmona M, Facchiano F, Burger KN, Mironov A, Luini A, Corda D (1999) CtBP/BARS induces fission of Golgi membranes by acylating lysophosphatidic acid. Nature 402:429–433

Wong AK, Ormonde PA, Pero R, Chen Y, Lian L, Salada G, Berry S, Lawrence Q, Dayananth P, Ha P, Tavtigian SV, Teng DH, Bartel PL (1998) Characterization of a carboxy-terminal BRCA1 interacting protein. Oncogene 17:2279–2285

Yang X.-J, Ogrysko VV, Nishikawa J, Howard BH, Nakatani Y (1996) A p300/CBP-associated factor that competes with adenoviral oncoprotein E1A. Nature 382:319–324

Yu X, Wu LC, Bowcock AM, Aronheim A, Baer R (1998) The C-terminal (BRCT). domains of BRCA1 interact in vivo with CtIP, a protein implicated in the CtBP pathway of transcriptional repression. J Biol Chem 273:25388–25392

Zerler B, Moran B, Maruyama K, Moomaw J, Grodzicker T, Ruley HE (1986) Adenovirus E1A coding sequences that enable ras and pmt oncogenes to transform cultured primary cells. Mol Cell Biol 6:887–899

Zhang CL, McKinsey TA, Lu JR Olson EN (2001a). Association of COOH-terminal-binding protein (CtBP) and MEF2-interacting transcription repressor (MITR) contributes to transcriptional repression of the MEF2 transcription factor. J Biol Chem 276: 35–39

Zhang H, Levine M (1999) Groucho and dCtBP mediate separate pathways of transcriptional repression in the Drosophila embryo. Proc Natl Acad Sci USA 96:535–540

Zhang Q, Piston DW, Goodman RH (2002) Regulation of corepressor function by nuclear NADH. Science 295:1895–1897

Zhang Q, Yao H, Vo N, Goodman RH (2000) Acetylation of adenovirus E1A regulates binding of the transcriptional corepressor CtBP. Proc Natl Acad Sci USA 97:14323–14328

Zhang Z, Smith MM, Mymryk JS (2001b) Interaction of the E1A oncoprotein with Yak1p, a novel regulator of yeast pseudohyphal differentiation, and related mammalian kinases. Mol Biol Cell 12:699–710

Cell Transformation by Human Adenoviruses

C. Endter · T. Dobner
Institut für Medizinische Mikrobiologie und Hygiene, Universität Regensburg, Landshuterstr. 22, 93047 Regensburg, Germany
E-mail: thomas.dobner@klinik.uni-regensburg.de

Abstract The last 40 years of molecular biological investigations into human adenoviruses have contributed enormously to our understanding of the basic principles of normal and malignant cell growth. Much of this knowledge stems from analyses of their productive infection cycle in permissive host cells. Also, initial observations concerning the carcinogenic potential of human adenoviruses subsequently revealed decisive insights into the molecular mechanisms of the origins of cancer, and established adenoviruses as a model system for explaining virus-mediated transformation processes. Today it is well established that cell transformation by human adenoviruses is a multistep process involving several gene products encoded in early transcription units 1A (E1A) and 1B (E1B). Moreover, a large body of evidence now indicates that alternative or additional mechanisms are engaged in adenovirus-mediated oncogenic transformation involving gene products encoded in early region 4 (E4) as well as epigenetic changes resulting from viral DNA integration. In particular, detailed studies on the tumorigenic potential of subgroup D adenovirus type 9 (Ad9) E4 have now revealed a new pathway that points to a novel, general mechanism of virus-mediated oncogenesis. In this chapter, we summarize the current state of knowledge about the oncogenes and oncogene products of human adenoviruses, focusing particularly on recent findings concerning the transforming and oncogenic properties of viral proteins encoded in the E1B and E4 transcription units.

1 Introduction

Almost exactly 40 years ago, Trentin, Yabe, and Taylor reported for the first time that human adenovirus type 12 (Ad12) can induce malignant tumors following injection into newborn hamsters (Trentin et al. 1962). Their seminal discovery showing that a pathogenic human virus has oncogenic potential inspired a period of intense research into adenovirus biology that continues to this day. Following the initial discovery with Ad12, it soon became apparent that adenoviruses provide an excellent experimental tool to analyze fundamental events of normal and malignant cell growth. In particular, their ability to oncogenically transform primary mammalian cells in culture, which are then capable of initiating tumor growth in a susceptible host animal, has been of immense value in elucidating events in cellular and viral growth control. Much of our current understanding of the molecular mechanisms underlying virus-mediated oncogenic transformation has derived from the study of human adenoviruses and the viral gene products involved in transformation. These studies led to the identification of the adenovirus E1A and E1B oncoproteins, which in turn have served as molecular probes to resolve many fundamental aspects of cellular gene expression, cell-cycle control, and programmed cell death.

Today, it is well established that adenovirus-mediated oncogenic transformation is a multistep process that involves a complex interplay between viral and cellular proteins, as well as insertional mutagenesis by viral DNA integration and presumably epigenetic changes. Most of the viral genes involved in adenovirus transformation are now known, and nearly all of their products have been identified, yet much remains a mystery. Even four decades later it is still unclear why human adenoviruses are oncogenic in rodents but not, as far as we know, in humans. It still remains to be determined why transformation of primary human cells with cloned adenovirus oncogenes is extremely inefficient compared to primary rodent cells, and why some types of adenoviruses and/or virus-transformed cells vary considerably in their oncogenic potential after direct injection into different rodents. A series of recent findings, however, have given new insights into adenovirus-mediated oncogenesis. Collectively, these data demonstrate that alternative or additional mechanisms are engaged in adenovirus-mediated oncogenic transformation involving gene products encoded in early regions 1B (E1B) and 4 (E4). In particular, detailed studies on the tumorigenic potential of subgroup D adenovirus type 9 (Ad9) E4 have revealed a new pathway that points to a novel, general mechanism of virus-mediated oncogenesis. This chapter reviews the current state of knowledge about the oncogenes and oncogene products of human adenoviruses, focusing particularly on recent findings concerning the transforming and oncogenic properties of viral proteins encoded in the E1B and E4 transcription units.

2 Oncogenic Potential of Adenoviruses

2.1 Tumor Induction in Animals

Human adenoviruses are classified as DNA tumor viruses by virtue of their ability to cause tumors when inoculated into newborn rodents (Graham 1984; Branton et al. 1985). On the basis of their oncogenicity in animals, the human serotypes can be subdivided into three classes: highly oncogenic adenoviruses, producing tumors at a high frequency within a few months; weakly oncogenic viruses, inducing tumors inconsistently and after long incubation times; and nononcogenic viruses (Table 1).

Table 1 Oncogenic and nononcogenic human adenoviruses

Subgroup	Serotypes	Oncogenicity in animals	Principle type of tumors
A	12, 18, 31	High	Undifferentiated sarcomas
D	9, 10	High	Fibroadenomas
B	3, 7	Moderate	Undifferentiated sarcomas
C–F	All others	None	None

Tumors in rodents induced by highly oncogenic subgroup A and weakly oncogenic subgroup B adenoviruses develop at the site of injection and, depending on the route of administration, vary greatly in type, including neurogenic, neuroepithelial, medulloblastic, and adenocarcinomatous tumors (Graham 1984). Moreover, Ad12, the prototype of the highly oncogenic subgroup A serotypes, can induce retinal tumors in newborn baboons (Mukai and Murao 1975) that are indistinguishable from human retinoblastomas (Mukai et al. 1980). By contrast, subgroup D Ad9 and Ad10 elicit exclusively estrogen-dependent mammary tumors at 100% frequency within three months of subcutaneous or intraperitoneal injection in female Wistar-Furth rats (Ankerst et al. 1974; Jonsson and Ankerst 1977, 1989; Javier et al. 1991). The differences in the oncogenicity among different adenoviruses are related to the genetic background of the host animal, as well as the susceptibility of the transformed cell to the destructive effect of the host's immune system (van der Eb and Zantema 1992; Williams et al. 1995). Also, as will be discussed in Sect. 3.3.1, the unique oncogenicity of subgroup D Ad9 apparently coincides with a novel molecular mechanism that differs fundamentally from the oncogenic subgroup A and B viruses.

2.2 Transformation of Rodent Cells in Culture

Despite the remarkable pattern of differential oncogenicity exhibited by the human adenoviruses in animals, the members of all subgroups tested so far are able to stably transform a variety of rodent cells in vitro with broadly similar efficiencies. Likewise, rodent cells, e.g., from rat, mouse, or hamster, can be efficiently transformed in culture by transfecting or microinjecting noninfectious subgenomic viral DNA fragments containing the adenovirus transforming genes (Sect. 3). Cells transformed in vitro are morphologically similar to cells explanted from virus-induced tumors and display many characteristic properties of an oncogenic cellular phenotype (Graham 1984). Their tumorigenicity varies in relation to the virus type and immune status of the test animal. Analogous to the situation following virus infection in animals, most cells transformed by highly oncogenic group A virus Ad12 are capable of initiating tumor growth in syngeneic immunocompetent rats, whereas most cells transformed with nononcogenic subgroup C Ad2 or Ad5 generate no tumors. However, some Ad2-transformed rat cell lines can induce tumors in immunosuppressed rats (Gallimore 1972), and both Ad12- and Ad5-transformed cells can produce tumors in thymus-deficient nude mice (Bernards et al. 1983) lacking major histocompatibility complex (MHC) class I-restricted cytotoxic T lymphocytes (CTLs). Also, tumorigenicity can vary greatly even among cell lines transformed by nononcogenic adenoviruses. For example, Ad2- or Ad5-transformed rat cells are not tumorigenic in rats, whereas similarly transformed hamster cells are able to induce tumors in newborn animals from the same species, although oncogenicity is lost in weanling animals that have developed their thymic response (Cook and Lewis 1979; Lewis and Cook 1982).

Collectively, these results indicate that tumorigenicity of adenovirus-transformed cells is dependent on the host animal used and is greatly influenced by the thymus-dependent CTL components of the immune system. In support of this, it has been shown that transformation of rat cells with Ad12 leads to a considerable reduction in MHC class I antigens on the cell surface (Schrier et al. 1983), and that some Ad12 transformants are more resistant to recognition by allogeneic CTLs in vitro than Ad5-transformed cells (Bernards et al. 1983; Yewdell et al. 1988; and Sect. 3.2.3). In addition, natural killer (NK) cells appear to play a role in determining the tumorigenicity of adenovirus-transformed cells, because Ad2 transformants are more susceptible to NK-mediated cell lysis than their Ad12 counterparts (Raska and Gallimore 1982; Cook et al. 1987). Finally, more recent data hint at the possibility that there are additional mechanisms governing tumor formation in animals (Sects. 3.3 and 5).

2.3
Transformation of Human Cells in Culture

Despite their oncogenic potential in animals and their ability to completely transform primary rodent cells in culture, extensive screens in the past have failed to correlate the presence of adenoviruses with oncogenesis in humans (Sect. 6.2). Also, many attempts to transform primary human cells in culture with adenovirus have been unsuccessful, indicating that abortive infection is one of the factors associated with highly efficient transformation of nonpermissive rodent cells. Interestingly, however, transformation of human cells with noninfectious subgenomic viral DNA fragments is extraordinarily inefficient compared to rodent cells, arguing that differences in permissivity to viral growth may not be a principle determining factor in transformation efficiency (Shenk 1996; Hahn et al. 1999). To date, only a few primary human cells have been successfully transformed by Ad12 or Ad5 DNA fragments in culture, including human embryo kidney (HEK) cells (Graham et al. 1977; Whittaker et al. 1984), human embryonic lung (HEL) cells (van den Heuvel et al. 1992), human embryo retinoblasts (HERs) (Byrd et al. 1982; Gallimore et al. 1986; Fallaux et al. 1996, 1998), and amniocytes (Schiedner et al. 2000). Among these only HERs and amniocytes can be transformed reproducibly, although less efficiently than rodent embryo or kidney cultures. The molecular basis underlying the differences in transformation efficiencies among various human cell types is unknown. However, recent work demonstrates that most transformed human cell lines derived from cultures of HEK and HER cells exhibit a pattern of intermediate filament expression similar to that seen in early differentiating neurons (Shaw et al. 2002). Since HER cell cultures and to a much lesser extent HEK cell cultures contain predominantly neuronal lineage cells, it has been proposed that human neuronal cells are a favored target for adenovirus transformation (Shaw et al. 2002). Whether transformed cells from transfections of amniocytes display a similar pattern remains to be seen.

3 Adenovirus Transforming Genes

3.1 Persistence of Viral Information

Most adenovirus tumors, tumor cell lines, and transformed cell lines are characterized by the persistence of viral DNA in a chromosomally integrated form and the expression of virus-specific antigens (reviewed by Graham 1984). Thus, adenovirus transformation follows the classical concept of viral oncogenesis where viral genes persist within the transformed cells. In a few cases, however, some Ad12-transformed cells established from hamster tumors gradually lost all viral DNA sequences after several passages in culture but continued to induce tumors after reinjection into hamsters (Kuhlmann et al. 1982; Pfeffer et al. 1999). Also, no viral DNA was detected in some Ad12-transformed rat cell lines (Paraskeva and Gallimore 1980; Paraskeva et al. 1982). These findings indicate that persistence of adenovirus DNA is not absolutely necessary for maintenance of the transformed state and suggest that a minority of oncogenic transformations by human adenoviruses can occur through a "hit-and-run" mechanism (Sect. 5).

While the majority of Ad12 and Ad9 tumor and transformed cells contain one or more copies of the entire genome (Graham 1984; Brusca et al. 1984; Javier et al. 1991), most rodent cells transformed by subgroup C Ad2 or Ad5 harbor only a fraction of the viral chromosome, although exceptions to this rule are numerous (Graham 1984). The only sequences of viral DNA that are almost invariably retained in virus-transformed cells are those derived from the left end of the viral genome, encompassing all of early region 1 (E1; Fig. 1). Likewise, in most completely transformed cell lines established following transfection with subgenomic viral DNA fragments, the E1 sequences are retained in an integrated form, although less of the viral genome is needed for complete transformation of primary cells in these assays (Fig. 1). Another frequently observed rearrangement of viral DNA in adenovirus-transformed cells is where left and right ends of the genome are joined together. Nevertheless, since the E1 region is the only one that is consistently retained in virus-transformed cells, and because transfection of cultured cells with cloned E1 genes can lead to oncogenic transformation, the prevailing view is that E1 gene products mediate the initiation of transformation and apparent-

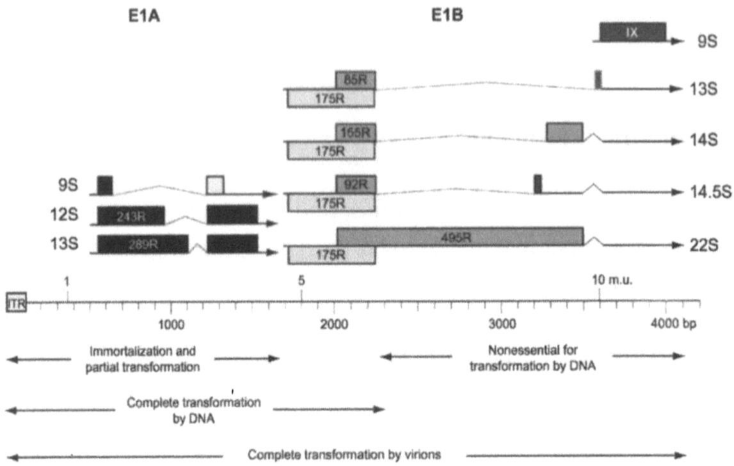

Fig. 1 Genomic organization and function of Ad2 E1. The left 4,000 bp of the linear double-stranded genome are depicted in the *center* of the figure. *Top:* Location of the E1A and E1B transcription units. The E1A- and E1B-derived mRNAs are shown as *solid lines*, the protein coding sequences as *boxes* containing the number of amino acid residues (*R*) in each of the different E1A and E1B polypeptides. The IX gene encodes for a structural polypeptide expressed at the intermediate stage of infection. *Bottom:* Regions identified as being required for transformation under various conditions. See text for details and references. *ITR*, inverted terminal repeat; *m.u.*, map units. (Adapted from Graham 1984)

ly provide functions required to maintain some of the phenotypic characteristics of the transformed cell. While this concept is applicable to most human adenoviruses examined so far, subgroup D Ad9, at least, does not conform to this generally accepted rule. Ad9 is unique among human adenoviruses because its primary oncogenic determinant is a function encoded in early region 4 (E4) rather than E1 (Sect. 3.3.1). In addition, recent data with plasmid-based genes derived from Ad5 E4 support the view that the transforming potential of subgroup C adenoviruses does not exclusively lie within the E1 region (Sect. 3.3.2).

3.2
Early Region 1

Detailed insight into the function of adenovirus E1 in the transformation process comes from extensive genetic and molecular analyses carried out largely with subgroup A and C serotypes and plasmid-based genes derived from these viruses. Collectively, these studies led to the conclusion that adenovirus-mediated cell transformation is a multistep process requiring the cooperation of E1 gene products encoded in two separate transcription units termed E1A and E1B (Fig. 1), both of which are necessary and perhaps sufficient for oncogenic transformation of primary rodent cells by most adenoviruses. In Ad2/5 the E1A and E1B genes are located between map units 1.3 and 11.2 on the adenovirus genome and are transcribed in a rightward direction. Sequence comparisons with other serotypes show that all human adenoviruses possess similarly organized E1A and E1B genes and express nearly the same complement of gene products.

E1A (map units 1.3 to 4.5) is the first transcription unit to be expressed after the viral chromosome reaches the nucleus in infected cells (Boulanger and Blair 1991). During the early phase of infection it encodes two mRNAs (13S and 12S, according to their sedimentation coefficients), which are generated by alternative splicing from a common E1A RNA precursor. Three additional E1A mRNA species (11S, 10S, and 9S) accumulate later in the infectious cycle, but no definitive function has been described for their products. The 13S mRNA produces a polypeptide of 289 amino acids (aa) in the case of Ad2/5 and 266 aa in Ad12, while Ad2/5 and Ad12 12S mRNAs encode polypeptides of 243 aa and 235 aa, respectively. The E1B gene (map units 4.6 to 11.2) lies adjacent to E1A and generates two major mRNA species of 22S and 13S with common 5′ and 3′ termini (Fig. 1). Both 22S and 13S mRNAs are derived from a common mRNA precursor by alternative splicing. Translation of the 22S mRNA from Ad2 used discrete, but overlapping reading frames to produce two unrelated polypeptides of 175 aa (E1B 19-kDa) and 495 aa (E1B 55-kDa) (Anderson et al. 1984; Takayesu et al. 1994). The 13S E1B mRNA also encodes the 19-kDa protein and a smaller species (85 aa) containing the amino-terminal 75 residues of the 55-kDa product. Two minor mRNAs of 14.5S and 14S (Virtanen and Pettersson 1985) coding for E1B 19-kDa and two 55-kDa related proteins (92 aa and 155 aa) have also been described, but as yet no function has been attributed to either the 92-aa or 155-aa E1B species in infection or transformation.

3.2.1 Role of E1A Proteins in Transformation

Following the initial description of the E1A transcription unit and its gene products in the late 1970s, it very soon became clear that E1A proteins from subgroup A and C viruses are essential for transformation of rodent cells and, in the case of the oncogenic serotypes, e.g., Ad12, provide functions for direct tumor induction in rats and for the tumorigenicity of transformants produced in culture (reviewed by Graham 1984; Williams et al. 1995; Ricciardi 1999). As understood at present, E1A proteins mediate the most critical step in cell transformation. In transfection experiments, expression of E1A alone is sufficient for immortalization and partial transformation of primary rodent cells. Both the 289-aa and 243-aa E1A proteins from Ad2/5 are equally effective in this process (Haley et al. 1984; Moran et al. 1986). Although the majority of E1A-immortalized cells can be subcultivated continuously, transformation is incomplete, in that the transformants do not grow to high cell densities and are rarely tumorigenic in animals (Gallimore et al. 1984). Full manifestation of the transformed phenotype requires the coexpression of E1B or other cooperating oncogenes such as activated Ha-*ras* (Ruley 1983). Again, each of the major E1A products is equally effective in completely transforming primary rat cells in such cooperation with E1B or the activated Ha-*ras* oncogene (Zerler et al. 1986; Jochemsen et al. 1986).

The transforming and oncogenic properties of E1A in nonpermissive cells derive from their ability to both deregulate normal cell growth (in an effort to establish productive infection) and their ability to modulate the infected cell's interaction with the immune system of the host. For example, E1A proteins stimulate entry into the S phase of the cell cycle to create optimal conditions for productive virus infection in nonproliferating human cells. In the course of an abortive infection in nonpermissive rodent cells, however, these same growth deregulatory functions can lead to immortalization or transformation. Also, E1A proteins from the oncogenic subtype Ad12 provide functions which down-regulate expression of MHC class I genes. This activity may be used by the virus to escape the host's immune defense mechanism in order to establish a persistent infection in humans (Pääbo et al. 1989), while the same activity may contribute to the observed tumorigenicity of Ad12-transformed rodent cells in immunocompetent animals (Ricciardi 1999).

E1A proteins accomplish deregulation of cell growth and immune evasion by binding to and modulating the normal function of key regulators that control transcription, cell-cycle progression, programmed cell death (apoptosis), and presumably protein turnover (for a recent comprehensive review, see Gallimore and Turnell 2001). To date, over 40 different cellular E1A binding proteins have been identified, including components of the general transcriptional machinery, transcriptional coactivators, and repressors as well as proteins directly involved in cell-cycle regulation, most notably the retinoblastoma tumor suppressor gene product (pRb) and the related pocket-proteins p107 and p130 (see below).

Extensive mutant analyses combined with the availability of E1A sequences from a number of different human and simian adenovirus serotypes have defined four evolutionarily conserved regions (CRs) (Kimelman 1986; Moran 1994; Avvakumov et al. 2002). These subdomains mediate nearly all of the biological properties of E1A and are the sites which interact with most cellular E1A-binding partners (Fig. 2). While CR1, CR2, and CR4 are common to both major E1A products, CR3 is unique to the larger 13S-derived E1A proteins. In addition, E1A products from the oncogenic serotypes contain a nonconserved alanine-rich domain between CR2 and CR3 that appears to play a vital role in tumor induction by these viruses in animals (Sect. 3.2.3). The regions in E1A required for immortalization and complete transformation in cooperation with E1B or activated *ras* oncogene map to CR1, CR2, CR4, and segments in the nonconserved amino-terminal region (NTR) of E1A. CR3 is dispensable for these processes, although it may enhance transformation efficiency in some way. In general, exon 1-encoded functions comprising CR1, CR2, and a short region mapping to the extreme NTR are sufficient for complete transformation in cooperation with activated *ras* oncogene. By contrast, immortalization and efficient cotransformation with E1B additionally require functions encoded within exon 2 CR4 (Subramanian et al. 1991; Quinlan and Douglas 1992; Douglas and Quinlan 1995).

As well as being required for transformation, the CR1 and CR2 domains together with residues from the NTR are responsible for the progression of quiescent cells to S-phase cellular DNA synthesis. Numerous studies indicate that stimulation of cell-cycle progression occurs through two independent pathways, which may act synergistically. The most extensively studied pathway involves CR1 and CR2 and their interaction with the pocket-family proteins pRb and related members p107 and p130, as well as cellular cyclin and kinases associated with them. Binding to pRb, p107, and p130 involves a highly conserved LXCXE motif located in CR2 (Fig. 2). It is now well established that these interactions displace the pocket-proteins from complexes with the E2F family of transcription factors (reviewed by Ben-Israel and Kleinberger 2002). E2F activity is regulated in a cell cycle-dependent manner, with interactions involving pRb, p107, and p130 repressing E2F activities (reviewed by Cress and Nevins 1996). Thus, binding of E1A proteins to Rb family members, causing their dissociation from E2Fs (Kovesdi et al. 1986a,b, 1987; Cress and Nevins 1996), results in activation of cellular genes containing E2F-binding sites and induction of cell-cycle progression (Cress and Nevins 1996). The second pathway maps to the NTR and a segment within CR1 that is distinct from the region required for pRb binding. Several lines of evidence indicate that these regions act as a transcription repression domain and operate through binding to the histone-directed acetyltransferases p300/CBP and the p300/CBP-associated factor (P/CAF), as well as the TATA-binding protein (TBP) and other factors involved in chromatin remodeling such as the SWI/SNF family member p400. There is good reason to believe that the amino-terminal pathway contributes to transformation by transcriptionally repressing cellular genes involved in cell proliferation and terminal cell differentiation.

Although exon 1-encoded functions are necessary for immortalization, these activities are not sufficient to maintain the cells in a proliferative mode and extend their lifespan in culture. This function is provided by properties encoded in exon 2 CR4, which apparently modulates the transforming activity of E1A exon 1 (Quinlan and Douglas 1992; see the chapter by Chinnadurai, this volume). As described above, CR4 is apparently dispensable for transformation in combination with activated *ras*, but plays a central role in plasmid-induced transformation in cooperation with E1B and in transformation by the virus (Subramanian et al. 1991). In fact, CR4 suppresses transformation by E1A/*ras*, since deletions in this subdomain enhance *ras*-combined transformation, resulting in a "hyper-transformed" phenotype (Subramanian et al. 1989; Douglas et al. 1991; Boyd et al. 1993). The ability of CR4 to repress transformation by *ras* resides in part in its ability to interact with the C-terminal binding protein (CtBP), since mutants lacking the conserved CtBP binding motif PXDLS (Fig. 2) enhance transformation (Boyd et al. 1993; Schaeper et al. 1995).

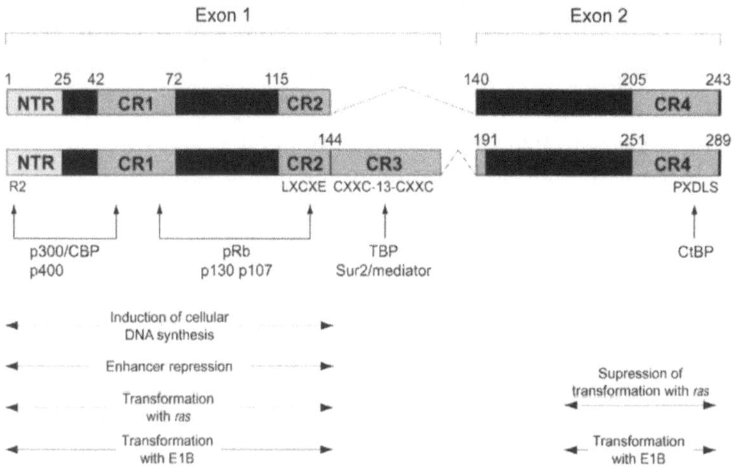

Fig. 2 Schematic representation of the two major E1A proteins of Ad5. E1A 289 aa and 243 aa differ by only a 46-aa internal cysteine-rich segment (*CR3*) unique to the 289-aa protein produced by differential splicing of the E1A transcripts. *Top:* Location of the aminoterminal region (*NTR*) and the regions conserved among adenovirus serotypes (*CR*). Numbers denote amino acid residues at the boundaries of the conserved regions and were derived from Avvakumov et al. (2002). The invariant arginine residue at position 2 (*R2*), the LXCXE motif in CR2, and the PXDLS motif in CR4 are indicated *below.* CXXC-13-CXXC denotes a four-cysteine zinc finger structure in CR3 that is required for binding to TBP and the Sur2 subunit of the mammalian mediator complex (Stevens et al. 2002; Wang and Berk 2002). R2 is absolutely required for E1A immortalizing activity and for E1A binding to p300. *Middle:* Important E1A-interacting proteins. Mapped binding sites are indicated by *lines* and *arrows.* For the sake of clarity, most E1A-binding partners are not shown (for detailed information, see Gallimore and Turnel 2001). *Bottom:* Functional domains identified as being required for transformation with activated *ras* and E1B, induction of cellular DNA synthesis and enhancer repression. CR3 is not essential for transformation but is crucial for efficient transcriptional activation of viral promoters from E1B, E2, E3, and E4. (Adapted from Boulanger and Blair 1991; Shenk 1996; Gallimore and Turnel 2001)

Taken together, these findings indicate that E1A proteins are essential for oncogenic transformation by most adenoviruses and mediate the most critical step in this process. Almost all of their transforming activities are mediated through multifunctional subdomains, which bind to and modulate the normal function of cellular key regulators of cell growth control. The consequence of these interactions is that E1A proteins induce unscheduled DNA synthesis and cell proliferation, so that primary cells with a limited lifespan can become immortal and can escape senescence (Moran 1994; Shenk 1996). Ironically, however, these very activities required for transformation also stimulate apoptosis and growth arrest, almost abrogating the transforming activity of E1A. As understood at present, most of these antiproliferative activities are due to E1A-induced metabolic stabilization of the tumor suppressor protein p53 (Debbas and White 1993; Lowe and Ruley 1993; Grand et al. 1994; Sabbatini et al. 1995). The ability of E1A to increase the half-life of p53 correlates in part with its ability to bind pRb and p300/CBP (Chiou and White 1997; Querido et al. 1997; Grossman et al. 1998; de Stanchina et al. 1998; Honda and Yasuda 1999). In addition, recent studies indicate that E1A can also inhibit proteasomal-mediated degradation of p53 through binding of its NTR to the 19S regulatory complex ATPases of the 26S proteasome (Turnell et al. 2000) and/or through interaction of CR2 with E2 enzymes of the ubiquitin/SUMO-1 conjugation machinery (Hateboer et al. 1996). As will be described in the next section, these proapoptotic and growth-arresting activities are efficiently counteracted by the E1B gene products, which have no transforming capability of their own, but markedly contribute to the transformation frequency of E1A, as well as the overall transformed and malignant phenotype.

3.2.2
Role of E1B Proteins in Transformation

Compared to the E1A proteins, considerably less is known about the role and the function(s) of the adenovirus E1B gene products in the complete transformation of rodent cells in culture and tumor induction in animals. In the past, numerous studies with virus mutants carrying lesions in the E1B unit gave rise to the hypothesis that both major E1B gene products, E1B 19-kDa and E1B 55-kDa, are required for complete transformation (reviewed by van der Eb and Zantema 1992). In contrast, a different conclusion was drawn from other genetic analyses showing that the 19-kDa product from Ad12 is not required for either efficient transformation of baby rat kidney (BRK) cells and baby mouse kidney (BMK) cells in culture or for direct induction of tumors by the virus in rats (Edbauer et al. 1988; Zhang et al. 1992). Similar results were obtained for transformation of a variety of rodent cells by Ad5 E1B 19-kDa mutant viruses in vitro (Telling and Williams 1993). Plasmid-based transformation assays show that both E1B 19-kDa and E1B 55-kDa proteins of Ad5 are individually capable of cooperating with E1A to completely transform BRK cells, albeit at lower efficiencies than when both proteins are expressed (Bernards et al. 1986; White and Cipriani 1990; McLorie et al. 1991; Rao et al. 1992). Several observations suggest that both E1B proteins contribute to complete cell transformation at least in part by antagonizing apoptosis and growth arrest, which, as described, primarily result from the induction and metabolic stabilization of p53 by E1A (Debbas and White 1993; Lowe and Ruley 1993). The fact that E1B 19-kDa and E1B 55-kDa cooperate with E1A in an additive fashion further implies that cell transformation may be the result of suppression of apoptosis induced by both p53-dependent and p53-independent mechanisms (Teodoro et al. 1995).

Although the molecular mechanism(s) by which E1B 19-kDa inhibits apoptosis in the course of the transformation process are not completely understood, how it functions is distinct from that of E1B 55-kDa. It is thought that E1B 19-kDa inhibits apoptosis induced by E1A, p53, TNF-α, Fas ligand, TRAIL, and many other stimuli by mechanisms analogous to the Bcl-2 family of apoptosis regulators (recently reviewed by White 2001). Within the context of modulating apoptosis during transformation, this adenovirus protein appears to block nearly all of the mitochondrial signaling events. These activities may involve: (1) inhibition of cytochrome *c* and Smac/DIABLO release from mitochondria (Henry et al. 2002) through direct binding to and modulation of the proapoptotic proteins Bax, Bak, and possibly related BH3-containing members of the Bcl-2 family (Boyd et al. 1994; Farrow et al. 1995; Han et al. 1996); (2) suppression of caspase-9 and caspase-3 activation downstream of mitochondria (Henry et al. 2002); (3) blocking of transcriptional repression by p53 mediated through the carboxy-terminal domain of the tumor suppressor protein (Shen and Shenk 1994; Horikoshi et al. 1995; Sabbatini et al. 1995); and (4) restoration of mdm-2 transcription, causing down-regulation of p53 levels and function (Thomas and White 1998).

As opposed to the 19-kDa product, the transforming potential of Ad5 E1B 55-kDa correlates with its ability to act directly as a transcriptional repressor (Yew and Berk 1992) targeted to p53-responsive promoters (Yew et al. 1994) by binding to the tumor suppressor protein (Sarnow et al. 1982). It is believed that these activities may antagonize p53-induced apoptosis (Sabbatini et al. 1995; Teodoro and Branton 1997) and/or cell-cycle arrest (Shepherd et al. 1993; Hutton et al. 2000) in the course of the transformation process. The larger E1B protein from Ad12 (E1B 54-kDa), although it does not form a stable complex with p53, also inhibits transcriptional activation by the tumor suppressor (Yew and Berk 1992), which provides further support for the view that blocking this activity is an integral part of the larger E1B gene product's contribution to complete transformation. The regions required for transformation map to several segments in the Ad5 protein, including a SUMO-1 conjugation motif (ΨKXE) at position 104 (Endter et al. 2001), the p53-binding domains located in the amino-terminal and in the central part (Yew et al. 1990; Yew and Berk 1992; Grand et al. 1999), as well as two regions at the carboxy terminus that mediate inhibition of p53-dependent and p53-independent transactivation (Yew et al. 1994; Teodoro et al. 1994; Teodoro and Branton 1997). The silencing activity of Ad5 E1B 55-kDa requires a cellular corepressor that copurifies with RNA polymerase II (Martin and Berk 1999) and the interaction with cellular factors known to be involved in transcriptional repression, such as histone-directed deacetylase 1 (HDAC1) and mSin3A (Punga and Akusjärvi 2000). In addition, consistent with its role in blocking p53-mediated transactivation, the Ad5 protein inhibits p53 acetylation by binding to both p53 and P/CAF (Liu et al. 2000). Moreover, the mode of action of Ad5 E1B 55-kDa during transformation may involve further functions and other protein interactions (Harada et al. 2002). For example, the Ad12 E1B 54-kDa protein significantly extends the lifespan of normal mammalian cells in culture, probably due to its ability to interfere with telomere metabolism during cell division (Gallimore et al. 1997). Such an activity would be consistent with the finding that E1B gene products cooperate with E1A CR4 to escape senescence of E1A-immortalized cells (see above). Whether this activity involves ablation of p53 functions remains to be seen. Also, in Ad2/5 but not Ad12 E1-transformed cells the majority of E1B 55-kDa and p53 proteins are excluded from the nucleus and colocalize in cytoplasmic dense bodies (Fig. 3) that also contain the tumor suppressor protein WT1 and microfilaments (Zantema et al. 1985; Maheswaran et al. 1998). This cytoplasmic restriction imposed upon p53 (see below) and WT1, which likely inactivates both factors, may include continuous nucleocytoplasmic export mediated through a leucine-rich nuclear export signal (NES) (Krätzer et al. 2000; Endter et al. 2001). Moreover, the E1B-associated protein 5 (E1B-AP5) and the nuclear body (NB)-associated promyelocytic leukemia (PML) protein can both individually suppress transformation of BRK cells by E1A plus E1B (Gabler

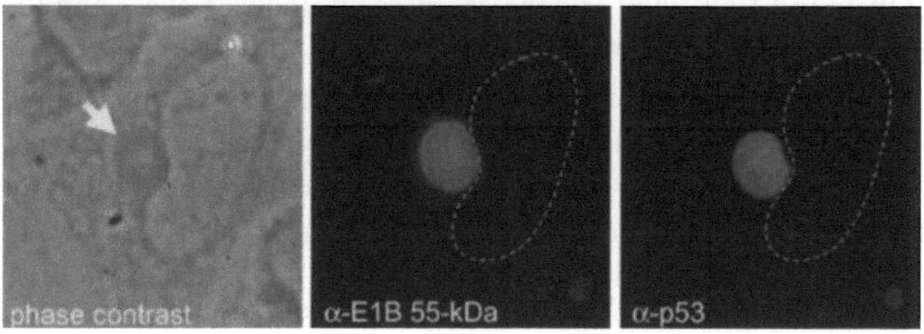

Fig. 3 Cytoplasmic accumulation of E1B 55-kDa and p53 in Ad5 E1-transformed BRK cells. The dense perinuclear body is indicated by an *arrow* and was stained with anti-E1B 55-kDa (2A6) and anti-p53 (PAb 421) monoclonal antibodies. (The images were kindly provided by Michael Nevels)

et al. 1998; Nevels et al. 1999a; Kzhyshkowska et al. 2001), indicating that modulation of these cellular factors by E1B proteins is required to enhance transformation. Finally, studies in lytically infected human cells have shown that Ad5 E1B 55-kDa relieves growth restrictions imposed on viral replication by the cell cycle through mechanisms independent of p53 (Goodrum and Ornelles 1997, 1998).

3.2.3
Role of E1A and E1B in Tumorigenesis

As mentioned (Sects. 2.1, 2.2) the differential oncogenicity exhibited by subgroup A and C adenoviruses and the tumorigenicity of their respective transformants is related to the animal host's ability to mount an immune response against tumor cells transformed by Ad2/5, but not against cells transformed by the oncogenic Ad12. The resistance of Ad12-transformed cells to the destructive effect of the host's immune system is dependent on a series of events that modulate pathways involved in immune regulation and presumably other, as yet unknown functions. Most of these events are guided by Ad12 E1A, but are evidently influenced by E1B and probably other viral gene products encoded outside of E1 (Sect. 3.3). Nearly all of this information has been chronicled in reviews over the years and will be summarized only briefly here. For detailed information on E1A from the oncogenic subgroup A Ad12, and its essential role in tumorigenesis, we refer to several excellent earlier reviews (Boulanger and Blair 1991; Ricciardi 1995, 1999; Williams et al. 1995; Blair and Hall 1998) and the chapters by Hohlweg et al. and J.F. Williams et al. in this volume.

E1A from oncogenic serotypes is thought to provide several unique functions allowing the transformed cell to evade NK cell- and CTL-mediated immune surveillance (Cook et al. 1982; Raska and Gallimore 1982; Bernards et al. 1983; Schrier et al. 1983; Eager et al. 1985; Sawada et al. 1985). Although the detailed mechanism(s) by which Ad12 E1A confers resistance to NK cell and CTL action is still not understood, the later activity has been shown to involve interference with MHC class I presentation, including transcriptional repression of MHC class I enhancer sites, as well as down-regulation of other genes in the MHC complex engaged in antigen processing and presentation (reviewed in Blair and Hall 1998). At present, several domains on E1A have been implicated in contributing to Ad12 tumorigenesis and MHC class I down-regulation: the alanine-rich spacer segment between CR2 and CR3 (Jelinek et al. 1994; Telling and Williams 1994), which is not present in Ad2/5 E1A; exon 1-encoded sequences between the NTR and CR2 (Pereira et al. 1995; Smirnov et al. 2001); and segments in the CR3-transactivating domain (Meijer et al. 1989; Huvent et al. 1996). Also, E1A CR4 may play some role in the differential oncogenicity between subgroup A and C viruses (see also the chapter by Chinnadurai, this volume). As described in Sect. 3.2.1, CR4 subdomain mutations that abolish CtBP-binding enhance transformation efficiency of Ad2/5 E1A in cooperation with *ras*, and, interestingly, the resulting transformants induce rapidly growing tumors in syngeneic rats and athymic mice (Subramanian et al. 1989; Boyd et al. 1993). These observations have led to the hypothesis that E1A CR4 may in fact suppress oncogenicity of subgroup C viruses (Williams et al. 1995). If this is true, it would be interesting to assess the role of this region in the highly oncogenic group A viruses as regards the transformed cell phenotype.

In addition to the central role of E1A in determining viral oncogenicity, it is clear that E1B gene products from subgroup A and C viruses provide functions which elevate the tumorigenic phenotype of the transformed cell (Bernards et al. 1982, 1983; Sawada et al. 1988). Several studies employing rat cells transformed by Ad5/12 E1A/E1B hybrid plasmids and viruses indicate that E1B gene products from these viruses may operate through different pathways in oncogenic transformation (Bernards et al. 1982; van den Heuvel et al. 1990, 1993). For example, virus-transformed rat cell lines containing the E1A and E1B genes from Ad12 are highly tumorigenic in syngeneic animals, while cell lines transformed with virus producing E1A proteins from Ad12 and E1B proteins from Ad5 exhibit reduced tumorigenicity in the same animals (Sawada et al. 1988). Furthermore, transfection experiments with Ad5/12 hybrid plasmids suggest that the difference in oncogenicity between Ad5- and Ad12-transformed BRK cells in athymic nude mice is probably specified by the E1B gene products (Bernards et al. 1982) and, more specifically, by the larger E1B protein (Bernards et al. 1983). In this context it has been shown that transformed cells producing the Ad12 E1B 54-kDa protein, but not transformants expressing the Ad5 55-kDa E1B gene, are highly tumorigenic in athymic nude mice (Bernards et al. 1982).

Other studies indicate that the oncogenicity of Ad5 E1-transformed 3Y1 rat cells in athymic nude mice inversely correlates with both the expression levels of E1B 55-kDa and the subcellular localization of p53 in these cells. In other words, 3Y1 cells which produce low levels of the 55-kDa protein have nuclearly located p53 and are highly oncogenic in nude mice, while high levels of the E1B protein cause the cytoplasmic restriction of p53 and, significantly, these cells only form tumors after very long latency periods (van den Heuvel et al. 1990). Based on these observations it has been proposed that Ad5 E1B-dependent cytoplasmic sequestration of p53 reduces the intrinsic ability of these cells to grow in a neoplastic state in vivo (van den Heuvel et al. 1990).

The molecular basis underlying these observations is still unknown. Apparently, E1B proteins from Ad5 and Ad12 do not affect susceptibility to NK cell action (Sawada et al. 1985; Kast et al. 1989), and both forms can inhibit p53-transactivating activity in transient transfection assays (Yew and Berk 1992). However, the large E1B proteins from both viruses differ substantially from each other with respect to their subcellular localization in transformed cells, their ability to form a stable complex with p53 (see Sect. 3.2.2), and their ability to induce chromosome fragility at four specific human loci that each contain tandemly repeated genes for an abundant small RNA (Durnam et al. 1988; Li et al. 1998; Liao et al. 1999; Yu et al. 2000). While the Ad12 product and p53 are predominantly localized in the nucleus, the majority of the Ad5 E1B protein accumulates with the tumor suppressor protein in a large cytoplasmic, perinuclear body (Fig. 3; Zantema et al. 1985a,b; Blair-Zajdel and Blair 1988). As mentioned above, this cytoplasmic restriction imposed on p53 most likely involves continuous nucleocytoplasmic shuttling of the Ad5 E1B 55-kDa product mediated via an NES and a SUMO-conjugation motif. Both motifs are absent in the Ad12 E1B 54-kDa polypeptide (Krätzer et al. 2000; Endter et al. 2001). Also, Ad12 E1B 54-kDa but not the large Ad5 E1B protein can induce metaphase fragility, a process that requires p53 functions (Li et al. 1998; Liao et al. 1999).

Although clearly speculative at this point, assuming high levels of nuclear p53 contribute to the tumorigenicity of adenovirus-transformed cells (van den Heuvel et al. 1990), it seems possible that the differences between Ad5- and Ad12-transformed cells are due at least in part to the large Ad12 E1B protein's inability to shuttle between the nuclear and cytoplasmic compartments. Such a model would be compatible with the finding that nuclear localization of Ad12 E1B 54-kDa correlates with its ability to induce metaphase fragility in combination with p53 (Li et al. 1998; Liao et al. 1999; Yu et al. 2000). Conversely, the nucleocytoplasmic trafficking of Ad5 E1B 55-kDa may abolish induction of chromosome fragility through cytoplasmic sequestration of p53 and probably other cellular factors. Thus, by analogy to Ad2/5 E1A CR4, such shuttling may play a negative role in determining the oncogenicity of the subgroup C viruses. Such a model would fit well with the observation that transformed rat cells expressing E4orf3 and/or E4orf6 in addition to E1A plus E1B exhibit increased oncogenicity in nude mice: both E4 proteins negatively regulate nuclear export of the large E1B protein (Ornelles and Shenk 1991; Goodrum et al. 1996; König et al. 1999; Leppard and Everett 1999; Lethbridge et al. 2003; Sect. 3.3.2). Additionally, Ad5 E1B 55-kDa directly or indirectly interacts with components of dot-like matrix-associated multiprotein complexes known as NBs or PML oncogenic domains (PODs) (Carvalho et al. 1995; Doucas et al. 1996). NBs are nuclear substructures closely attached to the nuclear matrix and have been associated with various aspects of cell growth regulation (reviewed by Sternsdorf et al. 1997) including the control of genes devoted to MHC class I antigen presentation (Zheng et al. 1998). Clearly, further work is needed to fully understand the mechanisms by which both large E1B proteins contribute to the tumorigenic phenotype of adenovirus transformants.

3.3
Early Region 4

The E4 region is located at the right-hand end of the virus genome and is transcribed in a leftward direction (Fig. 4). In Ad2/5, the E4 region

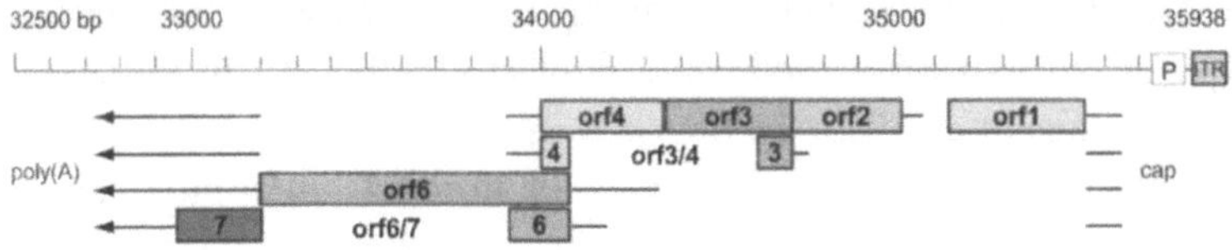

Fig. 4 Ad2/5 E4 transcription unit. The E4 unit is controlled by the E4 promoter (*P*) and generates a primary transcript of approximately 2,800 nucleotides in length. The various E4 transcripts are derived from one precursor RNA by alternative splicing and share common 5′ and 3′ termini (Herisse et al. 1981; Freyer et al. 1984; Virtanen et al. 1984). The processing steps are temporally regulated so that certain transcripts are produced in the early, and others in the late phase of virus infection (Tigges and Raskas 1984; Ross and Ziff 1992; Dix and Leppard 1993). In addition, a further gene product (E4orf3/4) is thought to exist (Virtanen et al. 1984). E4 mRNAs are indicated by *lines* and *arrows*; introns are indicated by *spaces*, and coding regions are indicated by *rectangles*. *ITR*, inverted terminal repeat; *cap*, cap-site; poly(A), polyadenylation signal. (Adapted from Dobner and Kzhyshkowska 2001)

codes for at least six different polypeptides named according to the arrangement of their open reading frames (ORFs): E4orf1, E4orf2, E4orf3, E4orf4, E4orf6, and E4orf6/7. Analyses of other serotypes indicate that all human adenoviruses characterized so far have a homologous E4 region with similar sequence organization. The significance of individual E4 gene products for lytic infection has long been unclear. Only in the last 10 years have a series of observations shown that the E4 gene products encompass a diverse collection of functions required for efficient viral replication. These functions operate through a complex network of protein interactions with key viral and cellular regulatory components involved in transcription, apoptosis, cell-cycle control, and DNA repair, as well as host cell factors that regulate cell signaling, posttranslational modifications, and the integrity of NBs discussed above. A recent and comprehensive review of the regulation and function of human adenovirus E4 proteins was given by Täuber and Dobner (2001).

Although expression of E1A and E1B oncogenes from subgroup A and C adenoviruses is apparently sufficient for full transformation of primary cells in culture, it has long been discussed whether the E4 region gene products are also implicated in the process of virus-mediated oncogenesis (Bernards et al. 1984; Graham 1984; Zalmanzon 1987). Evidence for involvement of the E4 region in transformation events stems from: (1) the observation that transformed cells often contain viral DNA structures in which E1- and E4-end segments are joined together (Graham 1984); (2) earlier studies describing the presence of E4-specific transcripts in virus-transformed cells (Flint et al. 1975, 1976; Flint and Sharp 1976; Esche et al. 1979; Esche 1982; Esche and Siegman 1982); (3) detection of antibodies directed against E4 products in tumor serum from hamsters (Brackmann et al. 1980; Sarnow et al. 1982; Downey et al. 1983); (4) the finding that tumor production by adenovirus virions is not determined by E1 gene products alone (Bernards et al. 1984; Zalmanzon 1987); and finally, (5) demonstrating that coexpression of E1 and E4 from the highly oncogenic Ad12 in rat 3Y1 cells resulted in a completely transformed phenotype as assessed by colony formation in soft-agar and increased tumorigenic capacity (Shiroki et al. 1984). Similarly, the entire E4 region from Ad2 was found to alter the morphology and stimulate anchorage-independent growth of Ad2 E1-transformed rat embryo fibroblasts (CREFs) (Öhman et al. 1995). In addition, studies with single E4 genes from Ad5 hint at the possibility that the transforming potential of subgroup C adenoviruses does not exclusively stem from the E1 region (Moore et al. 1996; Nevels et al. 1997, 1999a). Based on these results it has been suggested that, although not essential for adenovirus transformation, some E4 proteins may have partial transforming activities, stimulating E1 transformation and contributing to the transformed cell phenotype when present (Öhman et al. 1995; Moore et al. 1996; Nevels et al. 1997, 1999a). While this conclusion is probably valid for E4 products from subgroup A and C adenoviruses, subgroup D Ad9 E4 does not conform to this idea.

3.3.1 Ad9 E4orf1

As mentioned briefly in the beginning of this chapter, the oncogenicity of subgroup D Ad9 differs considerably from those of the subgroup A and B viruses (Sect. 2.1). First, subgroup A and B adenoviruses generally induce undifferentiated sarcomas at the site of virus inoculation in both female and male animals. In contrast, Ad9 elicits exclusively mammary tumors in female Wistar-Furth rats and no tumors of any type in males (Ankerst et al. 1974; Jonsson and Ankerst 1977; Ankerst and Jonsson 1989; Javier et al. 1991). Tumors that form in the female rats are predominantly mammary fibroadenomas (Javier et al. 1991), the most common type of benign breast tumor in young women (Cotran et al. 1994). Second, the mammary tumors elicited by Ad9 require estrogen for growth and maintenance (Ankerst and Jonsson 1989; Javier et al. 1991). Third, Ad9 is able to transform an established rat fibroblast cell line (3Y1 cells) in culture (Brusca et al. 1984) but not primary rodent cells, such as BRK cells (Javier et al. 1991). Fourth, Ad9 tumor induction in newborn female Wistar-Furth rats is apparently independent of E1A and E1B. Instead, the oncogenic potential of Ad9 is mediated by the gene product of orf 1 from E4 (E4orf1) (Javier et al. 1991, 1992; Javier 1994; Thomas et al. 1999; Thomas et al. 2001). Last, as will be discussed in more detail below, E4orf1-mediated transformation differs fundamentally from the classical pathway used by E1A and E1B oncoproteins from Ad12 and Ad2/5.

A variety of genetic and molecular analyses conducted by Ron Javier and coworkers have now clearly shown that E4orf1 is essential for tumorigenesis by Ad9, and that the 11-kDa product from E4orf3 and a noncoding segment between E4orf1 and E4orf2 may modulate this process (reviewed by Täuber and Dobner 2001). The Ad9 E4orf1 gene encodes a 14-kDa protein that localizes predominantly in the cellular cytoplasm (Weiss et al. 1996). Mutations in Ad2/5 E4orf1 have minimal effects on viral replication in HeLa cells (Halbert et al. 1985), and the gene product's function in viral lytic growth is unknown. Alone, E4orf1 exhibits cellular growth transforming activity in several different rodent cell lines, including CREF (Javier 1994; Weiss et al. 1996), as well as in the human cell line TE85 (Weiss et al. 1997a). Moreover, in TE85 cells, the related E4orf1 proteins from Ad5 and Ad12 demonstrate transforming activity similar to that of Ad9 E4orf1 (Weiss et al. 1997a). Three separate regions of this polypeptide are essential for CREF cell transformation (Weiss et al. 1997b). One region located at the extreme carboxy terminus is involved in the proper cytoplasmic localization of Ad9 E4orf1 (Weiss and Javier 1997) and represents a PDZ domain-binding motif (see below) that mediates interactions between E4orf1 products from Ad9, Ad5, or Ad12 and a specific group of cellular PDZ domain-containing proteins, such as MUPP1 and the membrane-associated guanylate kinase (MAGUK) homology proteins DLG and MAGI-1 (Lee et al. 1997, 2000; Weiss and Javier 1997; Glaunsinger et al. 2000). PDZ domains are protein interaction modules originally identified as a conserved element present in three structurally related proteins PSD-95, DLG, ZO-1. They comprise 80- to 90-aa-long segments that bind to short carboxy-terminal peptide sequences X(S/T)X(V/I/L)-COOH of target proteins (Fanning and Anderson 1999). Most proteins contain more than one

PDZ domain, each interacting with different target proteins. PDZ domain-containing proteins comprise a heterogeneous family of cellular factors mostly involved in signal transduction and cell signaling (Craven and Bredt 1998; Fanning and Anderson 1999). Mutational analyses show that the transforming properties of Ad9 E4orf1 correlate with binding to MUPP1 and MAGI-1 and aberrant sequestration of these proteins in the cell (Glaunsinger et al. 2000; Lee et al. 2000). Also, DLG, the functional mammalian orthologue of *Drosophila* disc large (dlg) tumor suppressor protein, can block progression into S phase of the cell cycle in NIH 3T3 fibroblasts (Ishidate et al. 2000). Since the viral oncoprotein Tax from HTLV-1 (Lee et al. 1997) and E6 proteins from the high-risk human papillomaviruses HPV16 and HPV18 also interact with these cellular factors (Kiyono et al. 1997; Gardiol et al. 1999), it has been proposed that DLG, MUPP1, MAGI-1, and possibly other Ad9 E4orf1-associated proteins function to suppress abnormal cell growth and that Ad9 E4orf1's task is to target these host factors for inactivation (Glaunsinger et al. 2001).

However, since E4orf1 products from nontumorigenic adenoviruses target the same proteins, interactions with these cellular factors cannot account for the additional, unique properties of Ad9 E4orf1. Recently, a further PDZ domain-containing protein, ZO-2, was identified that binds to Ad9 E4orf1 but not to E4orf1 proteins from Ad5 and Ad12 (Glaunsinger et al. 2001). The MAGUK protein ZO-2 is a candidate tumor suppressor whose expression is impaired in most breast cancer cell lines and primary breast adenocarcinomas (Chlenski et al. 2000 and references therein). As expected, complex formation is mediated by the carboxy-terminal PDZ domain-binding motif, and, similar to MUPP1 and MAGI-1, ZO-2 is aberrantly sequestered by the Ad9 E4orf1 protein in CREFs. Furthermore, transformation-defective E4orf1 mutant proteins exhibit impaired binding to and sequestration of ZO-2, and significantly, overexpression of ZO-2 in CREF cells inhibits Ad9 E4orf1-induced transformation (Glaunsinger et al. 2001). Collectively, these data suggest that the selective capacity to complex with the tumor suppressor protein ZO-2 is the key to defining the unique transforming and tumorigenic properties of the Ad9 E4orf1 oncoprotein, possibly pointing to a novel general mechanism of virus-mediated oncogenesis.

3.3.2 Ad5 E4orf3 and E4orf6

Several studies have now shown that two gene products from Ad5 E4, i.e., E4orf3 and E4orf6, possess transforming and oncogenic potential (Moore et al. 1996; Nevels et al. 1997, 1999a). Both E4 proteins substantially enhance focus formation in primary BRK cells in cooperation with Ad5 E1 proteins, acting synergistically in the presence of E1B (Moore et al. 1996; Nevels et al. 1997, 1999a). Transformed BRK cells that stably express E4orf6 and/or E4orf3 in addition to E1A and E1B, compared to cells expressing only E1A and E1B, exhibit multiple additional properties commonly associated with advanced oncogenic transformation including profound morphological alterations and, in the case of E4orf6, dramatically accelerated tumor growth in nude mice (Moore et al. 1996; Nevels et al. 1999a,b). Ad5 E4orf6 and E4orf3 also initiate focal transformation of BRK cells in an E1B-like fashion in cooperation with E1A. Curiously, however, whereas established E1A/E1B-transformed cells consistently maintain and express the viral genes, the majority of BRK cells transformed by E1A plus either E4orf6 or E4orf3 lack both E4- and E1A-specific DNA sequences (Nevels et al. 2001).

Despite this, some of these cells exhibit fully oncogenic phenotypes and can form tumors in nude mice, apparently in contradiction to conventional concepts of classical virus-induced oncogenesis, which consider the continuous expression of viral proteins necessary to sustain the transformed phenotype. Such observations spawned the "hit-and-run" model of viral transformation, which claims that viral genes are necessary to initiate but not to maintain cellular transformation (Sect. 5). The fact that transient expression of E1A plus E4orf6 or E4orf3 proved to be mutagenic (Nevels et al. 2001) suggests that the viral genes mediate such "hit-and-run" transformation by inducing oncogenic mutations in cellular genes.

As understood at present the oncogenic and mutagenic properties of both E4 proteins are related in part to their overlapping roles in lytic infection (reviewed by Dobner and Kzhyshkowska 2001; Täuber and Dobner 2001). Both E4 proteins independently augment viral DNA replication, late viral protein synthesis, and shut-off of host protein synthesis or production of progeny virions, and they inhibit concatemerization of viral genomes (Halbert et al. 1985; Bridge and Ketner 1989, 1990; Huang and Hearing 1989; Weiden and Ginsberg 1994). More recent evidence suggests that some of the redundant functions shared by the two E4 proteins are linked to their ability to interact with common cellular and viral factors. For example, E4orf3 and E4orf6 gene products physically associate with E1B 55-kDa (Leppard and Everett 1999; Nevels et al. 1999a), causing nuclear accumulation of 55-kDa and directing the E1B protein to different virus-induced structures in the nucleus (Ornelles and Shenk 1991; Goodrum et al. 1996; König et al. 1999; Leppard and Everett 1999; Lethbridge et al. 2003). Since, as described above, high levels of nuclear Ad12 E1B 54-kDa may contribute to enhanced tumorigenicity of transformed rat cells in nude mice (Sect. 3.2.2), it seems possible that nuclear targeting of Ad5 E1B 55-kDa by both E4 proteins may play some role in the accelerated oncogenic properties of E1/E4orf3- or E1/E4orf6-transformed rat cells. Also, the highly transformed E4orf6-associated phenotypes in E1-transformed rat cells correlate with a dramatic reduction of p53 steady-state levels that in turn correlate inversely with E4orf6 expression (Nevels et al. 1999b). Apparently, the Ad5 E4orf6 protein, in combination with E1A and E1B proteins, enhances the intrinsic ability of E1-transformed rat cells to grow in a neoplastic state by completely inactivating p53 tumor suppressor function (Nevels et al. 1999b). It seems that E4orf6 and E1B 55-kDa proteins cooperate to counteract E1A-induced stabilization of p53 through accelerated p53 proteolytic degradation via 26S proteasomes (Grand et al. 1994; Moore et al. 1996; Nevels et al. 1997, 1999b; Querido et al. 1997, 2001; Roth et al. 1998; Steegenga et al. 1998; Boyer and Ketner 2000; Cathomen and Weitzman 2000; Wienzek et al. 2000). This p53 degradation is dependent on 55-kDa interacting with both p53 and E4orf6, but occurs independently of MDM2 and p19ARF, regulators of p53 stability in mammalian cells (Roth et al. 1998; Nevels et al. 2000; Querido et al. 2001). A recent study shows that E4orf6 is part of a multiprotein complex that contains a novel Cullin-containing E3 ubiquitin ligase and is composed of Cullin family member Cul5, Elongins B and C, and the RING-H2 finger protein Rbx1 (Querido et al. 2001). Remarkably, this complex is similar to the von Hippel-Lindau (VHL) tumor suppressor and SCF (Skp1-Cul1/Cdc53-F-box) E3 ubiquitin ligase complexes and is capable of stimulating ubiquitination of p53 in vitro in the presence of E1/E2 ubiquitin activating and conjugating enzymes (Querido et al. 2001). Although still to be formally demonstrated, the same E3 ubiquitin ligase function may be involved in the recently reported E1B 55-kDa/E4orf6-dependent degradation of the cellular Mre11/Rad50/NBS1 complex (Stracker et al. 2002), a major component of double-strand break repair (DSBR), meiotic recombination, and telomere maintenance (reviewed by Hopfner et al. 2002). Apparently, inactivation of this complex by E1B 55-kDa and E4orf6 plays a major role in the inhibition of concatemerization of viral chromosomes in productively infected cells (Weiden and Ginsberg 1994; Stracker et al. 2002). This activity also involves interactions of E4orf6 and E4orf3 with the DNA-dependent protein kinase (DNA PK) (Boyer et al. 1999), another essential element involved in controlling the end-joining mechanism in the DSBR system (reviewed by Smith and Jackson 1999). Functional analyses show that E4orf6, although not E4orf3, can inhibit V(D)J-joining (Boyer et al. 1999), a process that requires DNA PK, and that both E4orf3 and E4orf6 can inhibit the repair of specific double-strand breaks induced by the *Saccharomyces cerevisiae* HO endonuclease in coinfected mammalian cells (Nicolás et al. 2000).

In contrast to E4orf6, the ability of E4orf3 to promote mutations and oncogenic cell growth is probably not linked to p53 function and stability. Rather, recent studies indicate that its activities may be due to combined effects involving binding to Ad5 E1B 55-kDa protein (Nevels et al. 1999a), mislocalization of the Mre11/Rad50/NBS1 complex (Stracker et al. 2002), as well as interactions with DNA PK (Boyer et al. 1999) and components of NBs (Carvalho et al. 1995; Doucas et al. 1996; Nevels et al. 1999a). The latter activity would fit well with a model suggesting that E4orf3-induced reorganization of NBs is involved in mislocalization of the Mre11/Rad50/NBS1 complex. This could trigger a cascade of processes causing uncontrolled cell proliferation, particularly since NBs have also been implicated in genomic stability (Zhong et al. 1999). Also, there are indications that NB reorganization is linked to modulation of the MHC class I-mediated immune response (Terris et al. 1995; Zheng et al. 1998) and cell-cycle-independent virus growth, the latter also involving E4orf6 and E1B 55-kDa (Goodrum and Ornelles 1999).

3.3.3
Ad5 E4orf6/7

The proximal E4 gene product uses open reading frames 6 and 7 to produce a fusion protein, E4orf6/7 (reviewed by Täuber and Dobner 2001), thought to act as a viral transactivator complementing the function of E1A proteins and promoting expression of viral (E2) and cellular genes controlled by the E2F family of transcription factors. In contrast to E1A

(Sect. 3.2.1), the E4orf6/7 gene product binds free E2F, resulting in E2F dimerization (Obert et al. 1994) and its increased binding stability on the two inverted E2F-binding sites in the Ad5 E2 early promoter (Hardy and Shenk 1989; Hardy et al. 1989; Huang and Hearing 1989; Raychaudhuri et al. 1989; Marton et al. 1990) as well as the cellular E2F-1 promoter (Schaley et al. 2000). Induction of stabilized E2F-binding in vitro correlates well with transcriptional activation of both cellular and viral promoters in vivo (Reichel et al. 1989; Neill et al. 1990; Neill and Nevins 1991; O'Connor and Hearing 1991; Obert et al. 1994; Schaley et al. 2000). Moreover, a recent study suggests that E4orf6/7 alone is sufficient to displace pRb and p107 from E2F complexes, which then activates expression of E2 by inducing binding of E2F to the E2 early promoter, and results in significantly enhanced replication of an E1A-defective adenovirus in HeLa cells (O'Connor and Hearing 2000).

Considering the complementary, almost redundant functions of E1A and E4orf6/7 proteins in modulating E2F-mediated transcription, it is interesting that Ad5 E4orf6/7 can both positively and negatively affect transformation. In the presence of an E1A mutant protein unable to bind pRb and thus release E2Fs from pRb complexes, Ad5 E4orf6/7 can partly promote immortalization and morphological transformation of primary BRK cells (Yamano et al. 1999). Conversely, Ad5 E4orf6/7 suppresses growth of untransformed rat cells through p53-dependent mechanisms (Yamano et al. 1999), and, if coexpressed with wild-type E1A or E1A plus E1B, reduces the transformation frequency of primary BRK cells (Nevels et al. 1997). Growth-stimulatory activities in BRK cells involve an E4orf6 amino-terminal region and a domain in the carboxy terminus, while growth-inhibitory activities map to the carboxy-terminal region of E4orf6 (Yamano et al. 1999). Since the latter domain contains regions involved in E2F-interaction and possibly E2F-1 release from pRb complexes, both positive and negative effects may be linked to E2F-1, which can confer either oncogenic or tumor suppressor activities depending on the cellular context (Johnson et al. 1994; Singh et al. 1994; Xu et al. 1995; Field et al. 1996).

3.4 Early Regions 2 and 3

The E2 unit is one of the most important early regions for completion of a productive infection, since it encodes proteins essential for viral DNA replication: the 72-kDa DNA binding protein (DBP), an 80-kDa precursor terminal protein (pTP), and the 140-kDa adenovirus DNA polymerase (reviewed by Shenk 1996). It seems that none of these proteins plays a major role in adenovirus transformation, although DBP has been detected in some transformed cell lines (Gilead et al. 1975; Levinson et al. 1976), and studies with Ad5 mutant viruses containing temperature-sensitive mutations in the 140-kDa polymerase and DBP genes indicate that both proteins may modulate transformation efficiency (Williams et al. 1974, 1979; Ginsberg et al. 1975; Logan et al. 1981; Fisher et al. 1982; Miller and Williams 1987; Rice et al. 1987). DBP may exert its influence on transformation via the E1 and E2 regions (Carter and Blanton 1978a,b; Blanton and Carter 1979; Babich and Nevins 1981) since it enhances expression of the E1A and E2 genes (Chang and Shenk 1990). By contrast, it has been hypothesized that the adenovirus DNA polymerase may play some role in integrating the viral genome into host cell DNA during transformation (Graham 1984).

The E3 transcription unit in Ad2/5 encodes at least seven proteins, four of which (E3 14.7-kDa, RIDα, RIDβ, and E3 19-kDa) are known to prevent killing of infected cells by the host immune system (reviewed by Wold and Tollefson 1998). While E3 14.7-kDa and the RIDα/β complex independently prevent TNF-induced apoptosis, E3 19-kDa inhibits killing by CTLs by blocking transport of MHC class I antigens to the cell surface.

Despite their vital role in suppressing immuno-induced death of infected cells, it is still unclear whether these same functions also play a role in adenovirus transformation and oncogenesis. Studies on the organization and expression of viral genes in adenovirus-transformed cells show that in Ad5 transformants, E3 is hardly ever present or expressed (Flint et al. 1976; Flint 1982), although the E3 19-kDa glycoprotein has been detected in some transformed cell lines (Jeng et al. 1978; Persson et al. 1979). Moreover, the Ad12 E3 region lacks an ORF corresponding to the Ad2/5 E3 19-kDa product (Sprengel et al. 1995), the E3 19-kDa function (downregulation of MHC class I molecules) in Ad12 tumor and transformed cells being provided at least in part by E1A. Nevertheless, given their importance in immune evasion it is certainly possible that E3 gene products have profound effects on the phenotype of transformed cells that express them, and that they may play some role in determining the oncogenicity of adenovirus-transformed cells. This, however, remains to be shown.

4 Viral DNA Integration

In addition to the persistence of viral DNA and expression of viral genes, other mechanisms in adenovirus oncogenesis can be considered, such as genetic and epigenetic changes possibly arising as a consequence of viral DNA integration (reviewed by Dörfler 1996; the chapter by Hohlweg et al., this volume). DNA sequence analyses at the sites of integration in a number of adenovirus tumor and transformed cells suggest that viral DNA does not integrate at only one or several specific sites in the host genome. However, short patch homologies between chromosomal and viral DNA may play a role in facilitating integrative recombination in a mechanism similar to nonhomologous recombination. Nevertheless, initial steps in viral malignant transformation could well be associated with insertional mutagenesis at various cellular sites, often followed by further genetic alterations such as deletions and translocations, as well as extensive alterations of methylation patterns in viral and adjacent host DNA sequences. It is conceivable that the later, epigenetic changes may account for some of the oncogenic properties of adenovirus-transformed cells and tumors, since they can alter cellular gene expression and influence genome stability far removed from the primary site(s) of viral DNA integration (reviewed by Jones 1999; Nakao 2001; Bird 2002).

5 "Hit-and-Run" Transformation

As early as the late 1970s and early 1980s, the classic principle of virus-mediated transformation based on continuous expression of viral oncogenic proteins had already been challenged by a hypothetical "hit-and-run" mechanism. In other words, the viral oncogenes were essential for initiation, but not maintenance of the transformed state (Skinner 1976; Schlehofer and zur Hausen 1982; Galloway and McDougall 1983). This hypothesis stems from several observations where infections with herpes simplex virus (HSV) (Waubke et al. 1968; Schlehofer and zur Hausen 1982; Hwang and Shilitoe 1990), Ad2 (Marengo et al. 1981; Paraskeva et al. 1983), Ad12 (zur Hausen 1967; Durnam et al. 1988), and the murine Abelson leukemia virus (ALV) (Mushinski et al. 1983) cause chromosomal aberrations and other mutations in the host genome, plus the fact that in HSV- and ALV-transformed cells, viral DNA sequences could never be detected with any consistency (Skinner 1976; Galloway and McDougall 1983; Mushinski et al. 1983). Similar observations were reported for some Ad12-transformed cells established from hamster tumors (Kuhlmann et al. 1982; Pfeffer et al. 1999) and rat cell lines transformed by Ad12 in culture (Paraskeva and Gallimore 1980; Paraskeva et al. 1982). Moreover, plasmid-based transfection assays employing Ad5 E1A in combination with human cytomegalovirus (HCMV) IE1/IE2 (Shen et al. 1997) or Ad5 E4orf3 and E4orf6 (Nevels et al. 2001) demonstrate that certain viral gene products possess mutagenic potential that correlates with their ability to transform primary rodent cells by a "hit-and-run" mechanism (Shen et al. 1997; Nevels et al. 2001). Altogether, these findings indicate that transient expression and/or integration of viral genes can be sufficient to induce oncogenic mutations ("hit"), which, in rare cases, can lead to cellular transformation. Manifestation of the transformed cell phenotype at the genetic level is then compatible with the loss of viral genetic information ("run").

The molecular origin of such mutagenic activities remains unclear, but is, as one might predict from the information summarized in this review, very likely related to the functions of viral gene products in viral replication. In the case of Ad12, induction of specific and random chromosomal damage is dependent on E1B 54-kDa (Schramayr et al. 1990) and apparently requires p53 functions (Li et al. 1998; Liao et al. 1999; Yu et al. 2000). For Ad5 E1A/E4orf3 and E1A/E4orf6, there is good reason to believe that accumulation of mutations is probably triggered by the cooperation of various factors, including the unscheduled induction of cellular DNA synthesis by E1A (Sect. 3.2.1), as well as the modulation of p53 and NBs by E4orf6 and E4orf3, respectively (Dobner et al. 1996; Nevels et al. 1999a, 2001). The latter possibility is intriguing because other viral oncoproteins similarly implicated in "hit-and-run" transformation, such as SV40 Tag (Ewald et al. 1996) and HCMV IE1/IE2 (Shen et al. 1997), also modulate the function of p53 (Bonin and McDougall 1997) as well as the integrity of NBs (Carvalho et al. 1995; Doucas et al. 1996; Ahn and Hayward 1997; Nevels et al. 1999a). Perhaps more significant, the reported inhibition of DNA double-strand break repair by E4orf3 and E4orf6 through binding to DNA-PK (Boyer et al. 1999; Nicolás et al. 2000) and/or modulation of the Mre11/Rad50/NBS1 complex (Stracker et al. 2002) could potentially contribute to the induction of mutations and genetic instability, thus forming the basis for "hit-and-run" transformation as observed with both E4 gene products.

Finally, it is interesting to note that E4orf3 and E4orf6 are stably expressed in the presence of E1B and confer multiple properties associated with a high grade of malignant transformation (see above). Perhaps E1B 19-kDa and/or E1B 55-kDa neutralize cytotoxic and/or mutagenic activities of both E4 proteins, where E1B 55-kDa could certainly contribute to this through its binding to E4orf3 and E4orf6. If true, this model could provide an explanation for why E4-specific mRNAs occasionally exist in adenovirus-transformed cells, but were always detected in connection with E1A plus E1B, but never with E1A alone (Flint et al. 1975, 1976; Flint and Sharp 1976; Esche et al. 1979; Esche 1982; Esche and Siegman 1982).

6 Conclusions

6.1 Adenovirus Transforming Genes—E1 plus E4

The last 40 years of molecular biological investigations into human adenoviruses have contributed enormously to our understanding of the basic principles of normal and malignant cell growth. Much of this knowledge stems from analyses of their productive infection cycle in permissive host cells. Also, initial observations concerning the carcinogenic potential of human adenoviruses subsequently revealed decisive insights into the molecular mechanisms of the origins of cancer, and established adenoviruses as a model system for explaining virus-mediated transformation processes. From the work summarized in this review it is evident that cell transformation by subgroup A and C adenoviruses is a multistep process involving the cooperation of several gene products encoded in E1 and presumably E4, as well as epigenetic changes resulting from viral DNA integration. While E1A is clearly the key player in subgroup A and C adenovirus-mediated transformation and oncogenesis, it is becoming increasingly clear that other gene products substantially affect the tumorigenic phenotype of the transformed cell. For example, downregulation of MHC class I molecules by Ad12 E1A provides strong evidence for the oncogenic nature of Ad12 but is clearly not the complete picture. Other mechanisms must exist, likely mediated by E1B proteins and probably E4 gene products. However, we are still far from fully understanding what function(s) E1B and E4 gene products provide in adenovirus oncogenesis.

Moreover, a large body of evidence now indicates that at least Ad5, in addition to the classical E1A and E1B oncoproteins, possesses two further gene products, which in combination with E1A can substitute for E1B in terms of initiating transformation. However, despite some functional similarities between these proteins, E1A/E4-mediated transformation differs fundamentally from the classical E1A/E1B pathway. While in the latter case the viral genes persist in the transformed cells, the coexpression of E1A and E4orf3 or E4orf6 leads to "hit-and-run" transformation. Finally, tumor induction by Ad9 in newborn female Wistar-Furth rats is independent of E1A and E1B but is mediated by the Ad9 E4orf1 gene product, apparently through a novel mechanism. Altogether, these observations strongly support the view that the oncogenic potential of adenoviruses does not exclusively lie within the E1 region. The transforming and oncogenic activities of Ad9 E4orf1, Ad5 E4orf6, and Ad5 E4orf3 are apparently based on novel molecular mechanisms mediated by interactions with key cellular regulators of transcription, apoptosis, DNA repair, and signal transduction. Without question, molecular analysis of these E4 gene products will further dissect intricate pathways involved in neoplastic cell growth, and at the same time lead to improved safety strategies for therapeutic adenovirus vectors.

6.2
Oncogenesis in Humans

To date, adenoviruses could never be convincingly associated with malignant diseases in humans. Although individual adenoviral genome regions have been detected in malignant tumors (Maitland et al. 1981; Ibelgaufts 1982; Ibelgaufts et al. 1982; Lawler et al. 1994; Kuwano et al. 1997a,b), and a possible connection with tumors of the urogenital tract has been discussed based on serological data (Csata et al. 1982a,b), no reliable epidemiological, serological, or genetic evidence exists as yet pointing to an involvement in the origin of particular tumor types (Mackey et al. 1976, 1979; Green et al. 1979, 1980; Chauvin et al. 1990; Fernandez-Soria et al. 2002). The capacity of Ad12 virions and Ad5 E1A/E4 genes to mediate "hit-and-run" transformation combined with the ability of Ad12 E1B 54-kDa to induce specific and random chromosomal damage (Sects. 3.2.3 and 5) implies, however, that adenoviruses could quite possibly intervene in tumorigenesis without the viral genes persisting in the transformed cells. It would then be almost impossible to prove a causal correlation between adenovirus infection and tumorigenesis. An analogous situation also exists for other human pathogenic viruses. For example, viral DNA was never consistently found in HSV-associated cervical carcinoma and HSV-transformed cells (Jones 1995; zur Hausen 1996). There are similar observations for other herpes viruses (Gelb and Dohner 1984; Karran et al. 1990; Jox et al. 1997; Legrand et al. 1997; Ambinder 2000) as well as human papillomaviruses (Iwasaka et al. 1992; zur Hausen 1996). The "hit-and-run" principle of virus-mediated oncogenesis could therefore be generally valid and thus form the basis of certain virus-dependent transformation events.

Apart from a possible "hit-and-run" etiology for human tumor diseases, it should be noted that the last comprehensive investigation into the association between adenoviruses and tumor diseases in humans dates back 20 years, and was mostly limited to the detection of DNA sequences and antigens from the E1 region of only six human serotypes. Today's group of human adenoviruses now includes 51 different serotypes, and comparable studies with the highly sensitive methods of PCR available today would be informative if not desirable. It cannot be completely excluded that like in the large family of human papillomaviruses only a few, as yet molecular biologically uncharacterized serotypes, may be connected with instigating malignant diseases.

Acknowledgements. We thank Birgitt Täuber and Michael Nevels for critical comments on the manuscript. Work in this laboratory was supported by grants from the Deutsche Forschungsgemeinschaft (DFG) and from the Fonds der Chemischen Industrie (FCI) to T.D.

References

Ahn J-H, Hayward GS (1997) The major immediate-early proteins IE1 and IE2 of human cytomegalovirus colocalize with and disrupt PML-associated nuclear bodies at very early times in infected permissive cells. J Virol 71:4599–4613

Ambinder RF (2000) Gammaherpesviruses and hit-and-run oncogenesis. Americ J Pathol 156:1–3

Anderson CW, Schmitt RC, Smart JE, Lewis JB (1984) Early region 1B of adenovirus 2 encodes two coterminal proteins of 495 and 155 amino acid residues. J Virol 50:387–396

Ankerst J, Jonsson N, Kjellen L, Norrby E, Sjogren HO (1974) Induction of mammary fibroadenomas in rats by adenovirus type 9. Int J Cancer 13:286–290
Ankerst J, Jonsson N (1989) Adenovirus type 9-induced tumorigenesis in the rat mammary gland related to sex hormonal state. J Natl Cancer Inst 81:294–298
Avvakumov N, Wheeler R, D'Halluin JC, Mymryk JS (2002) Comparative sequence analysis of the largest E1A proteins of human and simian adenoviruses. J Virol 76:7968–7975
Babich A, Nevins JR (1981) The stability of early adenovirus mRNA is controlled by the viral 72 kd DNA-binding protein. Cell 26:371–379
Ben-Israel H, Kleinberger T (2002) Adenovirus and cell cycle control. Front Biosci 7: D1369-D1395
Bernards R, Houweling A, Schrier PI, Bos JL, van der Eb AJ (1982) Characterization of cells transformed by Ad5/Ad12 hybrid early region. Virology 120:422–432
Bernards R, Schrier PI, Bos JL, Van der Eb AJ (1983) Role of adenovirus types 5 and 12 early region 1b tumor antigens in oncogenic transformation. Virology 127:45–53
Bernards R, Schrier PI, Houweling A, Bos JL, van der Eb AJ, Zijlstra M, Melief CJ (1983) Tumorigenicity of cells transformed by adenovirus type 12 by evasion of T-cell immunity. Nature 305:776–779
Bernards R, de Leeuw MG, Vaessen MJ, Houweling A, van der Eb AJ (1984) Oncogenicity by adenovirus is not determined by the transforming region only. J Virol 50:847–853
Bernards R, de Leeuw MG, Houweling A, van der Eb AJ (1986) Role of the adenovirus early region 1B tumor antigens in transformation and lytic infection. Virology 150:126–139
Bird A (2002) DNA methylation patterns and epigenetic memory. Genes Dev 16:6-21
Blair GE, Hall KT (1998) Human adenoviruses: evading detection by cytotoxic T lymphocytes. Semin Virol 8:387–397
Blair-Zajdel ME, Blair GE (1988) The intracellular distribution of the transformation-associated protein p53 in adenovirus-transformed rodent cells. Oncogene 2:579–584
Blanton RA, Carter TH (1979) Autoregulation of adenovirus type 5 early gene expression. III. Transcription studies in isolated nuclei. J Virol 29:458–465
Bonin LR, McDougall JK (1997) Human cytomegalovirus IE2 86-kilodalton protein binds p53 but does not abrogate G1 checkpoint function. J Virol 71:5861–5870
Boulanger PA, Blair GE (1991) Expression and interactions of human adenovirus oncoproteins. Biochem J 275:281–299
Boyd JM, Subramanian T, Schaeper U, La Regina M, Bayley S, Chinnadurai G (1993) A region in the C-terminus of adenovirus 2/5 E1a protein is required for association with a cellular phosphoprotein and important for the negative modulation of T24-ras mediated transformation, tumorigenesis and metastasis. EMBO J 12:469–478
Boyd JM, Malstrom S, Subramanian T, Venkatesh LK, Schaeper U, Elangovan B, D'Sa Eipper C, Chinnadurai G (1994) Adenovirus E1B 19 kDa and Bcl-2 proteins interact with a common set of cellular proteins. Cell 79:341–351

Boyer JL, Rohleder K, Ketner G (1999) Adenovirus E4 34 k and E4 11 k inhibit double strand break repair and are physically associated with the cellular DNA-dependent protein kinase. Virology 263: 307–312

Boyer JL, Ketner G (2000) Genetic analysis of a potential zinc-binding domain of the adenovirus E4 34 k protein. J Biol Chem 275:14969–14978

Brackmann KH, Green M, Wold WS, Cartas M, Matsuo T, Hashimoto S (1980) Identification and peptide mapping of human adenovirus type 2-induced early polypeptides isolated by twodimensional gel electrophoresis and immunoprecipitation. J Biol Chem 255:6772–6779

Branton PE, Bayley ST, Graham FL (1985) Transformation by human adenoviruses. Biochim Biophys Acta 780:67–94

Bridge E, Ketner G (1989) Redundant control of adenovirus late gene expression by early region 4. J Virol 63:631–638

Bridge E, Ketner G (1990) Interaction of adenoviral E4 and E1b products in late gene expression. Virology 174:345–353

Brusca JS, Jannun R, Chinnadurai G (1984) Efficient transformation of rat 3Y1 cells by human adenovirus type 9. Virology 136:328–337

Byrd P, Brown KW, Gallimore PH (1982) Malignant transformation of human embryo retinoblasts by cloned adenovirus 12 DNA. Nature 298:69–71

Carter TH, Blanton RA (1978a) Autoregulation of adenovirus type 5 early gene expression II. Effect of temperature-sensitive early mutations on virus RNA accumulation. J Virol 28:450–456

Carter TH, Blanton RA (1978b) Possible role of the 72,000 dalton DNA-binding protein in regulation of adenovirus type 5 early gene expression. J Virol 25:664–674

Carvalho T, Seeler JS, Öhman K, Jordan P, Pettersson U, Akusjärvi G, Carmo Fonseca M, Dejean A (1995) Targeting of adenovirus E1A and E4-ORF3 proteins to nuclear matrix-associated PML bodies. J Cell Biol 131:45–56

Cathomen T, Weitzman MD (2000) A functional complex of the adenovirus proteins E1B-55 kDa and E4orf6 is necessary to modulate the expression level of p53 but not its transcriptional activity. J Virol 74:11407–11412

Chang LS, Shenk T (1990) The adenovirus DNA-binding protein stimulates the rate of transcription directed by adenovirus and adeno-associated virus promoters. J Virol 64:2103–2109

Chauvin C, Suh M, Remy C, Benabid AL (1990) Failure to detect viral genomic sequences of three viruses (herpes simplex, simian virus 40 and adenovirus) in human and rat brain tumors. Ital J Neurol Sci 11:347–357

Chiou SK, White E (1997) p300 binding by E1A cosegregates with p53 induction but is dispensable for apoptosis. J Virol 71:3515–3525

Chlenski A, Ketels KV, Korovaitseva GI, Talamonti MS, Oyasu R, Scarpelli DG (2000) Organization and expression of the human zo-2 gene (tjp-2) in normal and neoplastic tissues. Biochim Biophys Acta 1493:319–324

Cook JL, Lewis AM, Jr. (1979) Host response to adenovirus 2-transformed hamster embryo cells. Cancer Res 39:1455–1461

Cook JL, Hibbs JB, Jr., Lewis AM, Jr. (1982) DNA virus-transformed hamster cell-host effector cell interactions: level of resistance to cytolysis correlated with tumorigenicity. Int J Cancer 30:795–803

Cook JL, May DL, Lewis AM, Jr., Walker TA (1987) Adenovirus E1A gene induction of susceptibility to lysis by natural killer cells and activated macrophages in infected rodent cells. J Virol 61:3510–3520

Cotran RS, Robbins SL, Kumar V (1994) Robbins pathology basis of disease, 5th ed. W. B. Saunders Co., Philadelphia, PA

Craven SE, Bredt DS (1998) PDZ proteins organize synaptic signaling pathways. Cell 93:495–498

Cress WD, Nevins JR (1996) A role for a bent DNA structure in E2F-mediated transcription activation. Mol Cell Biol 16:2119–2127

Cress WD, Nevins JR (1996) Use of the E2F transcription factor by DNA tumor virus regulatory proteins. Curr Top Microbiol Immunol 208:63–78

Csata S, Kulcsar G, Dan P, Horvath J, Nasz I, Ongradi J, Verebelyi A (1982a) Adenovirus antibodies in tumorous diseases of the urogenital system. Acta Chir Acad Sci Hung 23:15–22

Csata S, Kulcsar G, Horvath J, Nasz I, Ongradi J, Verebelyi A (1982b) Study of antibodies to adenoviruses in patients with tumors of the urogenital system. Int Urol Nephrol 14:115–119

de Stanchina E, McCurrach ME, Zindy F, Shieh SY, Ferbeyre G, Samuelson AV, Prives C, Roussel MF, Sherr CJ, Lowe SW (1998) E1A signaling to p53 involves the p19(ARF) tumor suppressor. Genes Dev 12:2434–2442

Debbas M, White E (1993) Wild-type p53 mediates apoptosis by E1A, which is inhibited by E1B. Genes Dev 7:546–554

Dix I, Leppard KN (1993) Regulated splicing of adenovirus type 5 E4 transcripts and regulated cytoplasmic accumulation of E4 mRNA. J Virol 67:3226–3231

Dobner T, Horikoshi N, Rubenwolf S, Shenk T (1996) Blockage by adenovirus E4orf6 of transcriptional activation by the p53 tumor suppressor. Science 272:1470–1473

Dobner T, Kzhyshkowska J (2001) Nuclear export of adenovirus RNA. Curr Top Microbiol Immunol 259:25–54

Dörfler W (1996) A new concept in adenoviral oncogenesis: integration of foreign DNA and its consequences. Biochim Biophys Acta 1288: F79-F99

Doucas V, Ishov AM, Romo A, Juguilon H, Weitzman MD, Evans RM, Maul GG (1996) Adenovirus replication is coupled with the dynamic properties of the PML nuclear structure. Genes Dev 10: 196–207

Douglas JL, Gopalakrishnan S, Quinlan MP (1991) Modulation of transformation of primary epithelial cells by the second exon of the Ad5 E1A12S gene. Oncogene 6:2093–2103

Douglas JL, Quinlan MP (1995) Efficient nuclear localization and immortalizing ability, two functions dependent on the adenovirus type 5 (Ad5) E1A second exon, are necessary for cotransformation with Ad5 E1B but not with T24ras. J Virol 69:8061–8065

Downey JF, Rowe DT, Bacchetti S, Graham FL, Bayley ST (1983) Mapping of a 14,000-dalton antigen to early region 4 of the human adenovirus 5 genome. J Virol 45:514–523

Durnam DM, Menninger JC, Chandler SH, Smith PP, McDougall JK (1988) A fragile site in the human U2 small nuclear RNA gene cluster is revealed by adenovirus type 12 infection. Mol Cell Biol 8:1863–1867

Eager KB, Williams J, Breiding D, Pan S, Knowles B, Appella E, Ricciardi RP (1985) Expression of histocompatibility antigens H-2 K, -D, and -L is reduced in adenovirus-12-transformed mouse cells and is restored by interferon gamma. Proc Natl Acad Sci USA 82:5525–5529

Edbauer C, Lamberti C, Tong J, Williams J (1988) Adenovirus type 12 E1B 19-kilodalton protein is not required for oncogenic transformation in rats. J Virol 62:3265–3273

Endter C, Kzhyshkowska J, Stauber R, Dobner T (2001) SUMO-1 modification required for transformation by adenovirus type 5 early region 1B 55-kDa oncoprotein. Proc Natl Acad Sci USA 98:11312–11317

Esche H, Schilling R, Dörfler W (1979) In vitro translation of adenovirus type 12-specific mRNA isolated from infected and transformed cells. J Virol 30:21–31

Esche H (1982) Viral gene products in adenovirus type 2-transformed hamster cells. J Virol 41: 1076–1082

Esche H, Siegman B (1982) Expression of early viral gene products in adenovirus type 12-infected and -transformed cells. J Gen Virol 60:99–113

Ewald D, Li M, Efrat S, Auer G, Wall RJ, Furth PA, Hennighausen L (1996) Time-sensitive reversal of hyperplasia in transgenic mice expressing SV40 T antigen. Science 273:1384–1386

Fallaux FJ, Kranenburg O, Cramer SJ, Houweling A, Van Ormondt H, Hoeben RC, Van Der Eb AJ (1996) Characterization of 911: a new helper cell line for the titration and propagation of early region 1-deleted adenoviral vectors. Hum Gene Ther 7:215–222

Fallaux FJ, Bout A, van der Velde I, van den Wollenberg DJ, Hehir KM, Keegan J, Auger C, Cramer SJ, van Ormondt H, van der Eb AJ, Valerio D, Hoeben RC (1998) New helper cells and matched early region 1-deleted adenovirus vectors prevent generation of replication-competent adenoviruses. Hum Gene Ther 9:1909–1917

Fanning AS, Anderson JM (1999) PDZ domains: fundamental building blocks in the organization of protein complexes at the plasma membrane. J Cin Invest 103:767–772

Farrow SN, White JH, Martinou I, Raven T, Pun KT, Grinham CJ, Martinou JC, Brown R (1995) Cloning of a bcl-2 homologue by interaction with adenovirus E1B 19 K. Nature 374:731–733

Fernandez-Soria V, Bornstein R, Forteza J, Parada C, Sanchez-Prieto R, Ramon y Cajal S (2002) Inconclusive presence of adenovirus sequences in human leukemias and lymphomas. Oncol Rep 9: 897–902

Field SJ, Tsai FY, Kuo F, Zubiaga AM, Kaelin WG, Jr., Livingston DM, Orkin SH, Greenberg ME (1996) E2F-1 functions in mice to promote apoptosis and suppress proliferation. Cell 85:549–561

Fisher PB, Babiss LE, Weinstein IB, Ginsberg HS (1982) Analysis of type 5 adenovirus transformation with a cloned rat embryo cell line (CREF). Proc Natl Acad Sci USA 79:3527–3531

Flint SJ, Gallimore PH, Sharp PA (1975) Comparison of viral RNA sequences in adenovirus 2-transformed and lytically infected cells. J Mol Biol 96:47–68

Flint SJ, Sambrook J, Williams JF, Sharp PA (1976) Viral nucleic acid sequences in transformed cells. IV. A study of the sequences of adenovirus 5 DNA and RNA in

four lines of adenovirus 5-transformed rodent cells using specific fragments of the viral genome. Virology 72:456–470
Flint SJ, Sharp PA (1976) Adenovirus transcription. V. Quantitation of viral RNA sequences in adenovirus 2-infected and transformed cells. J Mol Biol 106:749–771
Flint SJ (1982) Organization and expression of viral genes in adenovirus-transformed cells. Int Rev Cytol 76:47–65
Freyer GA, Katoh Y, Roberts RJ (1984) Characterization of the major mRNAs from adenovirus 2 early region 4 by cDNA cloning and sequencing. Nucl Acids Res 12:3503–3519
Gabler S, Schütt H, Groitl P, Wolf H, Shenk T, Dobner T (1998) E1B 55-kilodalton-associated protein: a cellular protein with RNA-binding activity implicated in nucleocytoplasmic transport of adenovirus and cellular mRNAs. J Virol 72:7960–7971
Gallimore PH (1972) Tumour production in immunosuppressed rats with cells transformed in vitro by adenovirus type 2. J Gen Virol 16:99–102
Gallimore PH, Byrd P, Grand RJ, Whittaker JL, Breiding D, Williams J (1984) An examination of the transforming and tumor-inducing capacity of a number of adenovirus type 12 early region 1, hostrange mutants and cells transformed by subgenomic fragments of Ad12 E1 region. Cancer Cells 2: 519–526
Gallimore PH, Grand RJ, Byrd PJ (1986) Transformation of human embryo retinoblasts with simian virus 40, adenovirus and ras oncogenes. Anticancer Res 6:499–508
Gallimore PH, Lecane PS, Roberts S, Rookes SM, Grand RJA, Parkhill J (1997) Adenovirus type 12 early region 1B 54 K protein significantly extends the life span of normal mammalian cells in culture. J Virol 71:6629–6640
Gallimore PH, Turnell AS (2001) Adenovirus E1A: remodelling the host cell, a life or death experience. Oncogene 20:7824–7835
Galloway DA, McDougall JK (1983) The oncogenic potential of herpes simplex viruses: evidence for a 'hit-and-run' mechanism. Nature 302:21–24
Gardiol D, Kuhne C, Glaunsinger BA, Lee SS, Javier RT, Banks L (1999) Oncogenic human papillomavirus E6 proteins target the discs large tumour suppressor for proteasome-mediated degradation. Oncogene 18:5487–5496
Gelb L, Dohner D (1984) Varicella-zoster virus-induced transformation of mammalian cells in vitro. J Invest Dermatol 83:77s-81 s
Gilead Z, Arens MQ, Bhaduri S, Shanmugam G, Green M (1975) Tumour antigen specificity of a DNA-binding protein from cells infected with adenovirus 2. Nature 254:533–536
Ginsberg HS, Ensinger MJ, Kauffman RS, Mayer AJ, Lundholm U (1975) Cell transformation: a study of regulation with types 5 and 12 adenovirus temperature-sensitive mutants. Cold Spring Harb Symp Quant Biol: 419–426
Glaunsinger BA, Lee SS, Thomas M, Banks L, Javier RT (2000) Interactions of the PDZ-protein MAGI-1 with adenovirus E4-ORF1 and high-risk papillomavirus E6 oncoproteins. Oncogene 19: 5270–5280
Glaunsinger BA, Weiss RS, Lee SS, Javier RT (2001) Link of the unique oncogenic properties of adenovirus type 9 E4-ORF1 to a select interaction with the candidate tumor suppressor protein ZO-2. EMBO J 20:5578–5586

Goodrum FD, Shenk T, Ornelles DA (1996) Adenovirus early region 4 34-kilodalton protein directs the nuclear localization of the early region 1B 55-kilodalton protein in primate cells. J Virol 70:6323–6335

Goodrum FD, Ornelles DA (1997) The early region 1B 55-kilodalton oncoprotein of adenovirus relieves growth restrictions imposed on viral replication by the cell cycle. J Virol 71:548–561

Goodrum FD, Ornelles DA (1998) p53 status does not determine outcome of E1B 55-Kilodalton mutant adenovirus lytic infection. J Virol 72:9479–9490

Goodrum FD, Ornelles DA (1999) Roles for the E4 orf6, orf3, and E1B 55-kilodalton proteins in cell cycle-independent adenovirus replication. J Virol 73:7474–7488

Graham FL, Smiley J, Russel WC, Nairn R (1977) Characteristics of a human cell line transformed by DNA from human adenovirus type 5. J Gen Virol 36:59–72

Graham FL (1984) Transformation by and oncogenicity of human adenoviruses. In: Ginsberg HS (ed) The adenoviruses. Plenum Press, New York, pp 339–398

Grand RJ, Grant ML, Gallimore PH (1994) Enhanced expression of p53 in human cells infected with mutant adenoviruses. Virology 203:229–240

Grand RJ, Parkhill J, Szestak T, Rookes SM, Roberts S, Gallimore PH (1999) Definition of a major p53 binding site on Ad2E1B58 K protein and a possible nuclear localization signal on the Ad12E1B54 K protein. Oncogene 18:955–965

Green M, Wold WSM, Mackey JK, Ridgen P (1979) Analysis of human tonsils and cancer DNAs and RNAs for DNA sequences of group C (serotypes 1,2,5, and 6) human adenoviruses. Proc Natl Acad Sci USA 76:6606–6610

Green M, Wold WSM, Brackmann KH (1980) Human adenoviruses transforming genes: group relationships, integration, expression in transformed cells and analysis of human cancers and tonsils. In: Essex M, Toardo G, zur Hausen H (eds) 7th Cold Spring Harbor conference on cell proliferation viruses in naturally occuring tumors. Cold Spring Harbor Laboratory, Cold Spring Harbor, New York, pp 373–397

Grossman SR, Perez M, Kung AL, Joseph M, Mansur C, Xiao Z-X, Kumar S, Howley PM, Livingston DM (1998) p300/MDM 2 complexes participate in MDM 2-mediated p53 degradation. Mol Cell 2:405–415

Hahn WC, Counter CM, Lundberg AS, Beijersbergen RL, Brooks MW, Weinberg RA (1999) Creation of human tumour cells with defined genetic elements. Nature 400:464–468

Halbert DN, Cutt JR, Shenk T (1985) Adenovirus early region 4 encodes functions required for efficient DNA replication, late gene expression, and host cell shutoff. J Virol 56:250–257

Haley KP, Overhauser J, Babiss LE, Ginsberg HS, Jones NC (1984) Transformation properties of type 5 adenovirus mutants that differentially express the E1A gene products. Proc Natl Acad Sci USA 81:5734–5738

Han J, Sabbatini P, Perez L, Modha D, White E (1996) The E1B 19 K protein blocks apoptosis by interacting with and inhibiting the p53-inducible and death-promoting Bax protein. Genes Dev: 461–477

Harada JN, Shevchenko A, Pallas DC, Berk AJ (2002) Analysis of the adenovirus E1B-55K-anchored proteome reveals its link to ubiquitination machinery. J Virol 76:9194–9206

Hardy S, Engel DA, Shenk T (1989) An adenovirus early region 4 gene product is required for induction of the infection-specific form of cellular E2F activity. Genes Dev 3:1062–1074

Hardy S, Shenk T (1989) E2F from adenovirus-infected cells binds cooperatively to DNA containing two properly oriented and spaced recognition sites. Mol Cell Biol 9:4495–4506

Hateboer G, Hijmans EM, Nooij JB, Schlenker S, Jentsch S, Bernards R (1996) mUBC9, a novel adenovirus E1A-interacting protein that complements a yeast cell cycle defect. J Biol Chem 271: 25906–25011

Henry H, Thomas A, Shen Y, White E (2002) Regulation of the mitochondrial checkpoint in p53-mediated apoptosis confers resistance to cell death. Oncogene 21:748–760

Herisse J, Rigolet M, de Dinechin SD, Galibert F (1981) Nucleotide sequence of adenovirus 2 DNA fragment encoding for the carboxylic region of the fiber protein and the entire E4 region. Nucl Acids Res 9:4023–4042

Honda R, Yasuda H (1999) Association of p19(ARF) with Mdm2 inhibits ubiquitin ligase activity of Mdm2 for tumor suppressor p53. EMBO J 18:22–27

Hopfner KP, Putnam CD, Tainer JA (2002) DNA double-strand break repair from head to tail. Curr Opin Struct Biol 12:115–122

Horikoshi N, Usheva A, Chen J, Levine AJ, Weinmann R, Shenk T (1995) Two domains of p53 interact with the TATA-binding protein, and the adenovirus 13S E1A protein disrupts the association, relieving p53-mediated transcriptional repression. Mol Cell Biol 15:227–234

Huang MM, Hearing P (1989) Adenovirus early region 4 encodes two gene products with redundant effects in lytic infection. J Virol 63:2605–2615

Huang MM, Hearing P (1989) The adenovirus early region 4 open reading frame 6/7 protein regulates the DNA binding activity of the cellular transcription factor, E2F, through a direct complex. Genes Dev 3:1699–1710

Hutton FG, Turnell AS, Gallimore PH, Grand RJ (2000) Consequences of disruption of the interaction between p53 and the larger adenovirus early region 1B protein in adenovirus E1 transformed human cells. Oncogene 19:452–462

Huvent I, Cousin C, Kiss A, Bernard C, D'Halluin JC (1996) Susceptibility to natural killer cells and down regulation of MHC class I expression in adenovirus 12 transformed cells are regulated by different E1A domains. Virus Res 45:123–134

Hwang CB, Shilitoe EJ (1990) DNA sequence of mutations induced in cells by herpes simplex virus type-1. Virology 178:180–188

Ibelgaufts H (1982) Are human DNA tumour viruses involved in the pathogenesis of human neurogenic tumours? Neurosurg Rev 5:3-24

Ibelgaufts H, Jones KW, Maitland N, Shaw JF (1982) Adenovirus-related RNA sequences in human neurogenic tumours. Acta Neuropathol Berl 56:113–117

Ishidate T, Matsumine A, Toyoshima K, Akiyama T (2000) The APC-hDLG complex negatively regulates cell cycle progression from the G0/G1 to S phase. Oncogene 19:365–372

Iwasaka T, Hayashi Y, Yokoyama M, Hara K, Matsuo N, Sugimori H (1992) 'Hit and run' oncogenesis by human papillomavirus type 18 DNA. Acta Obstet Gynecol Scand 71:219–223

Javier RT, Raska K, Jr., Macdonald GJ, Shenk T (1991) Human adenovirus type 9-induced rat mammary tumors. J Virol 65:3192–3202

Javier RT, Raska K, Jr., Shenk T (1992) Requirement for the adenovirus type 9 E4 region in production of mammary tumors. Science 257:1267–1271

Javier RT (1994) Adenovirus type 9 E4 open reading frame 1 encodes a transforming protein required for the production of mammary tumors in rats. J Virol 68:3917–3924

Jelinek T, Pereira DS, Graham FL (1994) Tumorigenicity of adenovirus-transformed rodent cells is influenced by at least two regions of adenovirus type 12 early region 1A. J Virol 68:888–896

Jeng YH, Wold WS, Green M (1978) Evidence for an adenovirus type 2-coded early glycoprotein. J Virol 28:314–323

Jochemsen AG, Bernards R, van Kranen HJ, Houweling A, Bos JL, van der Eb AJ (1986) Different activities of the adenovirus types 5 and 12 E1A regions in transformation with the EJ Ha-ras oncogene. J Virol 59:684–691

Johnson DG, Cress WD, Jakoi L, Nevins JR (1994) Oncogenic capacity of the E2F1 gene. Proc Natl Acad Sci USA 91:12823–12827

Jones C (1995) Cervical cancer: is herpes simplex virus type II a cofactor? Clin Microbiol Rev 8:549–556

Jones PA (1999) The DNA methylation paradox. Trends in Genetics 15:34–37

Jonsson N, Ankerst J (1977) Studies on adenovirus type 9-induced mammary fibroadenomas in rats and their malignant transformation. Cancer 39:2513–2519

Jox A, Rohen C, Belge G, Bartnitzke S, Pawlita M, Diehl V, Bullerdiek J, Wolf J (1997) Integration of Epstein-Barr virus in Burkitt's lymphoma cells leads to a region of enhanced chromosome instability. Ann Oncol 8:131–135

Karran L, Teo CG, King D, Hitt MM, Gao YN, Wedderburn N, Griffin BE (1990) Establishment of immortalized primate epithelial cells with sub-genomic EBV DNA. Int J Cancer 45:763–772

Kast WM, Offringa R, Peters PJ, Voordouw AC, Meloen RH, van der Eb AJ, Melief CJ (1989) Eradication of adenovirus E1-induced tumors by E1A-specific cytotoxic T lymphocytes. Cell 59: 603–614

Kimelman D (1986) A novel general approach to eucaryotic mutagenesis functionally identifies conserved regions within the adenovirus 13S E1A polypeptide. Mol Cell Biol 6:1487–1496

Kiyono T, Hiraiwa A, Fujita M, Hayashi Y, Akiyama T, Ishibashi M (1997) Binding of high-risk human papillomavirus E6 oncoproteins to the human homologue of the Drosophila discs large tumor suppressor protein. Proc Natl Acad Sci USA 94:11612–11616

König C, Roth J, Dobbelstein M (1999) Adenovirus type 5 E4orf3 protein relieves p53 inhibition by E1B-55-kilodalton protein. J Virol 73:2253–2262

Kovesdi I, Reichel R, Nevins JR (1986a) Identification of a cellular transcription factor involved in E1A trans-activation. Cell 45:219–228

Kovesdi I, Reichel R, Nevins JR (1986b) E1A transcription induction: enhanced binding of a factor to upstream promoter sequences. Science 231:719–722

Kovesdi I, Reichel R, Nevins JR (1987) Role of an adenovirus E2 promoter binding factor in E1Amediated coordinate gene control. Proc Natl Acad Sci USA 84:2180–2184

Krätzer F, Rosorius O, Heger P, Hirschmann N, Dobner T, Hauber J, Stauber RH (2000) The adenovirus type 5 E1B-55 k oncoprotein is a highly active shuttle protein and shuttling is independent of E4orf6, p53 and Mdm2. Oncogene 19:850–857

Kuhlmann I, Achten S, Rudolph R, Dörfler W (1982) Tumor induction by human adenovirus type 12 in hamsters: loss of the viral genome from adenovirus type 12-induced tumor cells is compatible with tumor formation. EMBO J 1:79–86

Kuwano K, Kawasaki M, Kunitake R, Hagimoto N, Nomoto Y, Matsuba T, Nakanishi Y, Hara N (1997a) Detection of group C adenovirus DNA in small-cell lung cancer with the nested polymerase chain reaction. J Cancer Res Clin Oncol 123:377–382

Kuwano K, Nomoto Y, Kunitake R, Hagimoto N, Matsuba T, Nakanishi Y, Hara N (1997b) Detection of adenovirus E1A DNA in pulmonary fibrosis using nested polymerase chain reaction. Eur Respir J 10:1445–1449

Kzhyshkowska J, Schütt H, Liss M, Kremmer E, Stauber R, Wolf H, Dobner T (2001) Heterogeneous nuclear ribonucleoprotein E1B-AP5 is methylated in its RGG-box and interacts with human arginine methyltransferase HRMT1L1. Biochem J 358:305–314

Lawler M, Humphries P, O'Farrelly C, Hoey H, Sheils O, Jeffers M, O'Briain DS, Kelleher D (1994) Adenovirus 12 E1A gene detection by polymerase chain reaction in both the normal and coeliac duodenum. Gut 35:1226–1232

Lee SS, Weiss RS, Javier RT (1997) Binding of human virus oncoproteins to hDlg/SAP97, a mammalian homolog of the Drosophila discs large tumor suppressor protein. Proc Natl Acad Sci USA 94:6670–6675

Lee SS, Glaunsinger BA, Mantovani F, Banks L, Javier RT (2000) Multi-PDZ domain protein MUPP1 is a cellular target for both adenovirus E4-ORF1 and high-risk papillomavirus type 18 E6 oncoproteins. J Virol 74:9680–9693

Legrand A, Mayer EP, Dalvi SS, Nachtigal M (1997) Transformation of rabbit vascular smooth muscle cells by human cytomegalovirus morphological transforming region I. Am J Pathol 151: 1387–1395

Leppard KN, Everett RD (1999) The adenovirus type 5 Elb 55 K and E4 Orf3 proteins associate in infected cells and affect ND10 components. J Gen Virol 80:997–1008

Lethbridge KJ, Scott GE, Leppard KN (2003) Nuclear matrix localization and SUMO-1 modification of adenovirus type 5 Elb 55 K protein are controlled by E4 Orf6 protein. J Gen Virol 84:259–268

Levinson A, Levine AJ, Anderson S, Osborn M, Rosenwirth B, Weber K (1976) The relationship between group C adenovirus tumor antigen and the adenovirus single-strand DNA-binding protein. Cell 7:575–584

Lewis AM, Jr., Cook JL (1982) Spectrum of tumorigenic phenotypes among adenovirus 2-, adenovirus 12-, and simian virus 40-transformed Syrian hamster cells defined by host cellular immune-tumor cell interactions. Cancer Res 42:939–944

Li Z, Yu A, Weiner AM (1998) Adenovirus type 12-induced fragility of the human RNU2 locus requires p53 function. J Virol 72:4183–4191

Liao D, Yu A, Weiner AM (1999) Coexpression of the adenovirus 12 E1B 55 kDa oncoprotein and cellular tumor suppressor p53 is sufficient to induce metaphase fragility of the human RNU2 locus. Virology 254:11–23

Liu Y, Colosimo AL, Yang XJ, Liao D (2000) Adenovirus E1B 55-kilodalton oncoprotein inhibits p53 acetylation by PCAF. Mol Cell Biol 20:5540–5553

Logan J, Nicolas JC, Topp WC, Girard M, Shenk T, Levine AJ (1981) Transformation by adenovirus early region 2A temperature-sensitive mutants and their revertants. Virology 115:419–422

Lowe SW, Ruley HE (1993) Stabilization of the p53 tumor suppressor is induced by adenovirus 5 E1A and accompanies apoptosis. Genes Dev 7:535–545

Mackey JK, Rigden PM, Green M (1976) Do highly oncogenic group A human adenoviruses cause human cancer? Analysis of human tumors for adenovirus 12 transforming DNA sequences. Proc Natl Acad Sci USA 73:4657–4661

Mackey JK, Green M, Wold WSM, Ridgen P (1979) Analysis of human cancer DNA for DNA sequences of human adenovirus type 4. J Natl Cancer Inst 62:23–26

Maheswaran S, Englert C, Lee SB, Ezzel RM, Settleman J, Haber DA (1998) E1B 55 K sequesters WT1 along with p53 within a cytoplasmic body in adenovirus-transformed kidney cells. Oncogene 16: 2041–2050

Maitland NJ, Kinross JH, Busuttil A, Ludgate SM, Smart GE, Jones KW (1981) The detection of DNA tumour virus-specific RNA sequences in abnormal human cervical biopsies by in situ hybridization. J Gen Virol 55:123–137

Marengo C, Mbikay M, Weber J, Thirion JP (1981) Adenovirus-induced mutations at the hypoxanthine phosphoribosyltransferase locus of Chinese hamster cells. J Virol 38:184–190

Martin ME, Berk AJ (1999) Corepressor required for adenovirus E1B 55,000-molecular-weight protein repression of basal transcription. Mol Cell Biol 19:3403–3414

Marton MJ, Baim SB, Ornelles DA, Shenk T (1990) The adenovirus E4 17-kilodalton protein complexes with the cellular transcription factor E2F, altering its DNA-binding properties and stimulating E1A-independent accumulation of E2 mRNA. J Virol 64:2345–2359

McLorie W, McGlade CJ, Takayesu D, Branton PE (1991) Individual adenovirus E1B proteins induce transformation independently but by additive pathways. J Gen Virol 72:1467–1471

Meijer I, Jochemsen AG, de Wit CM, Bos JL, Morello D, van der Eb AJ (1989) Adenovirus type 12 E1A down regulates expression of a transgene under control of a major histocompatibility complex class I promoter: evidence for transcriptional control. J Virol 63:4039–4042

Miller BW, Williams J (1987) Cellular transformation by adenovirus type 5 is influenced by the viral DNA polymerase. J Virol 61:3630–3634

Moore M, Horikoshi N, Shenk T (1996) Oncogenic potential of the adenovirus E4orf6 protein. Proc Natl Acad Sci USA 93:11295–11301

Moran E, Grodzicker T, Roberts RJ, Mathews MB, Zerler B (1986) Lytic and transforming functions of individual products of the adenovirus E1A gene. J Virol 57:765–775

Moran E (1994) Mammalian cell growth controls reflected through protein interactions with the adenovirus E1A gene products. Semin Virol 5:327–340

Mukai N, Murao T (1975) Retinal tumor induction by ocular inoculation of human adenovirus in 3 day old rats. J Neuropathol Exp Neurol 34:28–35

Mukai N, Kalter SS, Cummins LB, Matthews VA, Nishida T, Nakajima T (1980) Retinal tumor induced in the baboon by human adenovirus 12. Science 210:1023–1025

Mushinski JF, Potter M, Bauer SR, Reddy EP (1983) DNA rearrangement and altered RNA expression of the c-myb oncogene in mouse plasmacytoid lymphosarcomas. Science 220:795–798

Nakao M (2001) Epigenetics: interaction of DNA methylation and chromatin. Gene 278:25–31

Neill SD, Hemstrom C, Virtanen A, Nevins JR (1990) An adenovirus E4 gene product transactivates *E2* transcription and stimulates stable E2F binding through a direct association with E2F. Proc Natl Acad Sci USA 87:2008–2012

Neill SD, Nevins JR (1991) Genetic analysis of the adenovirus E4 6/7 trans activator: interaction with E2F and induction of a stable DNA-protein complex are critical for activity. J Virol 65:5364–5373

Nevels M, Rubenwolf S, Spruss T, Wolf H, Dobner T (1997) The adenovirus E4orf6 protein can promote E1A/E1B-induced focus formation by interfering with p53 tumor suppressor function. Proc Natl Acad Sci USA 94:1206–1211

Nevels M, Täuber B, Kremmer E, Spruss T, Wolf H, Dobner T (1999a) Transforming potential of the adenovirus type 5 E4orf3 protein. J Virol 73:1591–1600

Nevels M, Spruss T, Wolf H, Dobner T (1999b) The adenovirus E4orf6 protein contributes to malignant transformation by antagonizing E1A-induced accumulation of the tumor suppressor protein p53. Oncogene 18:9-17

Nevels M, Rubenwolf S, Spruss T, Wolf H, Dobner T (2000) Two distinct activities contribute to the oncogenic potential of the adenovirus type 5 E4orf6 protein. J Virol 74:5168–5181

Nevels M, Täuber B, Spruss T, Wolf H, Dobner T (2001) "Hit-and-run" transformation by adenovirus oncogenes. J Virol 75:3089–3094

Nicolás AL, Munz PL, Falck-Pedersen E, Young CSH (2000) Creation and repair of specific DNA double-strand breaks *in vivo* following infection with adenovirus vectors expressing Saccharomyces cerevisiae HO endonuclease. Virology 266:211–224

O'Connor RJ, Hearing P (1991) The C-terminal 70 amino acids of the adenovirus E4-ORF6/7 protein are essential and sufficient for E2F complex formation. Nucl Acids Res 19:6579–6586

O'Connor RJ, Hearing P (2000) The E4–6/7 protein functionally compensates for the loss of E1A expression in adenovirus infection. J Virol 74:5819–5824

Obert S, O'Connor RJ, Schmid S, Hearing P (1994) The adenovirus E4–6/7 protein transactivates the E2 promoter by inducing dimerization of a heteromeric E2F complex. Mol Cell Biol 14:1333–1346

Öhman K, Nordquist K, Linder S, Akusjärvi G (1995) Effect of adenovirus-2 early region 4 products on E1 transformation. Int J Oncol 6:663–668

Ornelles DA, Shenk T (1991) Localization of the adenovirus early region 1B 55-kilodalton protein during lytic infection: association with nuclear viral inclusions requires the early region 4 34-kilodalton protein. J Virol 65:424–429

Pääbo S, Severinsson L, Andersson M, Martens I, Nilsson T, Peterson PA (1989) Adenovirus proteins and MHC expression. Adv Cancer Res 52:151–163

Paraskeva C, Gallimore PH (1980) Tumorigenicity and in vitro characteristics of rat liver epithelial cells and their adenovirus-transformed derivatives. Int J Cancer 25:631–639

Paraskeva C, Brown KW, Dunn AR, Gallimore PH (1982) Adenovirus type 12-transformed rat embryo brain and rat liver epithelial cell lines: adenovirus type 12 genome content and viral protein expression. J Virol 44:759–764

Paraskeva C, Roberts C, Biggs P, Gallimore PH (1983) Human adenovirus type 2 but not adenovirus type 12 is mutagenic at the hypoxanthine phosphoribosyltransferase locus of cloned rat liver epithelial cells. J Virol 46:131–136

Pereira DS, Rosenthal KL, Graham FL (1995) Identification of adenovirus E1A regions which affect MHC class I expression and susceptibility to cytotoxic T lymphocytes. Virology 211:268–277

Persson H, Kvist S, Ostberg L, Peterson PA, Philipson L (1979) The early adenovirus glycoprotein E3-19 K and its association with transplantation antigens. Cold Spring Harb Symp Quant Biol 44: 509–514

Pfeffer A, Schubbert R, Orend G, Hilger-Eversheim K, Doerfler W (1999) Integrated viral genomes can be lost from adenovirus type 12-induced hamster tumor cells in a clone-specific, multistep process with retention of the oncogenic phenotype. Virus Res 59:113–127

Punga T, Akusjärvi G (2000) The adenovirus-2 E1B-55 K protein interacts with a mSin3A/histone deacetylase 1 complex. FEBS Lett 476:248–252

Querido E, Marcellus RC, Lai A, Rachel C, Teodoro JG, Ketner G, Branton PE (1997) Regulation of p53 levels by the E1B 55-kilodalton protein and E4orf6 in adenovirus-infected cells. J Virol 71:3788–3798

Querido E, Blanchette P, Yan Q, Kamura T, Morrison M, Boivin D, Kaelin WG, Conaway RC, Conaway JW, Branton PE (2001) Degradation of p53 by adenovirus E4orf6 and E1B55 K proteins occurs via a novel mechanism involving a Cullin-containing complex. Genes Dev 15:3104–3017

Querido E, Morisson MR, Chu-Pham-Dang H, Thirlwell SW, Boivin D, Branton PE (2001) Identification of three functions of the adenovirus E4orf6 protein that mediate p53 degradation by the E4orf6-E1B55 K complex. J Virol 75:699–709

Quinlan MP, Douglas JL (1992) Immortalization of primary epithelial cells requires first- and second- exon functions of adenovirus type 5 12S. J Virol 66:2020–2030

Rao L, Debbas M, Sabbatini P, Hockenbery D, Korsmeyer S, White E (1992) The adenovirus E1A proteins induce apoptosis, which is inhibited by the E1B 19-kDa and Bcl-2 proteins. Proc Natl Acad Sci USA 89:7742–7746

Raska K, Jr., Gallimore PH (1982) An inverse relation of the oncogenic potential of adenovirus-transformed cells and their sensitivity to killing by syngeneic natural killer cells. Virology 123:8-18

Raychaudhuri P, Bagchi S, Nevins JR (1989) DNA-binding activity of the adenovirus-induced E4F transcription factor is regulated by phosphorylation. Genes Dev 3:620–627

Reichel R, Neill SD, Kovesdi I, Simon MC, Raychaudhuri P, Nevins JR (1989) The adenovirus E4 gene, in addition to the E1A gene, is important for trans-activation of E2 transcription and for E2F activation. J Virol 63:3643–3650

Ricciardi RP (1995) Transformation and tumorigenesis mediated by the adenovirus E1A and E1B oncogenes. In: Barbanti-Brodano G (ed) DNA Tumor Viruses: Oncogenic Mechanisms. Plenum Press, New York, pp 195–210
Ricciardi RP (1999) Adenovirus transformation and tumorigenicity. In: Seth P (ed) Adenoviruses: Basic biology to gene therapy. RG Landes Co, Austin, pp 217–227
Rice SA, Klessig DF, Williams J (1987) Multiple effects of the 72-kDa, adenovirus-specified DNA binding protein on the efficiency of cellular transformation. Virology 156:366–376
Ross D, Ziff E (1992) Defective synthesis of early region 4 mRNAs during abortive adenovirus infections in monkey cells. J Virol 66:3110–3117
Roth J, König C, Wienzek S, Weigel S, Ristea S, Dobbelstein M (1998) Inactivation of p53 but not p73 by adenovirus type 5 E1B 55-Kilodalton and E4 34-Kilodalton oncoproteins. J Virol 72:8510–8516
Ruley HE (1983) Adenovirus early region 1A enables viral and cellular transforming genes to transform primary cells in culture. Nature 304:602–606
Sabbatini P, Chiou SK, Rao L, White E (1995) Modulation of p53-mediated transcriptional repression and apoptosis by the adenovirus E1B 19 K protein. Mol Cell Biol 15:1060–1070
Sabbatini P, Lin J, Levine AJ, White E (1995) Essential role for p53-mediated transcription in E1Ainduced apoptosis. Genes Dev 9:2184–2192
Sarnow P, Hearing P, Anderson CW, Reich N, Levine AJ (1982) Identification and characterization of an immunologically conserved adenovirus early region 11,000 Mr protein and its association with the nuclear matrix. J Mol Biol 162:565–583
Sarnow P, Ho YS, Williams J, Levine AJ (1982) Adenovirus E1b-58kd tumor antigen and SV40 large tumor antigen are physically associated with the same 54 kd cellular protein in transformed cells. Cell 28:387–394
Sawada Y, Fohring B, Shenk TE, Raska K, Jr. (1985) Tumorigenicity of adenovirus-transformed cells: region E1A of adenovirus 12 confers resistance to natural killer cells. Virology 147:413–421
Sawada Y, Raska K, Jr., Shenk T (1988) Adenovirus type 5 and adenovirus type 12 recombinant viruses containing heterologous E1 genes are viable, transform rat cells, but are not tumorigenic in rats. Virology 166:281–284
Schaeper U, Boyd JM, Verma S, Uhlmann E, Subramanian T, Chinnadurai G (1995) Molecular cloning and characterization of a cellular phosphoprotein that interacts with a conserved C-terminal domain of adenovirus E1A involved in negative modulation of oncogenic transformation. Proc Natl Acad Sci USA 92:10467–10471
Schaley J, O'Connor RJ, Taylor LJ, Bar-Sagi D, Hearing P (2000) Induction of the cellular E2F-1 promoter by the adenovirus E4-6/7 protein. J Virol 74:2084–2093
Schiedner G, Hertel S, Kochanek S (2000) Efficient transformation of primary human amniocytes by E1 functions of ad5: generation of new cell lines for adenoviral vector production. Hum Gene Ther 11:2105–2116
Schlehofer J, zur Hausen H (1982) Induction of mutations within the host cell genome by partially inactivated Herpes Simplex Virus type 1. Virology 122:471–475
Schramayr S, Caporossi D, Mak I, Jelinek T, Bacchetti S (1990) Chromosomal damage induced by human adenovirus type 12 requires expression of the E1B 55-kilodalton viral protein. J Virol 64: 2090–2095

Schrier PI, Bernards R, Vaessen RT, Houweling A, van der Eb AJ (1983) Expression of class I major histocompatibility antigens switched off by highly oncogenic adenovirus 12 in transformed rat cells. Nature 305:771–775

Shaw G, Morse S, Ararat M, Graham FL (2002) Preferential transformation of human neuronal cells by human adenoviruses and the origin of HEK 293 cells. Faseb J 16:869–871

Shen Y, Shenk T (1994) Relief of p53-mediated transcriptional repression by the adenovirus E1B 19-kDa protein or the cellular Bcl-2 protein. Proc Natl Acad Sci USA 91:8940–8944

Shen Y, Zhu H, Shenk T (1997) Human cytomegalovirus IE1 and IE2 proteins are mutagenic and mediate "hit-and-run" oncogenic transformation in cooperation with the adenovirus E1A proteins. Proc Natl Acad Sci USA 94:3341–3345

Shenk T (1996) Adenoviridae: the viruses and their replication. In: Fields BN, Knipe DM, Howley PM (eds) Virology, Third ed, vol. 2. Lippincott-Raven, New York, pp 2111–2148

Shepherd SE, Howe JA, Mymryk JS, Bayley ST (1993) Induction of the cell cycle in baby rat kidney cells by adenovirus type 5 E1A in the absence of E1B and a possible influence of p53. J Virol 67: 2944–2949

Shiroki K, Hashimoto S, Saito I, Fukui Y, Kato H, Shimojo H (1984) Expression of the E4 gene is required for establishment of soft-agar colony-forming rat cell lines transformed by the adenovirus 12 E1 gene. J Virol 50:854–863

Singh P, Wong SH, Hong W (1994) Overexpression of E2F-1 in rat embryo fibroblasts leads to neoplastic transformation. EMBO J 13:3329–3338

Skinner GR (1976) Transformation of primary hamster embryo fibroblasts by type 2 simplex virus: evidence for a "hit and run" mechanism. Br J Exp Pathol 57:361–376

Smirnov DA, Hou S, Liu X, Claudio E, Siebenlist UK, Ricciardi RP (2001) COUP-TFII is upregulated in adenovirus type12 tumorigenic cells and is a repressor of MHC class I transcription. Virology 284:13–19

Smith GCM, Jackson SP (1999) The DNA-dependent protein kinase. Genes Dev 13:916–934

Sprengel J, Schmitz B, Heuss Neitzel D, Doerfler W (1995) The complete nucleotide sequence of the DNA of human adenovirus type 12. Curr Top Microbiol Immunol 199:189–274

Steegenga WT, Riteco N, Jochemsen AG, Fallaux FJ, Bos JL (1998) The large E1B protein together with the E4orf6 protein target p53 for active degradation in adenovirus infected cells. Oncogene 16: 349–357

Sternsdorf T, Grotzinger T, Jensen K, Will H (1997) Nuclear dots: actors on many stages. Immunobiology 198:307–331

Stevens JL, Cantin GT, Wang G, Shevchenko A, Berk AJ (2002) Transcription control by E1A and MAP kinase pathway via Sur2 mediator subunit. Science 296:755–758

Stracker TH, Carson CT, Weitzman MD (2002) Adenovirus oncoproteins inactivate the Mre11 Rad50 NBS1 DNA repair complex. Nature 418:348–352

Subramanian T, La Regina M, Chinnadurai G (1989) Enhanced ras oncogene mediated cell transformation and tumorigenesis by adenovirus 2 mutants lacking the C-terminal region of E1a protein. Oncogene 4:415–420

Subramanian T, Malstrom SE, Chinnadurai G (1991) Requirement of the C-terminal region of adenovirus E1a for cell transformation in cooperation with E1b. Oncogene 6:1171–1173

Takayesu D, Teodoro JG, Whalen SG, Branton PE (1994) Characterization of the 55 K adenovirus type 5 E1B product and related proteins. J Gen Virol 75:789–798

Täuber B, Dobner T (2001) Molecular regulation and biological function of adenovirus early genes: the E4 ORFs. Gene 278:1-23

Täuber B, Dobner T (2001) Adenovirus early E4 genes in viral oncogenesis. Oncogene 20:7847–7854

Telling GC, Williams J (1993) The E1B 19-kilodalton protein is not essential for transformation of rodent cells in vitro by adenovirus type 5. J Virol 67:1600–1611

Telling GC, Williams J (1994) Constructing chimeric type 12/type 5 adenovirus E1A genes and using them to identify an oncogenic determinant of adenovirus type 12. J Virol 68:877–887

Teodoro JG, Halliday T, Whalen SG, Takayesu D, Graham FL, Branton PE (1994) Phosphorylation at the carboxy terminus of the 55-kilodalton adenovirus type 5 E1B protein regulates transforming activity. J Virol 68:776–786

Teodoro JG, Shore GC, Branton PE (1995) Adenovirus E1A proteins induce apoptosis by both p53-dependent and p53-independent mechanisms. Oncogene 11:467–474

Teodoro JG, Branton PE (1997) Regulation of p53-dependent apoptosis, transcriptional repression, and cell transformation by phosphorylation of the 55-kilodalton E1B protein of human adenovirus type 5. J Virol 71:3620–3627

Terris B, Baldin V, Dubois S, Degott C, Flejou J-F, Hénin D, Dejean A (1995) PML nuclear bodies are general targets for inflammation and cell proliferation. Cancer Res 55:1590–1597

Thomas A, White E (1998) Suppression of the p300-dependent mdm2 negative-feedback loop induces the p53 apoptotic function. Genes Dev 12:1975–1985

Thomas DL, Shin S, Jiang BH, Vogel H, Ross MA, Kaplitt M, Shenk TE, Javier RT (1999) Early region 1 transforming functions are dispensable for mammary tumorigenesis by human adenovirus type 9. J Virol 73:3071–3079

Thomas DL, Schaack J, Vogel H, Javier RT (2001) Several E4 region functions influence mammary tumorigenesis by human adenovirus type 9. J Virol 75:557–568

Tigges MA, Raskas HJ (1984) Splice junctions in adenovirus 2 early region 4 mRNAs: multiple splice sites produce 18 to 24 RNAs. J Virol 50:106–117

Trentin JJ, Yabe Y, Taylor G (1962) The quest for human cancer viruses: a new approach to an old problem reveals cancer induction in hamster hy human adenoviruses. Science 137:835–849

Turnell AS, Grand RJ, Gorbea C, Zhang X, Wang W, Mymryk JS, Gallimore PH (2000) Regulation of the 26S proteasome by adenovirus E1A. EMBO J 19:4759–4773

van den Heuvel SJL, van Laar T, Kast WM, Melief CJM, Zantema A, van der Eb AJ (1990)

Association between the cellular p53 and the adenovirus 5 E1B-55kd proteins reduces the oncogenicity of Ad-transformed cells. EMBO J 9:2621–2629

van den Heuvel SJL, The SI, Klein B, Jochemsen AG, Zantema A, van der Eb AJ (1992) p53 shares an antigenic determinant with proteins of 92 and 150 kilodaltons that may be involved in senescence of human cells. J Virol 66:591–595

van den Heuvel SJL, van Laar T, The I, van der Eb AJ (1993) Large E1B proteins of adenovirus types 5 and 12 have different effects on p53 and distinct roles in cell transformation. J Virol 67:5226–5234
van der Eb AJ, Zantema A (1992) Adenovirus oncogenesis. In: Dörfler W, Böhm P (eds) Malignant transformation by DNA viruses. VCH, Weinheim, pp 115–140
Virtanen A, Gilardi P, Naslund A, LeMoullec JM, Pettersson U, Perricaudet M (1984) mRNAs from human adenovirus 2 early region 4. J Virol 51:822–831
Virtanen A, Pettersson U (1985) Organization of early region 1B of human adenovirus type 2: identification of four differentially spliced mRNAs. J Virol 54:383–391
Wang G, Berk AJ (2002) In vivo association of adenovirus large E1A protein with the human mediator complex in adenovirus-infected and -transformed cells. J Virol 76:9186–9193
Waubke R, zur Hausen H, Henle W (1968) Chromosomal and autoradiographic studies of cells infected with herpes simplex virus. J Virol 2:1047–1054
Weiden MD, Ginsberg HS (1994) Deletion of the E4 region of the genome produces adenovirus DNA concatemers. Proc Natl Acad Sci USA 91:153–157
Weiss RS, McArthur MJ, Javier RT (1996) Human adenovirus type 9 E4 open reading frame 1 encodes a cytoplasmic transforming protein capable of increasing the oncogenicity of CREF cells. J Virol 70:862–872
Weiss RS, Lee SS, Prasad BV, Javier RT (1997a) Human adenovirus early region 4 open reading frame 1 genes encode growth-transforming proteins that may be distantly related to dUTP pyrophosphatase enzymes. J Virol 71:1857–1870
Weiss RS, Gold MO, Vogel H, Javier RT (1997b) Mutant adenovirus type 9 E4 ORF1 genes define three protein regions required for transformation of CREF cells. J Virol 71:4385–4394
Weiss RS, Javier RT (1997) A carboxy-terminal region required by the adenovirus type 9 E4 ORF1 oncoprotein for transformation mediates direct binding to cellular polypeptides. J Virol 71:7873–7880
White E, Cipriani R (1990) Role of adenovirus E1B proteins in transformation: altered organization of intermediate filaments in transformed cells that express the 19-kilodalton protein. Mol Cell Biol 10:120–130
White E (2001) Regulation of the cell cycle and apoptosis by the oncogenes of adenovirus. Oncogene 20:7836–7846
Whittaker JL, Byrd PJ, Grand RJ, Gallimore PH (1984) Isolation and characterization of four adenovirus type 12-transformed human embryo kidney cell lines. Mol Cell Biol 4:110–116
Wienzek S, Roth J, Dobbelstein M (2000) E1B 55-kilodalton oncoproteins of adenovirus types 5 and 12 inactivate and relocalize p53, but not p51 or p73, and cooperate with E4orf6 proteins to destabilize p53. J Virol 74:193–202
Williams J, Williams M, Liu C, Telling G (1995) Assessing the role of E1A in the differential oncogenicity of group A and group C human adenoviruses. Curr Top Microbiol Immunol 199:149–175
Williams JF, Young CSH, Austin PE (1974) Genetic analysis of human adenovirus type 5 in permissive and nonpermissive cells. Cold Spring Harbor Symp Quant B 39:427–437

Williams JF, Galos RS, Binger MH, Flint SJ (1979) Location of additional early regios within the left quarter of the adenovirus genome. Cold Spring Harbor Symp Quant Biol 44:353–366

Wold WSM, Tollefson AE (1998) Adenovirus E3 proteins: 14.7 K, RID, and gp19 K inhibit immuneinduced cell death; adenovirus death protein promotes cell death. Semin Virol 8:515–523

Xu G, Livingston DM, Krek W (1995) Multiple members of the E2F transcription factor family are the products of oncogenes. Proc Natl Acad Sci USA 92:1357–1361

Yamano S, Tokino T, Yasuda M, Kaneuchi M, Takahashi M, Niitsu Y, Fujinaga K, Yamashita T (1999) Induction of transformation and p53-dependent apoptosis by adenovirus type 5 E4orf6/7 cDNA. J Virol 73:10095–10103

Yew PR, Kao CC, Berk AJ (1990) Dissection of functional domains in the adenovirus 2 early 1B 55 K polypeptide by suppressor-linker insertional mutagenesis. Virology 179:795–805

Yew PR, Berk AJ (1992) Inhibition of p53 transactivation required for transformation by adenovirus early 1B protein. Nature 357:82–85

Yew PR, Liu X, Berk AJ (1994) Adenovirus E1B oncoprotein tethers a transcriptional repression domain to p53. Genes Dev 8:190–202

Yewdell JW, Bennink JR, Eager KB, Ricciardi RP (1988) CTL recognition of adenovirus-transformed cells infected with influenza virus: lysis by anti-influenza CTL parallels adenovirus-12-induced suppression of class I MHC molecules. Virology 162:236–238

Yu A, Fan HY, Liao D, Bailey AD, Weiner AM (2000) Activation of p53 or loss of the Cockayne syndrome group B repair protein causes metaphase fragility of human U1, U2, and 5S genes. Mol Cell 5:801–810

Zalmanzon ES (1987) Transforming and oncogenic properties of the adenovirus genome. Eksp Onkol 9:3-8

Zantema A, Fransen JA, Davis OA, Ramaekers FC, Vooijs GP, DeLeys B, van der Eb AJ (1985a) Localization of the E1B proteins of adenovirus 5 in transformed cells, as revealed by interaction with monoclonal antibodies. Virology 142:44–58

Zantema A, Schrier PI, Davis OA, van Laar T, Vaessen RT, van der Eb AJ (1985b) Adenovirus serotype determines association and localization of the large E1B tumor antigen with cellular tumor antigen p53 in transformed cells. Mol Cell Biol 5:3084–3091

Zerler B, Moran B, Maruyama K, Moomaw J, Grodzicker T, Ruley HE (1986) Adenovirus E1A coding sequences that enable ras and pmt oncogenes to transform cultured primary cells. Mol Cell Biol 6:887–899

Zhang S, Mak S, Branton PE (1992) Overexpression of the E1B 55-kilodalton (482R) protein of human adenovirus type 12 appears to permit efficient transformation of primary baby rat kidney cells in the absence of the E1B 19-kilodalton protein. J Virol 66:2302–2309

Zheng P, Guo Y, Niu Q, Levy DE, Dyck JA, Lu S, Sheiman LA, Liu Y (1998) Proto-oncogene PML controls genes devoted to MHC class I antigen presentation. Nature 396:373–376

Zhong S, Hu P, Ye TZ, Stan R, Ellis NA, Pandolfi PP (1999) A role for PML and the nuclear body in genomic stability. Oncogene 18:7941–7947

zur Hausen H (1967) Induction of specific chromosomal aberrations by adenovirus type 12 in human embryonic kidney cells. J Virol 1:1174–1185
zur Hausen H (1996) Papillomavirus infections—a major cause of human cancers. Biochim Biophys Acta 1288: F55-F78
zur Hausen H (1996) Viruses in human tumors—reminiscences and perspectives. Adv Cancer Res 68:1-22

Tumorigenesis by Adenovirus Type 12 in Newborn Syrian Hamsters

U. Hohlweg[1,4] · A. Dorn[1,2] · M. Hösel[1,2] · D. Webb[1] · R. Buettner[3,5] · W. Doerfler[1,2]

[1] Institute of Genetics, University of Cologne, 50931 Köln, Germany
[2] Institute for Clinical and Molecular Virology, University of Erlangen-Nürnberg, 91054 Erlangen, Germany
[3] Institute of Pathology, Technische Hochschule Aachen, 52074 Aachen, Germany
[4] Universität Tübingen, 72076 Tübingen, Germany
[5] Universität Bonn, 53012 Bonn, Germany

Abstract Ad12 oncogenesis in hamsters has been studied in detail to provide the following new data in this tumor model. Cells in the Ad12-induced tumors, often thought to be of neuronal origin, actually exhibit mesenchymal and neuronal characteristics and are probably of an undifferentiated derivation. Their intraperitoneal spread upon intramuscular injection of Ad12 adds another important new aspect. Differences in the integration patterns among the tumors suggest clonal origins from individual transformation events. Ad12 gene expression in the tumors is determined, at least in part, by the patterns of DNA methylation imprinted de novo upon the integrated Ad12 genomes. Differential Ad12 gene expression patterns, which have previously not been described in tumors, are an important parameter in Ad12 oncogenesis. The availability of cellular DNA arrays has opened up unprecedented possibilities to document changes in cellular transcription patterns, particularly of cancer-specific genes. These patterns exhibit differences and similarities among the different Ad12-induced tumors. Among the cellular genes, which are expressed in the Ad12-induced tumors, many are cancer-specific. We pursue the hypothesis that these alterations in cellular transcription patterns as a consequence of viral DNA integration and expression play an essential role in Ad12 oncogenesis.

1 Introduction

Human adenovirus type 12 (Ad12) was the first human virus shown to be oncogenic in newborn rodents (Trentin et al. 1962; Huebner et al. 1962). Much work has been devoted to studies on the role of the early adenoviral E1 and E4 gene functions for the mechanism of the transformation of cells in culture by human adenoviruses (Flint and Shenk 1989; Nevins 1995; Zantema and van der Eb 1995; Russell 2000; Thomas et al. 2001). There may be additional factors decisive for viral oncogenesis. Hence, our group has concentrated on an alternate mechanism involved in Ad12 oncogenesis. The integration of Ad12 DNA into the host genome seems very intimately linked to the oncogenic transformation by Ad12. In Ad12-induced tumors as well as in Ad12-transformed cells in culture, each tumor or transformed cell carries multiple copies of the Ad12 genome covalently linked to the DNA of the cell (Groneberg et al. 1977; Sutter et al. 1978; Knoblauch et al. 1996; Hilger-Eversheim and Doerfler 1997). The chromosomal insertion site is identical in all cells of a given tumor but varies from tumor to tumor. Thus, insertional mutagenesis in the conventional sense of alterations of cellular nucleotide sequences at the insertion sites is not likely to account for the generation of the oncogenic phenotype. Therefore, the role of the integration of foreign DNA in altering the overall structure of a mammalian genome and thus contributing to the process of oncogenic transformation has been of great interest in our laboratory (Doerfler 1968, 1970, 1995, 2000). In Ad12-transformed cells in culture, changes in cellular methylation and transcription patterns have been observed (Heller et al. 1995; Remus et al. 1999; Müller et al. 2001). The perturbation in chromatin structure and alterations in cellular and viral transcription and methylation patterns in the wake of foreign DNA insertion are thought to render a decisive contribution to the transition from a normal mammalian cell to a tumor cell (Doerfler 1996; 2000). The question arises of how the insertion of foreign DNA might affect the regional structure and function at the site of insertion and of neighboring chromosomes which are in contact with the insertion site on the chromosome targeted by foreign DNA.

2
Intramuscular or Subcutaneous Injection of Different Amounts of Ad12 and Tumor Response

The high frequency of tumor induction by Ad12 in newborn hamsters was likely facilitated by their immature defense systems. Intramuscular injection of Ad12 virions led to the intraperitoneal dissemination of many tumors. In some animals, tumors were also observed in the injected, not the contralateral, muscle. This finding confirmed the muscle as primary site of Ad12 application. By injecting decreasing amounts of Ad12 at different sites, threshold levels for viral tumor induction were determined. Independent of the site of application, 4.5×10^6 to 4.5×10^7 plaque-forming units (PFUs) of Ad12 per animal induced tumors within 29–49 days in 71%–94% of the surviving hamsters. Multiple tumors were observed in most animals, measuring 0.2–2 cm upon subcutaneous (s.c.) injection and 0.2–3 cm upon intramuscular (i.m.) injection. More than 300 hamster tumors were investigated, about 66 of them by Southern blot hybridization, and details of tumor characteristics are summarized in Table 1. Intraperitoneal tumors, with a tough fibrous capsule and suspension-like tumor cells inside, were attached to the lining of the peritoneal cavity or of the abdominal organs. The injection of only 4.5×10^5 Ad12 virions elicited up to two tumors in 8%–20% of the animals after 63 to 96 days. When 4.5×10^4 or less virions were applied, tumors were not observed as late as 16 months after injection (Table 1).

Table 1 Injection of Ad12 into newborn Syrian hamsters and tumor incidence

Site of injection	Plaque forming units	Number of injected animals	Animals with tumor(s)	Tumor formation (days)	Tumors per animal	Tumor size (cm)	Site of tumor formation
s.c.	4.5×10^7	14 (11)	9 (82%)	35–48	1–3	0.2–1.7	Nuchal region
	4.5×10^6	15 (14)	10 (71%)	37–49, 56, 64	1–5	0.2–2	Nuchal region
	4.5×10^5	14 (12)	1 (8.3%)	63	2	1.5	Nuchal region
	4.5×10^4	14 (11)	0	–	–	–	–
	4.5×10^3	12 (11)	0	–	–	–	–
i.m.	4.5×10^7	17 (16)	15 (94%)	29–36	12–16	0.5–3	Peritoneal cavity[a]
	4.5×10^6	15 (11)	7 (75%)	32–39	10–12	0.2–3	Peritoneal cavity[a]
	4.5×10^5	20 (15)	3 (20%)	63, 96	1–2	2	Peritoneal cavity[a]
	4.5×10^4	11 (10)	0	–	–	–	–
	4.5×10^3	9 (8)	0	–	–	–	–
Liver	4.5×10^7	12 (8)	7 (88%)	38	10–15	0.5–3	Liver, peritoneal cavity

[a] Tumor formation also on surfaces of liver and diaphragm plus at site of injected muscle.

3 Histological and Immunohistochemical Properties of Ad12-Induced Hamster Tumors

Since the first description in 1962 of tumor induction by Ad12 in newborn hamsters (Trentin et al. 1962), several suggestions have been made about the cellular origins of these tumors. On the basis of tumor histology, the tumors have been designated as (1) undifferentiated mesenchymal neoplasms or sarcomas (Trentin et al. 1962; Rabson et al. 1964; Chino et al. 1967; Yohn et al. 1968; Mukai and Ishida 1971; Mukai and Kobayashi 1972), (2) undifferentiated neuroectodermal tumors (Hamaya 1966; Nakajima and Mukai 1979), (3) undifferentiated malignant tumors (Ogawa et al. 1966; Neiders et al. 1968), (4) poorly differentiated or well differentiated adenocarcinomas (Yabe et al. 1966), and (5) retinoblastomas, medulloblastic, or malignant neurogenic tumors (Ogawa et al. 1966; Albert et al. 1968). Studies on the histology of tumors induced by Ad12 in the peritoneal cavity or in the subcutaneous tissue (Fig. 1a) of newborn hamsters revealed small, rounded, uniformly stained cells with a large hyperchromatic nucleus and granular chromatin, numerous mitotic cells, and Homer-Wright rosette-like structures characteristic of primitive neuroectodermal tumors (PNET) (Fig. 1b). Immunohistochemical assays identified neuronal-specific tissue markers (synaptophysin, neuronal-specific enolase) as well as the expression of vimentin, typical for mesenchymal origin. Based on these results, the most reasonable classification of the Ad12-induced hamster tumors appeared to be that of primitive neuroectodermal tumors with both neuronal and mesenchymal tissue markers. The target cell for Ad12 transformation in hamsters is not yet known. The formation of PNETs and the expression of neuronal-specific tissue markers might indicate a neuronal origin of the target cells for Ad12 transformation.

4 Clonal Origin of Ad12-Induced Tumors

In Ad12-induced tumors that have been generated by the subcutaneous (s.c.) or intramuscular (i.m.) injection of the virus into newborn hamsters, multiple copies of viral DNA persist in each tumor cell in a chromosomally integrated form (Hilger-Eversheim and Doerfler 1997). Free viral DNA has never been observed in Ad12-induced hamster tumor

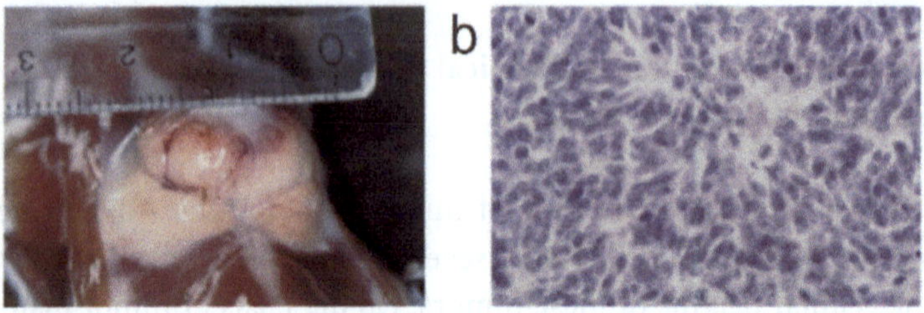

Fig. 1a, b Ad12-induced hamster tumor. **a** This tumor was induced by the s.c. injection of Ad12 into a newborn hamster. Hematoxylin/eosin staining: Histological examination reveals small, uniformly stained tumor cells with Homer-Wright rosette-like structures indicative of primitive neuroectodermal tumors (PNET) **b** Intraperitoneal spread of tumors after the intramuscular injection of Ad12

cells or in Ad12-transformed hamster cells. Usually, all of the integrated copies of Ad12 DNA are concentrated at one chromosomal site. Only in one case has viral DNA been found integrated on two chromosomes. Moreover, the chromosomal insertion site of Ad12 DNA is different in each tumor. There is no evidence for a specific Ad12 DNA insertion locus or a preferred cellular nucleotide sequence for the integrative recombination between cellular and viral DNA, except for short nucleotide sequence homologies between the reaction partners (Knoblauch et al. 1996; Wronka et al. 2002). Frequently, multiple tumors have been observed in one animal after the s.c. injection of Ad12 into newborn hamsters (Hilger-Eversheim and Doerfler 1997). Even in different tumors that have developed in one animal, the loci of viral DNA integration are different. Upon passage of the tumor cells in culture for up to 96 generations (Hilger-Eversheim and Doerfler 1997) or upon passing Ad12-transformed cell lines for decades (Sutter et al. 1978; Knoblauch et al. 1996), integration patterns or unique chromosomal locations have remained unaltered. Apparently, in the vast majority of cells, the Ad12 DNA is very stably integrated. In each individual tumor, all cells carry integrated Ad12 DNA at the same chromosomal site. The data demonstrate that Ad12-induced tumors are of clonal origin.

After the s.c. injection of Ad12 virions and tumor formation at the site of injection, metastases have not been observed, except for tumor cells in some of the local lymph vessels (Kuhlmann et al. 1982). In contrast, when Ad12 is administered into gluteal muscle tissue, extensive tumor formation spreads across the entire peritoneal cavity, involving the surface of many abdominal organs. Some of these tumors exhibit identical integration patterns, presumably due to a common clonal transformation event and subsequent spread.

The elucidation of the nucleotide sequence at a number of sites of linkage between Ad12 and hamster DNAs has not shown any sequence specificities at the insertion sites in a number of transformed and tumor cell lines (Deuring et al. 1981; Gahlmann et al. 1982, Gahlmann and Doerfler 1983; Stabel and Doerfler 1982; Doerfler et al. 1983; Knoblauch et al. 1996). However, the sites of linkage between the hamster cellular and the foreign (viral) DNA are characterized by the frequent occurrence of patch homologies between the recombination partners (summary in Knoblauch et al. 1996; Wronka et al. 2002). Frequently, the cellular sequences at or near the integration sites have been found to be transcriptionally active (Gahlmann and Doerfler 1983, Gahlmann et al. 1984; Schulz et al. 1987).

Since every tumor cell carries covalently linked Ad12 DNA, the integration of the foreign (Ad12) DNA into the host genome constitutes one of the important steps in the oncogenic transformation of hamster cells by Ad12. However, insertional mutagenesis in the conventional sense of alterations of cellular nucleotide sequences at the insertion site as a consequence of viral DNA insertion seems unlikely to account for the generation of the oncogenic phenotype, because each tumor is characterized by its individual Ad12 insertion site. On the other hand, the acquisition of many kilobases of inserted DNA might alter the chromatin topology and thus influence the function of specific parts of the genome (see below).

5
Integrated Ad12 Genomes

Previously published data identified Ad12-induced hamster tumors to be of clonal origin and to carry between a few and more than 20 copies of integrated Ad12 genomes at sites different from tumor to tumor, but at the same chromosomal location in all cells of one tumor (Hilger-Eversheim and Doerfler 1997). When the amounts of injected Ad12 were varied between 4.5×10^5 and 4.5×10^7 PFUs, the range of integrated copy numbers did not change. For several subcutaneously or intraperitoneally located Ad12 tumors, patterns of viral DNA integration were determined by Southern transfer hybridization using ^{32}P-labeled Ad12 DNA (Fig. 2a),

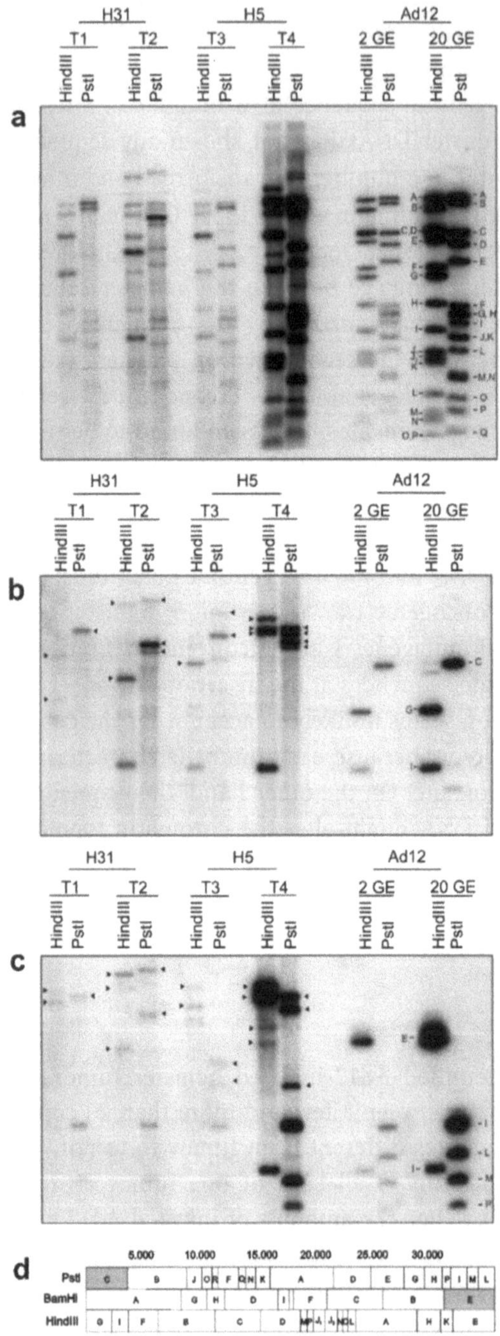

a
b
c
d
H31
H5
Ad12
T1
T2
T3
T4
2 GE
20 GE
HindIII
PstI
BamHI
5.000
10.000
15.000
20.000
25.000
30.000

the left (PstI-C) (Fig. 2b) or the right (BamHI-E) terminal fragment of Ad12 DNA (Fig. 2c) as hybridization probe (see Fig. 2d for the map of Ad12 DNA). The more than 120 tumors analyzed so far (Kuhlmann et al. 1982; Hilger-Eversheim and Doerfler 1997) exhibited individual patterns of off-size fragments which represented viral DNA covalently linked to cellular DNA or partly rearranged Ad12 DNA and did not correspond to any of the virion DNA fragments. Tumor T4 carried more than 20 genome equivalents, whereas the other tumors contained only one to a few copies of integrated Ad12 DNA (compare signal intensities in Fig. 2). Tumors T1 and T2 arose after injecting 4.5×10^7 PFUs, tumors T2 and T4 upon the injection of 4.5×10^5 PFU of Ad12. Only a few intraperitoneal tumors measuring 0.1–3 cm in diameter from one animal had similar integration patterns (not shown), presumably due to a clonal transformation event and the subsequent seeding of tumor cells in the peritoneal cavity and the growth of separate tumors.

6 Hit-and-Run Mechanism of Viral Oncogenesis

In Ad12-induced tumor cells, in Ad12-transformed cells, and in continuously passaged cell lines from these sources, the viral DNA is integrated in multiple copies, usually at a single chromosomal location. In most instances, the integrated Ad12 DNA resides very stably in the genomes of the Ad12-transformed cells or in the Ad12-induced hamster tumor cells.

Fig. 2a–d Patterns of Ad12 DNA integration into the DNA of different Ad12-induced hamster tumors. The DNA from four individual tumors, T1 to T4, isolated from two different hamsters (*H31*, *H5*) was extracted and cleaved with HindIII or PstI; the fragments were separated by electrophoreses on 0.8% agarose gels and transferred to positively charged nylon membranes. As size and quantity markers, 2 or 20 genome equivalents (*GE*) of authentic Ad12 virion DNA per diploid hamster genome were similarly treated and co-electrophoresed. The DNA was then hybridized to ^{32}P-labeled Ad12 DNA (**a**), the cloned left terminal PstI-C fragment (**b**), or the right terminal BamHI-E fragment of Ad12 DNA (**c**). Restriction maps of Ad12 DNA are shown in (**d**) with the hybridization probes shaded gray. The positions of the HindIII or PstI fragments of Ad12 virion marker DNA are indicated in the margins of the Ad12 DNA autoradiogram. Off-size fragments, which did not comigrate during electrophoresis with any of the virion DNA fragments, are designated by *arrowheads* (**a–c**). DNAs from tumors T5-T7 were not analyzed

However, upon continuous passage of such cell lines, the integrated viral DNA can be destabilized and lost (Groneberg et al. 1978; Groneberg and Doerfler 1979). At least in a few tumor cell lines, Ad12 DNA could no longer be detected by conventional Southern transfer hybridizations in the DNA of late-passage cell lines, but these revertants continued to induce tumors after reinjection into newborn hamsters (Kuhlmann et al. 1982; Pfeffer et al. 1999). Investigations by PCR on the presence of tiny segments of Ad12 DNA in some of the revertant-induced hamster tumor cells have revealed that individual cells from this initially cloned population had lost the Ad12 integrates, and others still contained tiny Ad12 DNA segments in highly variable distribution patterns. At present, it is unknown whether viral DNA excisions are due to random events or result from an active recognition and defense-excision mechanism against integrated foreign DNA in the mammalian genome. In any event, these data indicate that the presence of the viral genome and its products is not necessary for the maintenance of the transformed phenotype.

Furthermore, alterations in cellular DNA methylation patterns in the Ad12-transformed cell line T637 persist in the revertant TR3, in which Ad12 DNA cannot be detected any longer by Southern transfer hybridization experiments (Heller et al. 1995; Knoblauch et al. 1996). It is therefore likely that alterations in cellular transcription patterns and in cellular genome organization, which have been initiated by Ad12 DNA integration and/or by the expression of viral gene products, are permanently imprinted onto the cellular genome so that the oncogenic phenotype persists. These considerations are consistent with the hit-and-run mechanism of viral oncogenesis. Evidence for a hit-and-run mechanism in adenovirus transformation also comes from the findings of other laboratories (Nevels et al. 2001). Adenovirus infections in humans have thus far not been linked to human oncogenesis, because none of the analyzed human neoplasms contained adenoviral DNA or proteins (Green 1970, Green et al. 1979a,b; Graham 1984; Shenk 1996). However, since cells originally transformed by Ad12 to tumor cells can maintain their oncogenic phenotype despite the loss of the viral DNA from the cellular genome, the long-standing assumption that human adenoviruses have no role in human oncogenesis will have to be reconsidered.

7
Consequences of Foreign (Ad12) DNA Integration

7.1
De Novo Methylation of Integrated Ad12 DNA

From the study of different biological systems, there is ample evidence which supports the concept that sequence-specific patterns of promoter methylation are a signal for long-term gene inactivation (Doerfler 1983). Therefore, the de novo methylation of integrated foreign DNA can be viewed as an ancient cellular defense mechanism against the activity of foreign genes (Doerfler 1991; Yoder et al. 1997). An inverse correlation between DNA methylation and adenovirus gene activity was described for the first time in the early 1980s in adenovirus-transformed cells (Sutter and Doerfler 1980; Vardimon et al. 1980, 1982; Langner et al. 1984). which was later confirmed for a large number of mammalian promoters (for review, Munnes and Doerfler 1997). A particularly convincing case could be made for the late E2A promoter in Ad2-transformed hamster cells which carried partly deleted Ad2 genomes in an integrated state. In the cell line HE1, this promoter was methylated at all 5′-CCGG-3′sites and at additional 5′-CG-3′dinucleotides and was inactive. In contrast, in cell lines HE2 and HE3, the same promoter was unmethylated and active (Vardimon et al. 1980). In several Ad12-induced hamster tumors, the E1A and E1B promoters were hypo- or practically unmethylated, respectively, and the corresponding genes were transcribed in all tumors investigated by using the RT-PCR and DNA microarray methods (see Fig. 4). A similar inverse correlation was not observed in the tumors investigated for the major late promoter (MLP) of Ad12, which was silenced in all tumors. Previous studies documented a mitigator sequence in the first intron of the major late transcript of Ad12 DNA, which was thought to contribute to this efficient late transcriptional inactivation of Ad12 DNA in Syrian hamster cells (Zock and Doerfler 1990).

Virion DNA or free viral DNA in the nucleus of the host cell is not methylated (Günthert et al. 1976; Kämmer and Doerfler 1995), whereas chromosomally integrated viral DNA becomes de novo methylated in specific patterns (Sutter et al. 1978; Sutter and Doerfler 1980; Orend et al. 1995). By using methylation-sensitive restriction enzymes to analyze methylation patterns, it has been shown that de novo methylation is ini-

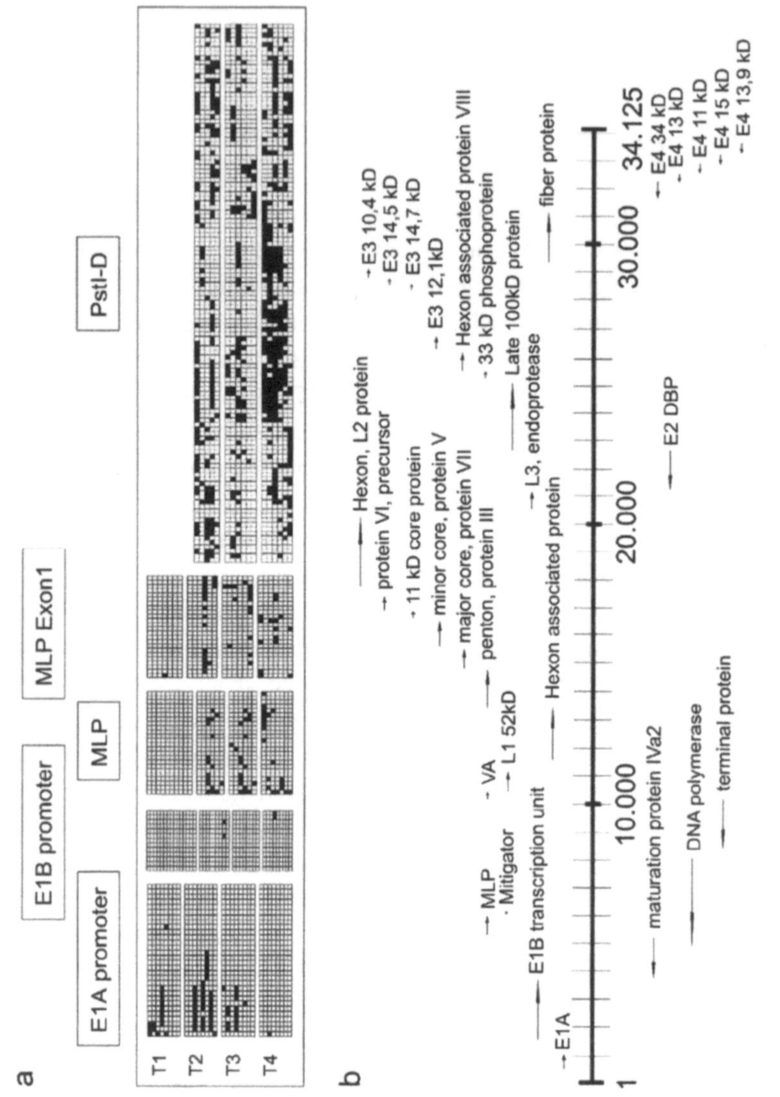

Fig. 3 **a** Methylation patterns of integrated Ad12 DNA in the Ad12-induced tumors T1, T2, T3, and T4. All 5′-CG-3′ dinucleotides of the E1A and E1B promoters, the MLP, the MLP exon 1, and the PstI-D fragment are represented by *squares*, the methylated 5′-CG-3′ dinucleotides as *filled squares*. The DNA sequence was determined in 6–14 independent clones after the genomic DNA had been treated with bisulfite, the fragments had been amplified by PCR, and individual PCR products had been molecularly cloned into the pGEM-T vector (Promega). In tumor T1, the PstI-D fragment was deleted as documented by restriction and Southern transfer hybridization

tiated at certain sites inside the integrated viral genome and spreads from there in both directions (Orend et al. 1995).

Recent results gleaned from analyses using the bisulfite protocol of the genomic sequencing technique indicate that de novo methylation is not initiated at a specific nucleotide or a narrowly restricted set of nucleotides, but rather in certain regions of the integrated genome (Fig. 3). Moreover, in terms of the site of initiation of de novo methylation, there is considerable variability from tumor to tumor. For some of the tumors, the extent of de novo methylation in the integrated Ad12 genomes correlates with tumor size, in other tumors it does not. The extent of de novo methylation might be determined by both the insertion site and the time after Ad12 DNA integration, which might be related to the time of malignant transformation of the target cell. From the site(s) of initiation of the de novo methylation, this DNA modification spreads gradually and progressively, but not uniformly, with tumor growth across major parts of the Ad12 genome. Certain parts of the Ad12 genome, particularly those that are actively transcribed in the tumor cell, such as the E1A and E1B regions, remain unmethylated or become only hypomethylated (Sutter and Doerfler 1980; Hohlweg et al. 2003).

It is unknown which factors in the mammalian genome or which signals in the integrated foreign viral DNA are decisive to direct and regulate the initiation and the extent of de novo methylation. Additional information about this complex problem came from studies with foreign nonviral DNA that had been integrated into the mouse genome by homologous or heterologous recombination. A set of experiments was devised in which part of the murine B lymphocyte tyrosine kinase (BLK) gene on chromosome 14 was reinserted by homologous recombination into one of its authentic allelic sites on the genome (Hertz et al. 1999). In this case, the previously unmethylated BLK gene, which had been cloned and propagated in a methylation-deficient bacterial host, became remethylated in exactly the same pattern which preexisted on the unmanipulated mouse allele. However, when the BLK gene was integrated by heterologous recombination at random sites somewhere in the mouse

experiments (Fig. 2a). **b** Functional map and nucleotide numbers (*top*) of the Ad12 genome with details of E1A, E1B, E2A, E2B, E3, and E4 regions. The MLP, the Ad12 mitigator (*Mi*), and the gene for the virus associated (*VA*) RNA are also designated genome, different patterns of de novo methylation were observed. Thus, position and neighborhood effects must be of considerable importance in determining the patterns of de novo methylation. Moreover, the nucleotide sequences of the transgenic foreign DNA also affected de novo methylation decisively. When foreign genes, like the luciferase gene under the control of the weak late E2A promoter of Ad2 or of the strong SV40 promoter, were attached to the BLK gene, hyper- or hypomethylation, respectively, was observed. These results indicate that sites of integration as well as the inserted nucleotide sequence, and possibly promoter strength, are factors which affect de novo methylation. We surmise that the chromatin-like structure across the integrated foreign DNA might have a decisive influence on the initiation and spreading of de novo methylation (Toth et al. 1989, 1990). Differences in the accessibility of individual 5′-CG-3′dinucleotides to the DNA methyltransferase system may be an important parameter. It is also conceivable that the chromatin structure in each genome segment provides an—as yet hypothetical—memory mechanism to preserve and reestablish site-specific methylation patterns.

7.2 Genome-Wide Perturbations in the Mammalian Genome upon Foreign DNA Insertion

The integration of viral DNA into mammalian genomes might play a decisive role in the generation of the malignant phenotype during viral oncogenesis. We therefore started to investigate the structural and functional consequences of the insertion of viral (foreign) DNA into established mammalian genomes. Alterations in the patterns of DNA methylation in the host genomes at the site of insertion and remote from it were of particular interest. By using different methods including the bisulfite protocol of the genomic sequencing technique (Frommer et al. 1992; Clark et al. 1994), we documented extensive changes in the patterns of DNA methylation at several cellular sites, some of which lay even remote from the loci of insertion of the DNA of Ad12. Lesser changes were found in cells transgenic for the DNA of bacteriophage λ (Heller et al. 1995; Remus et al. 1999). Because λ DNA was not detectably transcribed in transgenic mammalian cells, alterations of methylation patterns subsequent to foreign DNA insertion did not seem to depend on foreign gene transcription. Furthermore, it was shown that cellular

DNA sequences immediately abutting the integrated Ad12 DNA in hamster tumor cells also exhibited changes in DNA methylation (Lichtenberg et al. 1988). It is presently unknown by what mechanism the insertion of foreign DNA affects the organization and function of the recipient genome. Does the site of foreign gene integration determine where the remote effects occur, and is a critical size of integrated foreign DNA required? We surmise that the integration of foreign DNA with lengths close to one megabase can lead to perturbations in the chromatin structure at or remote from the site of insertion.

Potential structural and functional consequences of foreign DNA insertion were recently investigated by using the method of methylation-sensitive representational difference analysis (MS-RDA; Ushijima et al. 1997) and the novel method of methylation-sensitive amplicon subtraction (MS-AS) to detect cellular DNA segments with altered methylation patterns in transgenic cells (Müller and Doerfler 2000; Müller et al. 2001). These genome-wide scanning methods led to the isolation of several cellular genes and DNA segments from cells with integrated Ad12 or λ DNA which exhibited changes in DNA methylation or transcription patterns. In control experiments, similar differences in gene expression or DNA methylation patterns were not detectable among individual nontransgenic BHK21 cell clones. These data from extensive control experiments argued against the possibility that differences in methylation and transcription patterns had preexisted in the cells of the nontransgenic BHK21 line (Remus et al. 1999; Müller et al. 2001). Two mouse lines transgenic for an adenovirus promoter-indicator gene construct showed hypomethylation in the interleukin 10 locus. In another mouse line transgenic for the DNA of bacteriophage λ, hypermethylation was observed in the imprinted Igf2r gene in DNA from heart muscle. As the integrated λ DNA was not detectably transcribed in the transgenic cells, the possibility that products of the integrated foreign DNA might be involved in eliciting the observed alterations was unlikely. Furthermore, revertants of Ad12-transformed cell lines devoid of Ad12 genomes still exhibited the marked alterations in methylation patterns (Heller et al. 1995). Thus, factors other than viral gene products or, more specifically, other than transgene transcripts seemed to be decisive in causing alterations in cellular DNA methylation and transcription patterns.

We conclude that the insertion of foreign DNA into an established mammalian genome can lead to alterations in cellular DNA methylation and transcription patterns. It is conceivable that the locations of the genes and DNA segments affected by these alterations depend on the sites of foreign DNA insertion and on the loci adjacent to these sites. Since chromosomes exhibit unique topological relationships among each other in the interphase nucleus (Hubert and Bourgeois 1986; Haaf and Schmid 1991; Heun et al. 2001), structural perturbations caused by the insertion of foreign DNA at one site might be transmitted to and affect the structure and function of loci on adjacent chromosomes, even over considerable genetic distances. Probably, only a limited number of such alterations in cellular chromatin structure will be compatible with cell survival and normal growth regulation. This consideration is consistent with the finding that the integrated copies of Ad12 (foreign) DNA are generally concentrated at one chromosomal site, which is specific for each tumor. It will be interesting to investigate whether the insertion of foreign (Ad12) DNA per se can contribute to the mechanism of oncogenesis in virus-induced and/or "naturally occurring malignancies."

Evidence is accumulating from gene therapy trials in humans that the insertion of foreign (viral) DNA into the human genome can have catastrophic consequences and lead to human malignancies. In attempts to treat male infants suffering from X-linked severe combined immunodeficiency (X-SCID), the use of a retroviral vector system has led to leukemia-like disease in 2/10 treated boys (Marshall 2003). The notion that the activation of cellular oncogenes may be responsible for this iatrogenic oncogenesis conforms to currently popular ideas about the mechanism of oncogenesis. It is, however, more likely that the perturbations, which are caused in an established genome by foreign DNA insertion, play a decisive role in oncogenesis (Doerfler 1995, 2000).

Fig. 4 DNA array analyses of Ad12 gene transcription in the Ad12-induced tumors T1 to T4, in the Ad12-transformed hamster cell line T637, and in BHK21 mock-infected cells. The Ad12 DNA segments (map locations in Fig. 3b) fixed on each of the six arrays are indicated in the BHK21 mock panel. The Ad12 segments were PCR synthesized. Total RNAs from cells or tumors as indicated were reverse-transcribed in the presence of ^{32}P-dCTP

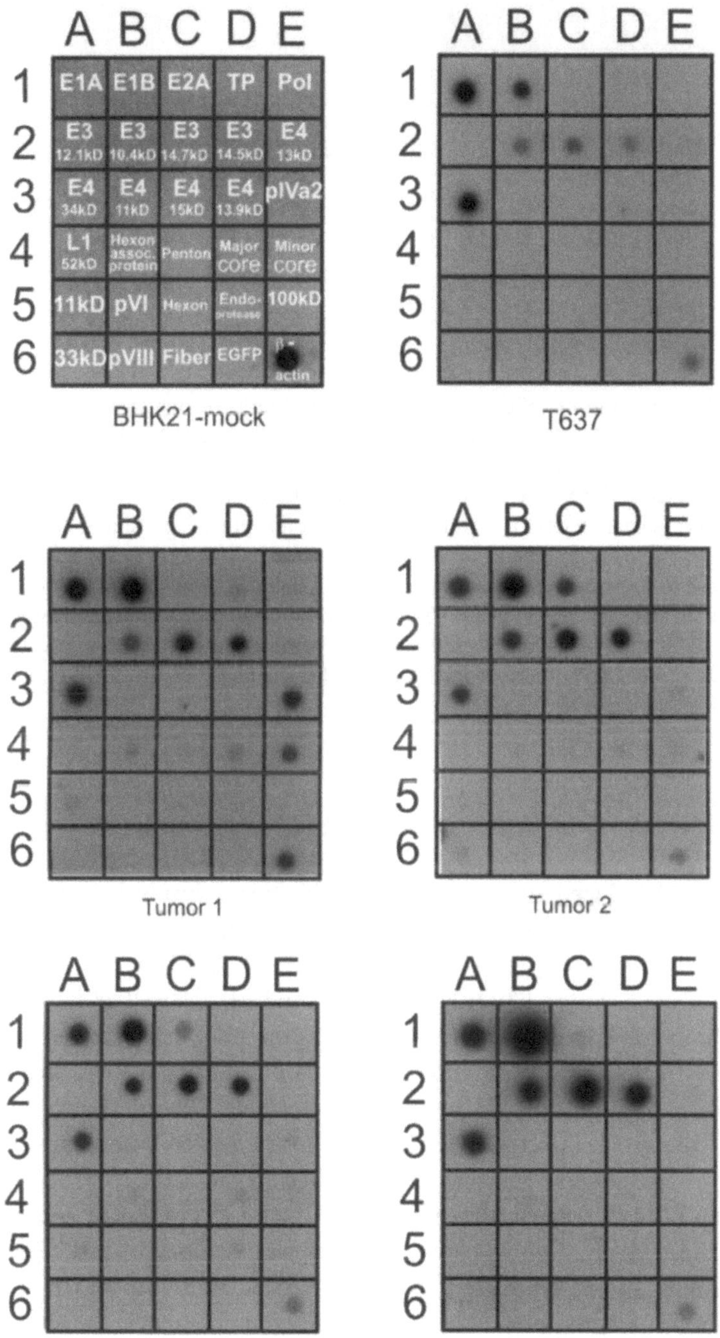
A B C D E
1 2 3 4 5 6
E1A
E1B
E2A
TP
Pol
E3 12.1kD
E3 10.4kD
E3 14.7kD
E3 14.5kD
E4 13kD
E4 34kD
E4 11kD
E4 15kD
E4 13.9kD
pIVa2
L1 52kD
Hexon assoc. protein
Penton
Major core
Minor core
11kD
pVI
Hexon
Endo-
100kD
33kD
pVIII
Fiber
EGFP
actin
BHK21-mock
T637
Tumor 1
Tumor 2

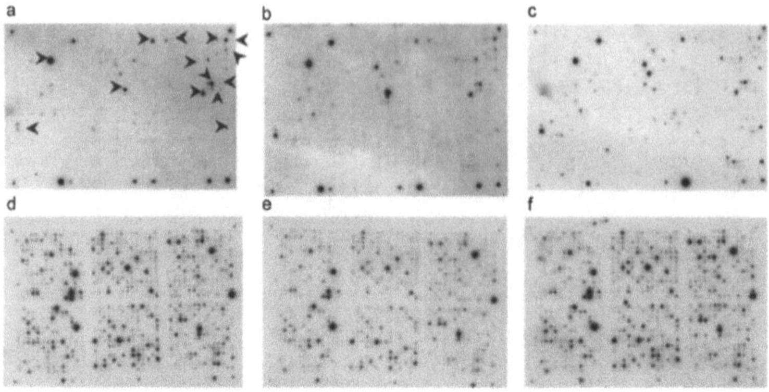

Fig. 5a–f Transcription of cellular genes in the Ad12-induced hamster tumors T1 (**a**), T2 (**b**), and T4-T7 (**c–f**). ^{32}P-labeled cDNAs from the six Ad12-induced tumors were synthesized and hybridized to the DNA on the arrays (Clontech). *Arrrowheads* in **a** indicate genes which were similarly transcribed in the three tumors T1, T2, and T4. Unmarked signals referred to transcripts unique for one of the tumors T1, T2, T4. The arrays in **a–c** correspond to the mouse gene arrays (section 1, Table 2), the arrays in **d–f** to the mouse cancer arrays (section 2, Table 2). The *bottom rows* of all panels carried internal control genes (*from left to right*) ubiquitin, phospholipase 2, HPRT, GAPDH, myosin 1 alpha, ornithine decarboxylase, beta actin, 45 kDa calcium-binding protein, and 40S ribosomal protein S29. The *upper left* and *right corners* on each panel show position markers

7.3 Viral and Cellular Gene Expression in Ad12-Induced Hamster Tumors

For the Ad12-induced tumors T1 to T4, the transcription patterns of Ad12 genes were analyzed by the DNA microarray technique (Fig. 4) and turned out to be very similar. Of the 28 Ad12-specific genes screened, E1A, E1B, the E3-ORFs 10.4, 14.5, and 14.7 as well as E4-ORF 34 were transcribed in all four tumors and in the Ad12-transformed cell line T637. In the tumors T1 and T3, signals of additional Ad12 genes (E2A, IVa2, and core protein V) were also detectable. Reverse transcripts of RNA from mock-infected BHK21 hamster cells lacked homologies to Ad12 genes. All reverse transcripts hybridized to the actin control, but not to the EGFP (green fluorescent protein) genes as a negative control (Fig. 5). After X-ray film exposure, the filter was stripped of the ^{32}P-labeled probes and rehybridized to ^{32}P-labeled Ad12 DNA, which hybridized to all positions carrying Ad12 DNA segments. Hence, the empty squares in Fig. 4 are not due to a lack of specific DNA fragments. The results of the DNA microarray experiments were confirmed by RT-PCR controls with positive transcription signals for the E1A 13S, E1B, E3 ORFs 10.4 and 14.7, and E4 ORF 34 genes, and by RNA (Northern) transfer experiments (data not shown).

In addition, for the analyses of cellular gene transcription patterns in three of the four Ad12-induced tumors and in the Ad12-transformed cell line T637, we initially employed DNA microarrays with 1,176 mouse genes, because hamster gene arrays were not available. Nucleotide sequence comparisons for some of the hamster genes expressed in the tumors revealed homologies between 85% and 96% to the corresponding mouse genes (Table 2). These data hence justified the use of mouse gene arrays for the analysis of cellular transcription patterns in hamster tumors. The hybridization results of the cDNAs from tumors T1, T2, and T4 (Fig. 5a–c) revealed similarities in the transcription of at least 25 genes (arrowheads in Fig. 5) and differences in the transcription patterns of numerous other cellular genes (unmarked signals in Fig. 5a). The patterns between the hamster cell line BHK21 and the Ad12-transformed BHK21 cell line T637 showed differences and similarities (data not shown). The T637 cell line, which was transformed by Ad12 in culture, differed markedly in its cellular transcription patterns from the Ad12-induced tumors in animals. Of course, a cell line cultured for many years would be expected to exhibit altered transcription patterns.

We have extended these analyses to 804 additional genes and used cDNA preparations from three further tumors: T5, T6, and T7. The 804 additional genes were cancer-specific genes which were tested for their possible involvement in Ad12 tumorigenesis. The hybridization results of the cDNAs from tumors T5 to T7 to the mouse gene arrays used (Table 2) were quite similar to those from tumors T1, T2, and T4. The annealing data of the cDNAs from tumors T5 to T7 to the arrays containing the cancer genes demonstrated that many of the cancer-specific genes were expressed in the Ad12-induced tumors (Fig. 5d–f).

8
Outlook

Much work has been devoted to studies on the mechanism of adenovirus transformation of cells in culture, and to the contributions of the viral E1 and E4 gene products (for reviews, see Doerfler 1983, 1984; Doerfler and Böhm 1995, 2003; see the chapter by J.F. Williams et al., this vol-

Table 2 Cellular genes transcribed in six different Ad12-induced tumors[a]

Gene name	Class	Homology to mouse genes
Section 1: mouse gene arrays		
SRY-box containing gene 6	Transcription activator and repressor	
High-mobility group protein 2	Transcription activator and repressor	
Heterogeneous nuclear ribonucleoprotein A1	RNA processing	
Heterogeneous nuclear ribonucleoprotein K	RNA processing	
Eukaryotic translation initiation factor 4A1	Translation factor	
Peptidylprolyl isomerase A	Posttranslational modification protein	
Ribosomal protein S5	Ribosomal protein	
Ribosomal protein L6	Housekeeping gene	93%
GAPDH	Housekeeping gene	
Ornithine decarboxylase[b]	Housekeeping gene	
Cytoplasmic beta-actin[b]	Housekeeping gene	
ATPase inhibitor	Energy metabolism	
Collagen I alpha 1 subunit (COL1A1)	Extracellular matrix protein	90%
Heatshock 84-kDa protein; tumor-specific transplantation 84-kDa antigen	Chaperones and heatshock proteins	
Heatshock protein cognate 70	Chaperones and heatshock proteins	
Tumor necrosis factor-induced protein 1		87%
Ewing sarcoma homolog	Oncogene and tumor suppressor	
Sperm-specific antigen 1		
GATA-binding protein	Transcription factor	
Transcription factor AP-2, gamma	Transcription factor	
Cyclin E	Kinase activator/inhibitor	
MAP kinase 7	Cell cycle-regulating kinase	
CDK inhibitor 1B (p27)	Oncogene	

Table 2 (continued)

Gene name	Class	Homology to mouse genes
Integrin beta 7	Cell adhesion protein	
Integrin alpha M (Cd 11b)	Cell–cell adhesion receptor	
Eph receptor B2	Cell–cell adhesion receptor	
Eph receptor B4	Cell–cell adhesion receptor	
CD 14 antigen	Cell–cell adhesion receptor	
Interferon gamma-inducing factor binding protein	Extracellular transport/carrier protein	85%
Abl proto-oncogene	Oncogene	
MAD homolog 1 (*Drosophila*)	Oncogene	
Peroxisome proliferator-activated receptor alpha	Transcription activator and repressor	92%
Gap junction membrane channel protein alpha	Cell–cell adhesion receptor	
Procollagen, type I, alpha 1	Extracellular matrix protein	
PKC substrate 80 K-H	Complex carbohydrate metabolism/ER	
Chaperonin subunit 8	Chaperone/ Heat shock protein	
Baculoviral IAP repeat-containing 2	Apoptosis-associated protein	
BH3 interacting (with Bcl family) domain, apoptosis agonist	Bcl2 family protein	
Granzyme A	Protein phosphatase	
CD28 antigen	Growth factor and chemokine receptor	
Interleukin 9 receptor		91%
Angiogenin-related protein	Growth factors, cytokines, chemokines	
Fibroblast growth factor 7	Growth factors, cytokines, chemokines	
GRO1 oncogene	Growth factors, cytokines, chemokines	
Vascular endothelial growth factor B	Growth factors, cytokines, chemokines	93%
Wingless-related MMTV integration site 2b	Oncogene	
Wingless-related MMTV integration site 8b	Oncogene	

Table 2 (continued)

Gene name	Class	Homology to mouse genes
Wingless-related MMTV integration site 5b	Oncogene	
Adenylate cyclase activating polypeptide 1	Extracellular communication protein	
Crystallin, zeta	Xenobiotic metabolism	93%
Integrin beta 1 binding protein 1	Cell–cell adhesion receptor	
MAPKKK 3	Kinase activator/inhibitor	
Proteinphosphatase 3, catalytic subunit, alpha isoform	Intracellular protein phosphatase	
Proteinphosphatase 3, catalytic ubunit, beta isoform	Intracellular protein phosphatase	
Proteinphosphatase 5, catalytic subunit	Intracellular protein phosphatase	
Protein tyrosine phosphatase, non-receptor type 8	Intracellular protein phosphatase	86%
Inositol 1,4,5-triphosphate receptor 5	Phospholipase and phosphoinositol kinases	
Protein tyrosine phosphatase, receptor-type, F interacting protein, binding protein 2	Other intracellular transducers/effectors/ modulators	
Alpha 1 microglobulin/bikunin	Inhibitor of proteases	85%
Alpha 2 macroglobulin	Inhibitor of proteases	
Plasminogen activator inhibitor, type I	Inhibitor of proteases	
Procollagen, type IX, alpha 2	Cytoskeleton/motility proteins	
Xeroderma pigmentosum, complementation group C	DNA damage signalling/repair proteins and DNA ligases	
Excision repair 3	DNA damage signalling/repair proteins and DNA ligases	

[a] Experimental details are described in the text and the legend to Fig. 5. Gene locations on the mouse arrays were given by Clontech. Two different arrays were used, each carrying 1,176 different mouse genes. The arrays in section 2 contain 804 cancer genes and 372 genes which were also part of the section 1 arrays. The section 1 arrays were hybridized to cDNAs from tumors T1, T2, T4; the section 2 arrays to cDNAs from tumors T5 to T7 only.

[b] Genes are not marked with arrowheads on the mouse array.

ume). One of the essential preconditions for the oncogenic potential of Ad12 in Syrian hamsters is the perfect nonpermissivity of hamster cells for infection with Ad12 and the ensuing abortive infection (Doerfler 1969; Hösel et al. 2001a,b, 2003). Ad12 cannot replicate in abortively infected hamster cells which are, therefore, able to survive and to undergo specific alterations which in some of the infected cells will eventually lead to oncogenesis. Viral oncogenesis is a highly complex biological event to which many different factors contribute. The integration of Ad12 DNA into the host genome is very intimately linked to the oncogenic transformation by Ad12. Since each tumor is characterized by its individual Ad12-integration site, insertional mutagenesis in the conventional sense pertaining to the alteration of cellular nucleotide sequences at the insertion site seems unlikely to be able to account for the generation of the oncogenic phenotype. Furthermore, the presence of the viral genome and its products is not necessary for the maintenance of the transformed phenotype. We surmise that the acquisition of many kilobases or even a megabase of inserted foreign DNA alters the regional structure and function of the genome at the site of foreign DNA insertion. These perturbations might also be transmitted to neighboring DNA sequences, to the chromatin structure, and possibly to adjacent chromosomes due to the close neighborhood among chromosomes in interphase nuclei. Although the mechanisms which connect genome perturbation due to foreign DNA insertions and alterations in gene transcription remain unknown, it will be interesting to pursue the possibility that these perturbations could be closely linked to viral oncogenic transformation and perhaps also to the generation of "spontaneously" occurring tumors.

The role of the integration of foreign DNA in changing the overall structure of a mammalian genome and thus contributing to the process of oncogenic transformation has been of particular interest in our laboratory (Doerfler 1968, 1970, 1995, 2000). Alterations in cellular DNA methylation and transcription patterns upon the integration of foreign (Ad12 bacteriophage λ) DNA have been documented previously (Heller et al. 1995; Remus et al. 1999; Müller et al. 2001) and again in this report. Could these changes have preexisted in all or some of the target cells for Ad12 oncogenesis? Although this possibility cannot be investigated prior to the identification of the target cells for Ad12 tumorigenesis, previous studies on BHK21 cells transgenic for Ad12 or bacteriophage λ DNA have not supported this notion (Heller et al. 1995; Remus et al. 1999;

Müller et al. 2001) but have not ruled it out completely. It will be of interest to determine whether and how the sites of viral DNA integration are topologically related to specific structures in the nuclear matrix of the cell.

The experimental findings on Ad12 insertion appear to be very similar or even identical for many types of foreign DNA which become integrated into mammalian genomes. Of course, viruses have evolved highly specialized mechanisms to transport their genomes into the nuclei of host cells. However, the main portal of entry of foreign DNA into mammalian organisms is the gastrointestinal tract, which is permanently exposed to foreign DNA by food ingestion. In the late 1980s, we started to investigate the fate of food-ingested foreign DNA in the mammalian organism, and have used the mouse as a model (Schubbert et al. 1994, 1997, 1998; Hohlweg and Doerfler 2001). Our results indicate that foreign DNA ingested with the daily food supply can survive the passage through the gastrointestinal tract of mice in fragmented form and at the level of a few percent of the DNA ingested. This DNA can reach peripheral white blood cells and spleen and liver cells via the intestinal epithelia. In rare instances, the food-ingested foreign DNA becomes integrated into the DNA of the mouse genome (Schubbert et al. 1997). These results on the fate of food-ingested DNA raise the question of what long-term biological effects food-ingested DNA might have in mammalian organisms.

The observation that the insertion of foreign DNA into established mammalian genomes can have structural and functional consequences for the recipient genome will be of interest to many fields of biomedical research: gene transfer studies, transgenic animals, and human somatic gene therapy. The findings reviewed here suggest that in transgenic animals, the foreign DNA could influence the activity of many more genes than merely that of the ones directly affected by the "knock-in" or the "knock-out" procedure. Thus, the interpretations of the results derived from these experiments must be viewed with caution. In addition, the stable insertion of foreign DNA into the human genome for purposes of human somatic gene therapy could have far-reaching consequences, with undesirable side effects which must be carefully weighed in all such therapeutic regimens. Recent reports on cases of leukemia apparently caused by the insertion of retroviral genomes (Marshall 2003) which have been used as gene transfer vectors add significance to the notion that structural genome alterations may also play a role in human oncogenesis.

Acknowledgements. This research was supported by the Deutsche Forschungsgemeinschaft through SFB 274, by the Bundesministerium für Bildung und Wissenschaft through a joint grant with Coley Pharmaceuticals Inc., by a grant from the Bayerisches Staatsministerium für Landesentwicklung und Umweltfragen, München, and by the Ministerium für Wissenschaft und Forschung, Nordrhein-Westfalen, Düsseldorf.

References

Albert DM, Rabson AS, Dalton AJ (1968) In vitro neoplastic transformation of uveal and retinal tissue by oncogenic DNA viruses. Invest Ophthalmol 7:357–365

Chino F, Tsuruhara T, Egashira Y (1967) Pathological studies on the oncogenesis of adenovirus type 12 in hamsters. J Med Sci Biol 20:483–500

Clark SJ, Harrison J, Paul CL, Frommer M (1994) High sensitivity mapping of methylated cytosines. Nucl Acids Res 22:2990–2997

Deuring R, Winterhoff U, Tamanoi F, Stabel S, Doerfler W (1981) Site of linkage between adenovirus type 12 and cell DNAs in hamster tumour line CLAC3. Nature 293:81–84

Doerfler W (1968) The fate of the DNA of adenovirus type 12 in baby hamster kidney cells. Proc Natl Acad Sci USA 60:636–643

Doerfler W (1969) Nonproductive infection of baby hamster kidney cells (BHK21) with adenovirus type 12. Virology 38:587–606

Doerfler W (1970) Integration of the deoxyribonucleic acid of adenovirus type 12 into the deoxyribonucleic acid of baby hamster kidney cells. J Virol 6:652–666

Doerfler W (1983) DNA methylation and gene activity. Ann Rev Biochem 52:93–124

Doerfler W, Gahlmann R, Stabel S, Deuring R, Lichtenberg U, Schulz M, Eick D, Leisten R (1983) On the mechanism of recombination between adenoviral and cellular DNAs: the structure of junction sites. Current Topics Microbiol Immunol 109:193–228

Doerfler W (1983, 1984) The molecular biology of adenoviruses. Current Topics in Microbiology and Immunology, vol. 109–111, Springer Verlag, Berlin, Heidelberg, New York, Tokyo.

Doerfler W (1991) Patterns of DNA methylation—evolutionary vestiges of foreign DNA inactivation as a host defense mechanism. A proposal. Biol Chem Hoppe-Seyler 372:557–564

Doerfler W (1995) The insertion of foreign DNA into mammalian genomes and its consequences: a concept in oncogenesis. Adv Cancer Res 66:313–344

Doerfler W, Böhm P, eds (1995) The molecular repertoire of adenoviruses. Current Topics in Microbiology and Immunology, vol 199/I-III, Springer Verlag, Berlin, Heidelberg, New York, Tokyo

Doerfler W (1996) A new concept in (adenoviral) oncogenesis: integration of foreign DNA and its consequences. BBA Reviews in Cancer 1288: F79-F99

Doerfler W (2000) Foreign DNA in mammalian systems. Wiley-VCH, Weinheim, New York, Chichester, Brisbane, Singapore, Toronto

Doerfler W, Böhm P, eds (2003) Adenoviruses: model and vectors in virus host interactions. Current Topics in Microbiology and Immunology, vols 272, 273, Springer Verlag, Heidelberg, Berlin, New York, Tokyo

Flint J, Shenk T (1989) Adenovirus E1A protein paradigm viral transactivator, Annu Rev Genet 23:141–161

Frommer M, Mc Donald LE, Millar DS, Collis CM, Watt F, Grigg GW, Molloy PL (1992) A genomic sequencing protocol that yields a positive display of 5-methylcytosine residues in individual DNA strands. Proc Natl Acad Sci USA 89:1827–1831

Gahlmann R, Leisten R, Vardimon L, Doerfler W (1982) Patch homologies and the integration of adenovirus DNA in mammalian cells. EMBO J. 1:1101–1104

Gahlmann R, Doerfler W (1983) Integration of viral DNA into the genome of the adenovirus type 2-transformed hamster cell line HE5 without loss or alteration of cellular nucleotides. Nucleic Acids Res 11:7347–7361

Gahlmann R, Schulz M, Doerfler W (1984) Low molecular weight RNAs with homologies to cellular DNA at sites of adenovirus DNA insertion in hamster or mouse cells. EMBO J 3:3263–3269

Graham FL (1984) Transformation by oncogenicity of human adenoviruses. In: H.S. Ginsberg (ed.) The adenoviruses. Plenum Press New York, 339–398

Green M (1970) Oncogenic viruses. Annu Rev Biochem 39:701–756

Green M, Mackey JK, Wold WSM, Ridgen P (1979a) Analysis of human cancer DNA for DNA sequences of human adenovirus serotypes 3, 7, 11, 14, 16 and 21 in group B. Cancer Res 39:3479–3484

Green M, Wold WSM, Mackey JK, Ridgen P (1979b) Analysis of human tonsil and cancer DNAs and RNAs for DNA sequences of group C (serotypes 1, 2, 5, and 6) human adenoviruses. Proc Natl Acad Sci 12:6606–6610

Groneberg J, Chardonnet Y, Doerfler W (1977) Integrated viral sequences in adenovirus type 12-transformed hamster cells. Cell 10:101–111

Groneberg J, Sutter D, Soboll H, Doerfler W (1978) Morphological revertants of adenovirus type 12-transformed hamster cells. J Gen Virol 40:635–645

Groneberg J, Doerfler W (1979) Revertants of adenovirus type 12-transformed hamster cells have lost part of the viral genomes. Int J Cancer 24:67–74

Günthert U, Schweiger M, Stupp M, Doerfler W (1976) DNA methylation in adenovirus, adenovirus-transformed cells, and host cells. Proc Natl Acad Sci USA 73:3923–3927

Haaf T, Schmid M (1991) Chromosome topology in mammalian interphase nuclei. Exp Cell Res 192:325–332

Hamaya K (1966) Pathological studies on carcinogenesis of hamster tumors induced by adenovirus type 12; on the intracranial virus inoculation of newborn hamsters. I. General description. J Karyopathol 11:23–34

Heller H, Kämmer C, Wilgenbus P, Doerfler W (1995) Chromosomal insertion of foreign (adenovirus type 12, plasmid, or bacteriophage λ) DNA is associated with enhanced methylation of cellular DNA segments. Proc Natl Acad Sci USA 92:5515–5519

Hertz J, Schell G, Doerfler W (1999) Factors affecting de novo methylation of foreign DNA in mouse embryonic stem cells. J Biol Chem 274:24232–24240

Heun P, Taddel A, Gasser SM (2001) From snapshots to moving pictures: new perspectives on nuclear organization. Trends Cell Biol 11:519–525

Hilger-Eversheim K, Doerfler W (1997) Clonal origin of adenovirus type 12-induced hamster tumors: nonspecific chromosomal integration sites of viral DNA. Cancer Research 57:3001–3009

Hösel M, Schröer J, Webb D, Jaroschevskaja E, Doerfler W (2001a) Cellular and early viral factors in the interaction of adenovirus type 12 with hamster cells: the abortive response. Virus Res 81:1-16

Hösel M, Webb D, Schröer J, Schmitz B, Doerfler W (2001b) The overexpression of the adenovirus type 12 pTP or E1A gene facilitates Ad12 DNA replication in nonpermissive BHK21 hamster cells. J Virol 75:10041–10053

Hösel M, Webb D, Schröer J, Doerfler W (2003) Abortive infection of hamster cells by adenovirus type 12. Current Topics Microbiol Immunol 272:415–440

Hohlweg U, Doerfler W (2001) On the fate of plant or other foreign genes upon the uptake in food or after intramuscular injection. Mol Gen Genomics 265:225–233

Hohlweg U, Hösel M, Dorn A, Webb D, Schramme A, Corzilius L, Hilger-Eversheim K, Remus R, Schmitz B, Buettner R, Doerfler W (2003) Intraperitoneal dissemination of Ad12-induced primitive neuroectodermal hamster tumors: De novo methylation and transcription patterns of integrated viral and of cellular genes. Submitted

Hubert J, Bourgeois CA (1986) The nuclear skeleton and the spatial arrangement of chromosomes in the interphase nucleus of vertebrate somatic cells. Hum Gent 74:1-15

Huebner RJ, Rowe WP, Lane WT (1962) Oncogenic effects in hamsters of human adenovirus types 12 and 18. Proc Natl Acad Sci USA 48:2051–2058

Kämmer C, Doerfler W (1995) Genomic sequencing reveals absence of DNA methylation in the major late promoter of adenovirus type 2 DNA in the virion and in productively infected cells. FEBS Letters 362:301–305

Knoblauch M, Schröer J, Schmitz B, Doerfler W (1996) The structure of adenovirus type 12 DNA integration sites in the hamster cell genome. J Virol 70:3788–3796

Kuhlmann I, Achten S, Rudolph R, Doerfler W (1982) Tumor induction by human adenovirus type 12 in hamsters: loss of the viral genome from adenovirus type 12-induced tumor cells is compatible with tumor formation. EMBO J 1:79–86

Langner K-D, Vardimon L, Renz D, Doerfler W (1984) DNA methylation of three 5' C-C-G-G 3' sites in the promoter and 5' region inactivates the E2a gene of adenovirus type 2. Proc Natl Acad Sci USA 81:2950–2954

Lichtenberg U, Zock C, Doerfler W (1988) Integration of foreign DNA into mammalian genome can be associated with hypomethylation at site of insertion. Virus Res 11:335–342

Marshall E (2003) Gene therapy: Second child in French trial is found to have leukemia. Science 299:320

Müller K, Doerfler W (2000) Methylation-sensitive amplicon subtraction: a novel method to isolate differentially methylated DNA sequences in complex genomes. Gene Funct Dis 1:154–160

Müller K, Heller H, Doerfler W (2001) Foreign DNA integration: genome-wide perturbations of methylation and transcription in the recipient genomes. J Biol Chem 276:14271–14278

Mukai N, Ishida Y (1971) Tumors derived from the nervous system. In: Pathology of Tumors, Edited by H.Sugano, H.Kobayashi, Tokyo, Asakura Co

Mukai N, Kobayashi S (1972) Undifferentiated intraperitoneal tumors induced by human adenovirus type 12 in hamsters. Am J Pathol 69:331–339

Munnes M, Doerfler W (1997) DNA methylation in mammalian genomes: promoter activity and genetic imprinting. In: Dulbecco, R. (Ed.) Encyclopedia of Human Biology, vol. 3, 435–446, Academic Press, San Diego, New York, Boston, London, Sydney, Tokyo, Toronto

Nakajima T, Mukai N (1979) Cell origin of human adenovirus type 12-induced subcutaneous tumors in Syrian hamsters. Acta Neuropathol. 15:187–194

Neiders ME, Weiss L, Yohn DS (1968) A morphologic comparison of tumors produced by type 12 adenovirus and by the HA-12-1T line of adeno-12 tumor cells. Cancer Res 28:577–584

Nevels M, Täuber B, Spruss T, Wolf H, Dobner T (2001) "Hit-and-run" transformation by adenovirus oncogenes. J Virol 75:3089–3094

Nevins JR (1995) Adenovirus E1A: transcription regulation and alteration of cell growth control. Current Topics Microbiol Immunol 199/III: 25–32

Ogawa K, Tsutsumi A, Iwata K, Fujii Y, Ohmori M, Taguchi K, Yabe Y (1966) Histogenesis of malignant neoplasm induced by adenovirus type 12. Gann 57:43–52

Orend G, Knoblauch M, Kämmer C, Tjia ST, Schmitz B, Linkwitz A, Meyer zu Altenschildesche G, Maas J, Doerfler W (1995) The initiation of de novo methylation of foreign DNA integrated into a mammalian genome is not exclusively targeted by nucleotide sequence. J.Virol 69:1226–1242

Pfeffer A, Schubbert R, Orend G, Hilger-Eversheim K, Doerfler W (1999) Integrated viral genomes can be lost from adenovirus type 12-induced hamster tumor cells in a clone-specific, multistep process with retention of the oncogenic phenotype. Virus Res 59:113–127

Rabson AS, Kirschstein RL, Paul FJ (1964) Tumors produced by adenovirus 12 in mastomys and mice. J Natl Cancer Inst 32:77–87

Remus R, Kämmer C, Heller H, Schmitz B, Schell G, Doerfler W (1999) Insertion of foreign DNA into an established mammalian genome can alter the methylation of cellular DNA sequences. J Virol 73:1010–1022

Russell WC (2000) Update on adenovirus and its vectors. J Gen Virol 81:2573–2604

Schubbert R, Lettmann C, Doerfler W (1994) Ingested foreign (phage M13) DNA survives transiently in the gastrointestinal tract and enters the bloodstream of mice. Mol Gen Genetics 242:495–504

Schubbert R, Renz D, Schmitz B, Doerfler W (1997) Foreign (M13) DNA ingested by mice reaches peripheral leukocytes, spleen and liver via the intestinal wall mucosa and can be covalently linked to mouse DNA. Proc Natl Acad Sci USA 94:961–966

Schubbert R, Hohlweg U, Doerfler W (1998) On the fate of food-ingested foreign DNA in mice: chromosomal association and placental transmission to the fetus. Mol Gen Genet 259:569–576

Schulz M, Freisem-Rabien U, Jessberger R, Doerfler W (1987) Transcriptional activities of mammalian genomes at sites of recombination with foreign DNA. J Virol 61:344–353

Shenk T (1996) Adenoviridae: the viruses and their replication, In: B.N. Fields, D.M. Knipe, P.M. Howley (ed.) Virology, 3rd ed. vol 2, Lippincott-Raven, New York, 2111–2148

Stabel S, Doerfler W (1982) Nucleotide sequence at the site of junction between adenovirus type 12 DNA and repetitive hamster cell DNA in transformed cell line CLAC1. Nucleic Acids Res 10:8007–8023

Sutter D, Westphal M, Doerfler W (1978) Patterns of integration of viral DNA sequences in the genomes of adenovirus type 12-transformed hamster cells. Cell 14:569–585

Sutter D, Doerfler W (1980) Methylation of integrated adenovirus type 12 DNA sequences in transformed cells is inversely correlated with viral gene expression. Proc Natl Acad Sci USA 77:253–256

Thomas DL, Schaak J, Vogel H, Javier R (2001) Several E4 region functions influence mammary tumorigenesis by adenovirus type 9. J Virol 75:557–568

Toth M, Lichtenberg U, Doerfler W (1989) Genomic sequencing reveals a 5-methylcytosine-free domain in active promoters and the spreading of preimposed methylation patterns. Proc Natl Acad Sci USA 86:3728–3732

Toth M, Müller U, Doerfler W (1990) Establishment of de novo DNA methylation patterns. Transcription factor binding and deoxycytidine methylation at CpG and non-CpG sequences in an integrated adenovirus promoter. J Mol Biol 214:673–683

Trentin JJ, Yabe Y, Taylor G (1962) The quest for human cancer viruses. Science 137:835–841

Ushijima T, Morimura K, Hosoya Y, Okonogi H, Tatematsu M, Sugimura T, Nagao M (1997) Establishment of methylation-sensitive-representational difference analysis and isolation of hypo- or hypermethylated genomic fragments in mouse liver tumors. Proc Natl Acad Sci USA 94:2284–2289

Vardimon L, Neumann R, Kuhlmann I, Sutter D, Doerfler W (1980) DNA methylation and viral gene expression in adenovirus-transformed and -infected cells. Nucleic Acids Res 8:2461–2473

Vardimon L, Kressmann A, Cedar H, Maechler M, Doerfler W (1982) Expression of a cloned adenovirus gene is inhibited by *in vitro* methylation. Proc Natl Acad Sci USA 79:1073–1077

Wronka G, Fechteler K, Schmitz B, Doerfler W. (2002) Recombination between adenovirus type 12 DNA and mammalian DNA: purification of a cell-free system and analyses of *in vitro* generated recombinants. Virus Res 90:225–242

Yabe Y, Ogawa K, Iwata K, Murakami S (1966) Effect of injection of adenovirus type 12 in adult hamsters. Acta Med Okayama 20:147–154

Yoder JA, Walsh CP, Bestor TH (1997) Cytosine methylation and the ecology of intragenomic parasites. Trend Genet 13:335–340

Yohn DS, Weiss L, Neiders ME (1968) A comparison of the distribution of tumors produced by intravenous injection of type 12 adenovirus and adeno-12 tumor cells. Cancer Res 28:571–576

Zantema A, van der Eb AJ (1995) Modulation of gene expression by adenovirus transformation. Current Topics Microbiol Immunol 199/III: 1–24

Zock C, Doerfler W (1990) A mitigator sequence in the downstream region of the major late promoter of adenovirus type 12 DNA. EMBO J 9:1615–1623

E1A-Based Determinants of Oncogenicity in Human Adenovirus Groups A and C

J. F. Williams[1] · Y. Zhang[1] · M. A. Williams[1] · S. Hou[2] · D. Kushner[3] · R. P. Ricciardi[2]

[1] Department of Biological Sciences, Carnegie Mellon University, 4400 Fifth Avenue, Pittsburgh, PA 15213, USA
E-mail: jfw1@andrew.cmu.edu

[2] Department of Microbiology, School of Dental Medicine, Department of Biochemistry and Molecular Biophysics, University of Pennsylvania, Philadelphia, PA 19104, USA

[3] Institute for Molecular Virology, University of Wisconsin, Madison, WI 53706, USA

Abstract A broad spectrum of genetic and molecular investigations carried out with group C, Ad2 and Ad5, and with group A, Ad12, have shown that early region1 (E1) gene products are sufficient for complete transformation of rodent cells in vitro by these viruses. During the past quarter century, the processes by which E1A proteins, in cooperation with E1B proteins, perturb the cell cycle and induce the transformed phenotype, have become well defined. Somewhat less understood is the basis for the differential oncogenicity of these two groups of viruses, and the processes by which the E1A proteins of Ad12 induce a tumorigenic phenotype in transformants resulting from infection of cells in vivo and in vitro. In this chapter we review previous findings and present new evidence which demonstrates that Ad12 E1A possesses two or more independent functions enabling it to induce tumors. One of these functions lies in its capacity to repress transcription of MHC class I genes, allowing the tumor cells to avoid lysis by cytotoxic T lymphocytes. We have shown that class I repression is mediated through increased binding of repressor COUP-TF and decreased binding of NF-kB to the class I enhancer. In addition to mediating immune escape, E1A also determines the susceptibility of transformants to Natural Killer (NK) cell lysis, and in this case, also, Ad12 transformants are not susceptible. By using Ad 12 mutants containing chimeric E1A Ad12-Ad5 genes, point mutations, or a specific deletion, we have shown that the unique spacer region of Ad12 E1A is an oncogenic determinant, but is not required for transformation in vitro. Given that the E1A regions responsible for class I repression are first exon encoded, we have examined a set of cell lines transformed by these altered viruses, and have found that while they display greatly reduced tumorigenicity, they maintain a wild-type capacity to repress class I transcription. Whether the spacer contributes to NK evasion remains unresolved. Lastly, we discuss the properties of the Ad2/Ad5 E1A C-terminal negative modulator of tumorigenicity, and examine the effects on transformation, tumor induction and transformant tumorigenicity, when the Ad5 negative modulator is placed by chimeric construction in Ad12 E1A.

1
Introduction: A Brief Overview of Human Adenovirus Oncogenicity

Some 40 years ago it was shown that human adenovirus (Ad) types 12 and 18 were capable of inducing rapidly growing, undifferentiated sarcomas in newborn Syrian hamsters (Huebner et al. 1962; Trentin et al. 1962). Subsequent to these initial findings it was shown that, using a number of different routes of inoculation, a variety of tumor types could be induced in hamsters and other rodents. Enigmatically, it was found that most of the other human Ad serotypes tested either failed to induce tumors in rodents, or did so very slowly and with low efficiencies. Based on these early findings, the group A viruses (Ad12, Ad18 and Ad31) were classified as highly oncogenic serotypes, and the group B viruses were categorized as weakly oncogenic. All of the remaining human Ad serotypes, comprising those of group C (Ad1, Ad2, and Ad5), groups D and E, and more recently the enteric viruses of group F have been collectively defined as essentially nononcogenic. A conspicuous and intriguing exception to this general rule is the group D virus Ad9, which was found to induce mammary fibroadenomas with high incidence in both neonatal and adult female Wistar/Firth rats (Jonsson and Ankerst 1977). It is of added interest that some of these slowly growing benign tumors progressed over a period of time into more malignant forms. In effect, these various tumor induction studies reveal that almost all of the 49 known human serotypes appear to be nononcogenic or at best weakly oncogenic, and only Ad12, Ad18, Ad31, and Ad9 are indeed highly oncogenic. However, it is not clear that each and every one of these serotypes has been as rigorously tested as have Ad5 and Ad12, and it is probable that some have not, in fact, been tested at all. Additional references to these earlier tumor induction studies can be found in previous reviews (Gallimore et al. 1984; Graham 1984; Mautner et al. 1995; Williams 1986).

Regardless of this clear-cut display of differential oncogenicity by the human adenoviruses, many representative members of all groups have been found to transform a variety of rodent cells in vitro (see Graham 1984 for some references), with efficiencies which are broadly similar, but dependent in some cases on cell type. Thus, the inability of an adenovirus to induce tumors in an animal host is most likely not due simply to its inability to infect and transform target cells per se. This may not be the case for the group F enteric viruses (or DNA from these) with which the transformation efficiency is extremely low, and/or the phenotype of transformants is apparently partial (see Mautner et al. 1995); similar exceptions may well emerge in the future. All these various cell transformants possess many similar properties in vitro, and are morphologically similar to Ad-induced tumor cells grown in culture. However, the cell transformants induced by different Ad groups differ conspicuously in tumorigenicity when tested in appropriate recipient animals, and generally follow the pattern of tumor induction exhibited by these viruses. For example, rat cells transformed by Ad12 in vitro are highly tumorigenic in both newborn and adult syngeneic rats, whereas similar Ad2 and Ad5 transformants generally are not (see some of our results below). However, it was reported that Ad2, Ad5, and Ad12 hamster cell transformants are tumorigenic in newborn hamster but not in weanlings, although only Ad12 induces tumors directly in these animals (Lewis and Cook 1982). In addition, it was also shown that Ad5 (temperature-sensitive late-function mutant) transformants will form tumors in newborn hamsters, that the resultant tumors can be repetitively transplanted even into adult hamsters, and that some of these transplanted tumors metastasize to lungs and liver (Williams 1973). Therefore, the tumorigenicity of Ad transformants may well depend on both the nature of the host used, and on the history of the transformed cell line used. Although Ad2 and Ad5 transformants are nontumorigenic in adult, immunocompetent, syngeneic rats, as mentioned above, it was shown that Ad2 transformants will form tumors when the recipient rats are immunosuppressed by treatment with antithymocytic serum (Gallimore 1972). Likewise, both Ad5 and Ad12 transformants are tumorigenic in immunodeficient nude mice (Bernards et al. 1983). These results suggested that thymus-dependent cytotoxic T lymphocytes (CTLs) of the recipient's immune system play a major role in rejecting cells transformed by nononcogenic group C viruses, but not those transformed by group A viruses. Support for this notion came from the discovery that Ad12, but not Ad5, mouse and rat transformants express greatly reduced levels of class I major histocompatibility (MHC) antigens on the cell surface (Eager et al. 1985; Schrier et al. 1983). However, this CTL-mediated immune attack may not be the only host response mobilized against the insurgent group C transformants. The host may also launch a nonadaptive, innate natural killer (NK) cell-mediated response, as it has been shown that Ad5 transformants are much more sensitive to NK cell killing than cells transformed by Ad12 (Cook et al. 1982; Raska and Gallimore; 1982). Whether or not either or both of these responses operate to prevent Ad2 and Ad5 virus-induced tumor growth in the host, or whether these responses must be suppressed to permit Ad12 tumor induction, is by no means certain. Moreover, these responses may not be the only potential factors which determine the capacity of target cells newly transformed by adenoviruses to form tumors in immunocompetent, syngeneic hosts; we have, in fact, found evidence that they most likely are not, as presented below.

2
Adenovirus Transforming Genes: E1A Is a Key Player in Viral Oncogenesis

A variety of genetic and molecular analyses have been conducted with group C, Ad2 and Ad5, and with group A, Ad12. Studies employing viral mutants, viral DNA fragments, and plasmids have shown that early region 1 (E1) genes and their products are necessary and evidently sufficient for complete transformation of rodent cells in culture. Most likely this finding will turn out to be applicable to the majority of human adenoviruses, although in the case of group D, Ad9, the E4 region is essential for complete transformation and for tumor induction (Javier 1994; Javier et al. 1992). The E1 region maps to the left end of the viral genome, and has a similar organization in all human serotypes examined. It consists of two separate transcription units, E1A (map units 1·3 to 4·5) and E1B (map units 4·6 to 11·2), each encoding multiple overlapping products. Although products encoded by each of these gene sequences are important for induction of tumors in rodent hosts and for transformation of rodent cells in vitro, we will focus largely upon E1A in this chapter, and will refer only peripherally to E1B (many references to topics outlined briefly above can be found in reviews by Boulanger and Blair 1991; Gallimore et al. 1984; Graham 1984; Ricciardi, 1999; Williams 1986; Williams et al. 1995). We will, in fact, concentrate upon evaluating the probable role(s) of the E1A gene products in tumor induction by group A and group C adenoviruses and the acquisition of tumorigenic capacity by cells transformed in vitro by these viruses.

2.1
Multiple Functions of E1A Operate Early During Infection of Human Cells

During viral infection, transcripts of the E1A gene are differentially spliced to generate five mRNA species with diverse size and common 5′ and 3′ termini. The two largest mRNAs (13 and 12S) are produced at early times during infection in permissive human cells and in nonper-

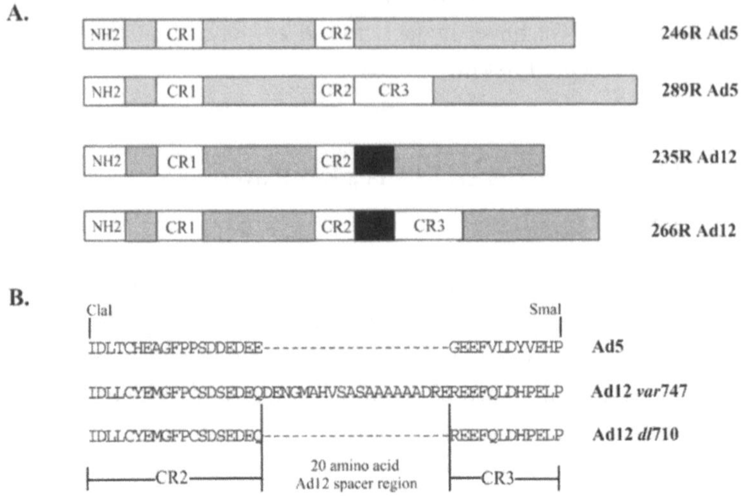

Fig. 1A, B The unique, nonconserved E1A spacer region of Ad12. **A** Comparison of the general structural features of the two largest E1A proteins of Ad5 and Ad12 showing the locations of the three major conserved regions and the location of the spacer in the Ad12 proteins. The region to the *left* of CR3 in each case is encoded by exon 1, and the region *right* of CR3 is exon 2 encoded. The spacer sequence is depicted as a *black block*. **B** A comparison of the amino acid sequences in and adjacent to the E1A spacer region of Ad12 (R124–143) with the corresponding regions in spacer deletion mutant dl710 E1A (Sect. 3.1) and Ad5 E1A. The sequences shown are those contained in the ClaI/SmaI (nts 814–964 in Ad12 and nts 917–1,007 in Ad5) cassettes used to generate the Ad12 and Ad5-based chimeric E1A genes and viruses previously described. (Telling and Williams; Williams et al. 1995)

missive rodent cells, whereas the three smaller species (11, 10, and 9S) appear only at later times in infection of permissive cells. Translation of the 13S and 12S species starts at a common AUG and uses a common reading frame throughout to generate polypeptides which are identical except for the internal amino acid residues (R) absent from the smaller protein and present only in the larger one. In Ad5, the translation products of the 13S and 12S mRNAs are 289R and 243R, respectively, whereas in Ad12 the corresponding products are 266R and 235R. In comparing the E1A amino acid sequences of these large E1A proteins of Ad5 and Ad12 and those of group B (Ad3, Ad7) and Simian adenovirus type 7 (SA7), it became clear that despite significant sequence variation between the serotypes, the proteins do retain three conserved regions (CRs). The relative positions of CR1, CR2, and CR3 in the Ad5 and Ad12 proteins are shown in Fig. 1. These E1A proteins are multifunctional; they play important roles in regulating the expression of viral early genes and some cellular genes during the initial stages of infection in human cells, as well as vital roles in immortalizing and transforming rodent cells. For example, the E1A protein activates the expression of viral early genes at the transcriptional initiation level, and can likewise activate certain cellular genes. CR3 is the main mediator of this function, but there is evidence that other regions of E1A, encoded by both exons 1 and 2, are also capable of doing so. In addition to this transactivational property, the E1A proteins also possess trans-repressive activity, a function which maps to CR1 and CR2. This repressive activity is also directed at some cellular genes, which in the case of Ad12 include the MHC1 genes. This particular suppressive activity maps to nonconserved regions of the first exon of E1A and involves cellular proteins of the MHC1 enhancer region that cause the production of MHC1 proteins to be dramatically reduced (see Sect. 4). In addition to these activities, the E1A proteins are required for induction of DNA synthesis in quiescent human (and rodent) cells in culture, and have been shown to form complexes with a variety of cellular proteins during infection of permissive cells with several human adenoviruses. One of these captured cellular proteins is p105, a protein product of the retinoblastoma gene, RB, which binds to regions of E1A adjacent to and within the N-terminal part of CR1 and to CR2. This interaction is likely to be of significance for the growth of the virus in nonproliferating cells, which may well be the type of target cell normally encountered in the natural human host. RB is the quintessential tumor suppressor gene. Its protein product pRb binds in normal, noninfected cells to E2F, a cellular transactivator of certain cell cycle-regulated genes, and in so doing blocks the cell cycle and hence cellular proliferation. pRb may have a higher affinity for E1A proteins than for E2F, and because of this preferential binding the E2F is released during the early stages of infection and made available to activate the promoters of these cellular genes. This results in the induction of cellular DNA synthesis and proliferation, and the support of efficient viral growth. Similar interactions involving Rb-E2F-E1A most likely contribute, either directly or indirectly, to the transformation of rodent cells by human adenoviruses. Readers who wish to expand on the various topics touched upon in this section will find substantial additional information and numerous relevant references in previous review articles

(Bayley and Mymryk 1994; Boulanger and Blair 1991; Jones 1992; Ricciardi 1999; Williams et al. 1995).

2.2
Functions of E1A Proteins Essential for Transformation of Rodent Cells

Using host-range (hr) mutants of serotypes 5 and 12, it was shown that complete transformation of primary baby rat kidney (BRK) cells and other rodent cells in vitro by these viruses requires cooperation of the E1A and E1B genes (Byrd et al. 1988; Graham et al. 1974). Experiments in which these cells were transfected with purified Ad5 and Ad12 viral DNA fragments or plasmids revealed that the E1A gene of each serotype is sufficient for immortalization and partial transformation, and that as with the viruses the cooperation of E1A and E1B is required for the fully transformed phenotype (Gallimore et al. 1984; Houweling et al. 1980; Shiroki et al. 1979; van der Eb et al. 1977). In addition, the E1A gene will transform such primary cells in cooperation with the activated Ha-*ras* oncogene (Ruley 1983) or the *src* gene (Fischer and Quinlan 1998). Using Ad2 and Ad5 mutants which synthesize only the 13S mRNA or only the 12S mRNA, it has been shown that both the 289R and the 243R proteins are required for complete transformation of rat cells (Montell et al. 1984; Winberg and Shenk 1984). Tests using similar mutants of Ad12 demonstrated that in the presence of the larger (266R) protein, the 235R product is not required for transformation of primary BRK, primary BMK (mouse) cells, and cells of the continuous rat line 3Y-1; conversely, in the absence of the 266R protein, the 235R protein is totally incapable of transforming any of these cell types (Lamberti and Williams 1990). Furthermore, only the 266R product is required for direct Ad12 tumor induction in rats and for the tumorigenicity of Ad12 transformants induced in vitro, and a mutant lacking the 266R protein and expressing only the 235R protein is completely negative for tumor induction. In addition to the use of viral mutants, Ad2 and Ad5 E1A plasmids have been employed to demonstrate that both the 289R and the 243R product can immortalize primary BRK cells and bring about their complete transformation in cooperation with the Ha-*ras* oncogene (Moran et al. 1986; Roberts et al. 1985). These findings formed the basis of a reductionist approach in which mutational analysis of Ad2/5 plasmids expressing only the 243R protein has provided a substantial storehouse of information concerning the potential roles of E1A in transformation and other cellular processes. The exon 1-encoded region of that protein is sufficient for complete transformation of BRK cells in cooperation with the *ras* oncogene, and within that region segments mapping to the extreme N-terminus, CR1, and CR2 are essential for this cooperative transformation (reviewed in Bayley and Mymryk 1994; Vasavada et al. 1986; Williams et al. 1995). These segments correspond quite closely with amino acid sequences deemed essential for other E1A functions. For example, the N-terminal and CR1 residues are apparently essential for DNA induction, CR2 is thought to be essential for cellular progression through the cell cycle, and all three are also necessary for immortalization of rodent cells. Similar plasmid studies have shown that the second exon of E1A also contributes important functions necessary to immortalization and transformation (see Sect. 5).

Although few, if any, of these particular plasmid studies directly address the issue of adenoviral oncogenicity or the tumorigenicity of bona-fide Ad transformants, it has been shown that the tumorigenicity in immunocompetent, syngeneic rats of BRK cells transformed by Ad5/Ad12 E1A/E1B hybrid plasmids and viruses is strongly correlated with expression of the Ad12 E1A gene, but is increased to some extent by expression of the E1B gene (Bernards et al. 1982; 1983; Sawada et al. 1988). Although the transformants expressing Ad5 E1A were not tumorigenic in syngeneic rats, they did form tumors in immunodeficient nude mice and rats, and in these the expression of E1B again elevates their tumorigenicity. These results suggest that E1A and E1B both contribute to the tumorigenic phenotype of Ad transformants, with E1A being the key player. As mentioned in Sect. 1, these results also suggested that the rat transformants expressing Ad12 E1A escape the host's CTL defense system and proceed to form tumors, whereas these expressing Ad5 E1A do not. The additional discovery that lysis of rat transformants by allogeneic CTLs in vitro is prevented by expression of Ad12 E1A but not Ad5 E1A supports this view (Bernards et al. 1983). As also mentioned in Sects. 1 and 2.1, MHC1 gene expression is reduced in rat and mouse cells transformed by Ad12 but not in those transformed by Ad5, and this downregulation is orchestrated by the E1A gene (Eager et al. 1985; Schrier et al. 1983), acting at the level of transcription of the MHC1 genes (Ackrill and Blair 1988; Freidman and Ricciardi 1988). Further discussion of this Ad12 E1A-mediated repression of MHC gene expression is contained in Sect. 4. It is worth mentioning here, however, that immunization with Ad12 (Sjögren et al. 1968; Trentin and Bryan 1966) and with Ad2 and Ad5 (Gallimore and Williams 1982) induces transplantation immunity against inoculated Ad12 and Ad2/5 tumorigenic transformants, respectively. This phenomenon is group-specific, in that Ad12 does not protect against Ad2 tumor cells or Ad2/5 protect against Ad12 tumors, and is ascribed to the Ad tumor-specific transplantation antigen (TSTA), against which CTLs react. Group C Ad TSTA was mapped to the E1 region of the Ad genome by analyzing the capacity of Ad5 mutants to induce immunity in Hooded Lister rats (haplotype RT-1^{u}) to a syngeneic, tumorigenic Ad2 transformant (Gallimore and Williams 1982). Subsequent studies using viable Ad5 and Ad12 recombinant viruses, in which the E1 regions carry heterologous E1A/E1B transcription units, showed that TSTA immunity (which as mentioned above is not cross-reactive between these two serotypes) of both viruses is an E1A gene function (Sawada et al. 1986). The regions of the E1A proteins which determine TSTA function in these rats (RT-1^{u}) were mapped using both a series of recombinant viruses mutated in the E1A gene of Ad5 (Urbanelli et al. 1989), and by viruses carrying chimeric Ad5/Ad12 E1A genes (Sawada et al. 1994; Raska 1995). In Ad5 this function maps to the second exon of E1A, whereas in Ad12 it is determined by the N-terminal 68-residue segment of E1A.

Viral antigens such as TSTA are presented to mature T cells as complexes between peptide fragments of the viral component and MHC class I antigens on the surface of infected and/or transformed cells. Antigen-specific recognition of the virally imprinted cells, activation of naive T lymphocytes by these cells, and the subsequent lysis of the antigen-presenting (target) cells by the activated T cells occurs only if the MHC haplotype of both the target and its attacker are identical; the process is known as MHC restriction (Zinkernagel and Doherty 1979). Using deletion mutants of Ad5, it was shown that the immunodominant Ad5 E1A epitope recognized by rat RT-1^{u}-restricted CTLs is encoded by the second exon of the E1A gene (Urbanelli et al. 1989). In Fischer rats (haplotype RT-1^{lvl}), however, preliminary results map the Ad5 CTL epitopes to the first exon of E1A (Routes et al. 1991). Mouse H-$2D^{b}$-restricted CTLs recognize a dominant epitope mapping within a sequence bounded by residues 232–247 of the type 5 E1A 289R protein (Kast et al. 1989). This finding was determined in vitro using a cytotoxicity assay in which uninfected syngeneic mouse embryo cells were sensitized for Ad5-specific CTL lysis by a set of overlapping 12-residue peptides covering the entire E1A amino acid sequence. In addition, these Ad5-specific CTLs, when injected along with recombinant interleukin-2, cause complete regression of large Ad5 tumors in nude mice. This finding supports the view that Ad5 transformants are nontumorigenic in immunocompetent syngeneic hosts because CTLs eradicate them. Ad12, on the other hand, suppresses MHC class I expression in transformed cells, and the resultant lack of the MHC self element means that TSTA presentation to CTLs is prohibited. As a result, these cells can most likely slip through the immune surveillance system of the host, and may progress to form tumors. This line of reasoning provides an explanation for the differential tumorigenicity between Ad12 and Ad5 transformants. Strong evidence in support of this view was obtained in experiments in which the tumorigenicity of an Ad12 mouse transformant was significantly reduced after functional H-2 genes resistant to repression by E1A were added to it (Tanaka et al. 1985). However, other investigators challenged the role of class I repression by concluding that transfection of MHC class I genes into Ad12-transformed cells enhances rather than reduces their tumorigenicity (Soddu and Lewis 1992), and that there is no correlation between level of MHCI antigen and tumorigenicity in a wide range of Ad (and SV40) transformants (Haddada et al. 1986, 1988; Mellow et al. 1984; Sawada et al. 1985). These differences may be related to differences in the methodologies for testing class I repression and inherent differences in the cell lines examined. In light of these inconsistencies, it may be unwise at present to make simple generalizations about the relationship between suppression of MHCI gene expression and Ad12 oncogenicity. The fact that both Ad5 and Ad12 induce robust, specific transplantation immunity in immunocompetent rats supports this view. It seems likely that, in addition to class I repression, other factors are also involved in the capacity of Ad transformants to grow in syngeneic hosts.

There is, as it happens, some evidence (mentioned in Sect. 1) that NK cells of the host defense system also play a role in deciding the fate of Ad tumor cells. It was shown that the level of tumorigenicity of Ad transformants correlates closely with their degree of resistance to NK cells, with Ad12 transformants being much more resistant to the cytolytic effect of these cells than Ad2 and Ad5 transformants (Cook et al. 1982; Kast et al. 1989; Raska and Gallimore 1982). Studies carried out with cells transformed by viruses or plasmids containing hybrid A5/Ad12 E1 regions showed that susceptibility to NK killing is determined by the E1A gene (Kast et al. 1989; Sawada et al. 1985). All cells expressing Ad5 E1A were highly susceptible to NK activity, whereas those expressing Ad12 E1A were not; the serotype specificity of E1B in these transformants had no apparent influence on NK sensitivity. It was proposed that E1A first and second exon functions are required to make these cells susceptible to NK cytolysis, and this alteration in the transformed cell may stem from regulatory action of the intact E1A proteins (Krantz et al. 1996). To date, little is known concerning either the mechanism(s) behind this phenotypic change that occurs during Ad transformation, or of the way in which NK cells recognize these target cells. It was suggested that increased susceptibility to NK cytolysis results from reduced levels or reduced inhibitory activity of class I antigens (Chadwick et al. 1992; Karre et al. 1986; Piontek et al. 1985; Storkus et al. 1992). Evidently, this is not the case for the Ad transformation system, where, for example, tumorigenic Ad12 transformants display low levels of surface MHCI antigen and low levels of NK susceptibility. There is little or no doubt that the capacity of Ad transformants to form tumors in immunocompetent, syngeneic hosts stems at least in part from the capacity of these cells to evade the host's defense mechanisms. E1A proteins clearly play a role in this evasion scenario, in addition to the many critical functions they provide for productive infection of human cells and transformation of rodent cells in vitro. It is likely that one (or more) of these additional E1A functions also plays a key role in viral oncogenicity.

3 The Unique Spacer Region of Ad12 E1A Is a Vital Oncogenic Determinant

In weighing the question of what other E1A functions could possibly influence the differential oncogenicity of human adenoviruses in rodents, we considered it logical to assess the potential roles of the nonconserved regions, which are strikingly different among the various serotypes. Our attention was drawn in particular to one Ad12 E1A region comprising a segment of 20 amino acid residues (R124–143) located between CR2 and CR3 (see Fig. 1). This unique sequence, which we term the spacer region, is completely missing in the Ad5 E1A proteins. A somewhat longer spacer is found in the E1A proteins of simian adenovirus 7 (SA7), and slightly shorter ones are present in the Ad3 and Ad7 proteins (Kimelman et al. 1985; Larsen and Tibbetts 1987; van Ormondt and Galibert 1984). Given that Ad12 and SA7 are both highly oncogenic, Ad3 and Ad7 are weakly oncogenic, and Ad5 is essentially nononcogenic (see Sect. 1), there is, therefore, a clear correlation between spacer length and level of viral oncogenicity, an observation first made by Larsen and Tibbetts in 1987. It thus seemed feasible to us that the Ad12 E1A spacer might contribute in some way to the oncogenicity of the virus and to the tumorigenicity of Ad12 transformants in syngeneic, immunocompetent hosts. We tested this hypothesis initially by reciprocally exchanging segments of the Ad12 E1A gene containing the spacer sequence and the equivalent Ad5 region, and examining the phenotypic effects of these genotypic rearrangements on the resultant Ad12- and Ad5-based chimeric viruses carrying chimeric type 12/type 5 E1A genes (Telling and Williams 1994). Two of these Ad12-based chimeric viruses, 12*ch*702 and 12*ch*704, (containing the substituted ClaI to SmaI segment of Ad5 E1A, see Fig. 1) displayed greatly reduced oncogenicity (lower incidence and longer latent period of tumors) in syngeneic Hooded Lister rats, and cells transformed in vitro by these chimerics were either less or non-tumorigenic in syngeneic adult rats. The Ad12 E1A-restriction site variant var747 used in the construction of these chimeric viruses was indistinguishable from Ad12 wild-type in terms of both viral oncogenicity and transformant tumorigenicity. Although defective for induction of tumors in rats and tumorigenic capacity of BRK cells transformed in vitro, these chimeric viruses were found to be phenotypically wild-type for all other relevant characteristics measured. For example, they grew and plaqued with high efficiency in both human A549 and A12E1-transformed human embryonic retinoblastoma 3 (HER3) cells, and they transformed BRK and BMK cells with wild-type frequencies over a wide range of input virus multiplicities. In addition, they displayed normal (wild-type) E1A gene expression and activation of E1B transcription in A549 cells, normal levels of E1A mRNAs in infected BRK and BMK cells, and normal synthesis of E1A proteins and E1B 55K protein in BRK transformants. Transformed foci induced in primary BRK monolayers were somewhat flatter, larger and less dense than those induced by Ad12 and *var*747 (see Telling and Williams 1994), but when isolated and passaged they grew vigorously, and their capacity to establish continuous transformed lines was similar to that of foci induced by Ad12 and var747. Once established they manifested phenotypic traits such as growth to high density, efficient growth in agar suspension, and colony morphology which were essentially identical to those displayed by wild-type Ad12 transformants. These results are consistent with the view that the spacer is an oncogenic determinant but is not essential for complete transformation in vitro.

3.1 Mutants Provide Additional Proof That the Spacer Sequence Is an Oncogenic Determinant

In making the chimeric genes described above, the cassettes used for reciprocal transfers contain flanking sequences from CR2 and CR3 (see Fig. 1). These are, of course, conserved sequences, but in Ad5 and Ad12 they do possess a number of different amino acids, which could perhaps contribute to the phenotypic changes found. To check for this possibility, and more precisely to map the oncogenic determinant, we constructed and phenotypically analyzed a site-specific spacer point mutant and a specific spacer deletion mutant. Point mutant *pm*715 carries a G-C transversion at nt 913 which results in the replacement of the fourth residue in the run of six alanines in the spacer sequence by a proline. The overall phenotype of this mutant is indistinguishable from that of the chimeric viruses in terms of viral growth, transformation of BRK and BMK cells in vitro, E1A and E1B expression in transformants, very poor oncogenicity in Hooded Lister rats, and very low tumorigenicity of transformed cells in syngeneic, immunocompetent, adult rats (Williams et al. 1995). In contrast, mutant *pm*714, with a C-G transversion at nt 931 which results in replacement of the arginine at the start of CR3 with glycine (the residue found at that position in Ad5) acts nearly the same as wild-type Ad12 in terms of viral growth, viral transformation in vitro, tumorigenicity, and tumor induction efficiency and latency. The deletion mutant *dl* 710 carries a precise deletion of the Ad12 spacer (see Fig. 1), generated by oligonucleotide mutagenesis (Chan and Smith 1984). The mutagenic oligodeoxyribonucleotide used contained 15 nucleotides flanking the 5′ side of the spacer contiguous with 15 nucleotides flanking the 3′ side. The method for the transfer of the segment containing the deletion from the M13 construct into the Ad12 E1A plasmid was described previously (Telling and Williams 1994). The E1A gene carrying the spacer deletion was transferred into the complete Ad12 genome by marker rescue (Lamberti and Williams 1990) of the Ad12 E1A deletion (bp993–1,385 deleted) mutant dl701 (D. Breiding, Ph.D. thesis, Carnegie Mellon University, 1985). This mutant also has an overall phenotype very similar to that of the two chimeric viruses and to the *pm*715 phenotype. It still

Table 1 Transformation of Hooded Lister baby rat kidney (BRK) and Balb/c baby mouse kidney (BMK) cells infected by Ad12 wild-type variant 747 and spacer deletion mutant *dl* 710

Cells	Input multiplicity of virus (pfu/cell)	No. of foci/dish[a]					
		var 747			*dl* 710		
BRK	100	3	5	4	2	3	4
BRK	30	11	9	12	8	8	10
BRK	10	16	21	19	15	13	13
BRK	3	8	10	9	7	5	5
BRK	1	3	2	3	1	2	2
BRK	None	0	0	0	0	0	0
BMK	10	24	22	21	24	27	23
BMK	None	0	0	0	0	0	0

[a] Procedures used for transformation were as described (Edbauer et al. 1988). Following infection of BRK cells, all cultures were grown at 36.5°C, and final counts of transformed foci were made at 32 days postinfection. Following infection of BMK cultures, cells were removed by trypsinization, seeded on new dishes at a density of 5×10^5 cells/dish, and grown at 36.5°C; final counts were made at 33 days postinfection. Values shown correspond to the number of foci on each of 3 replicate dishes. pfu, Plaque-forming units.

grows to wild-type levels and plaques with wild-type efficiencies on both A549 and HER3 cells (data not shown), and transforms BRK and BMK cells with frequencies similar to those of Ad12 wild-type var747 (Table 1). The *dl*710 mutant is, however, quite defective for tumor induction in Hooded Lister rats (Table 2). Moreover, cells transformed by dl710 display normal production of E1A and E1B proteins (Y. Zhang, S. Tuncer, and J. Williams, unpublished results), but are essentially nontumorigenic in syngeneic, immunocompetent, adult rats (Table 3). These results confirm that the spacer is an oncogenic determinant, but is not obligatory for complete transformation of both rat and mouse primary cells in vitro. In addition, the results with both *dl*710 and *pm*715 prove that the Ad5 CR2 and CR3 flanking regions present in ch702 and ch704 do not contribute to the oncogenic defects observed with these mutants. Furthermore, the results with *pm*715 suggest that it is unlikely that spacer removal simply causes spatial disturbance in the E1A protein molecule, resulting in the alteration of a functional domain elsewhere that is in turn responsible for reduced viral oncogenicity and transformant tumorigenicity. Were it to occur, such spatial perturbation could potential-

Table 2 Induction of tumors in Hooded Lister rats inoculated as newborns with Ad12 *dl* 710

Virus	Tumor-positive rats/rats inoculated[a]		
	6×10^8 pfu	3×10^8 pfu	1×10^8 pfu
var 747	N.D.	8/8	8/8
dl 710	5/19	4/18	6/18
Virus	Latency period (days)[b]		
	6×10^8 pfu	3×10^8 pfu	1×10^8 pfu
var 747	N.D.	50±3	52±4
dl 710	158±29	125±63	137±28

[a] In all cases, 24–36-h-old newborn rats were inoculated intraperitoneally with 0.05 ml of virus suspension.
[b] Average time in days after inoculation at which palpable tumors were first detected±standard deviation. Animals were killed when tumors reached a diameter of 3–4 cm, or 9 months after inoculation in animals in which no tumors developed.
N.D., not determined; pfu, plaque-forming unit.

Table 3 Tumorigenicity of transformed cell lines derived from BRK cells infected by the Ad12 E1A spacer deletion mutant *dl* 710

Cell line	Tumor incidence (no. of rats positive/total) by no. of cells inoculated[a]		
	1×10^7	3×10^6	1×10^6
RK 747A	6/6 (8, 25)	6/6 (8, 25)	6/6 (10, 31)
RK 747B	ND	6/6 (9, 25)	6/6 (13, 37)
RK 710A	2/6 (16, 60)	1/6 (20, 90)	0/6 (>90, 90)
RK 710B	0/6 (12, 90)	1/6 (17, 90)	0/6 (>90, 90)
RK 710C	2/6 (14, 44)	1/6 (14, 90)	1/6 (14, 90)

[a]As in previous experiments (Telling and Williams 1994; Williams et al. 1995), the appropriate number of cells contained in 0.5 ml of DMEM, was inoculated subcutaneously into 6-week-old syngeneic Hooded Lister rats. Following inoculation the recipients were examined every 2 days for evidence of tumor growth at the inoculation site. When tumors reached approx. 3 cm in diameter, the animals were killed. The numbers in parentheses refer respectively to the day on which palpable tumors first appeared and the day after inoculation on which animals with tumors were euthanized. In some cases (e.g., animals inoculated with 1×10^7 RK 710B), tumors appeared within a few weeks, grew slowly for some time, and then regressed completely; these animals were killed at 90 days postinoculation. In other cases (e.g., rats inoculated with 1×10^6 RK 710A and B), no tumors appeared within the 90-day period (shown as >90), and these animals were also killed at 90 days.
N.D., not determined.

ly alter the capacity of the E1A protein to bind with cellular proteins such as pRb. Preliminary co-immunoprecipitation analyses of extracts from a broad set of Ad12, chimeric, and spacer mutant-induced BRK transformants and tumor cells suggest that this is not the case (C. Lui, G. Telling, and J. Williams, unpublished results). Others have examined the binding capacity of GST-13S-E1A and spacer deletion fusion proteins to bind to pRb from human A549 cells and measured their ability to disrupt Rb/E2F complexes using gel shift analysis (Rumpf et al. 1999). In this system the hexa-alanine sequence of the Ad12 spacer is apparently required or involved in both processes, whereas the N-terminal half of the spacer is not. At this time we have no good explanation for the apparent difference in these results, but it may be related to the fact that we monitored protein interactions which had occurred in the milieu of the transformed cell, whereas those investigators measured these interactions by mixing components in vitro where conditions are likely to be quite different. That group also reported that the hexa-alanine sequence of the Ad12 spacer is necessary for the complete establishment of transformation in BMK cultures cotransfected with Ad12 E1A and Ad12 E1B plasmids (Rumpf et al. 1999). On the other hand, in all cases, whether using HRK or BMK cells, we have found no significant differences in the transforming efficiencies of Ad12 and a variety of spacer chimerics and mutants (see Table 1; Telling and Williams 1994). This apparent disagreement may be related in part to the fact that they used plasmid transfection whereas we used infection with virus in our experiments. It was previously found that transformation of primary rodent and human cells by Ad12 plasmids is considerably less efficient than that obtained using Ad5 plasmids (Gallimore et al. 1986; van den Elsen et al. 1982), although the proportion of established, continuous transformed cell lines from such cultures is similar, but possibly dependent on conditions used. Transformation efficiencies of Ad12 and Ad5 viruses in BRK and BMK cells are, however, quite the reverse, with Ad12-induced frequencies occurring at levels more than 20-fold higher than those with Ad5 (J. Williams, unpublished data).

3.2
Tumorigenicity of Transformants May Also Be Influenced by Other Regions of E1A

Whether or not such E1A/cellular protein interactions are influenced by the spacer region, it is likely that some other region(s) of E1A in addition to the spacer, and perhaps cooperating with it, also contributes to Ad12 oncogenicity. Evidence in favor of this view lies in the fact that all four of the spacer mutant viruses discussed above are not entirely non-oncogenic, and display low but measurable levels of tumor induction in rats, albeit with considerably longer latency periods than found with wild-type virus. Strong supportive evidence for this view also comes from studies in which the tumorigenicity of BRK cells, transformed in vitro by cotransfection with a series of plasmids containing Ad5-based Ad5/Ad12 chimeric E1A genes and Ad12 E1B plasmid, was measured in syngeneic newborn Hooded Lister rats (Jelinek et al. 1994). Transformants induced by transfection with chimerics containing Ad12 sequences from the N-terminus to the left edge of CR3 were found to be tumorigenic, whereas those transformed with chimerics containing sequences encoding only 80 residues of the Ad12 N-terminal region were not. Transformants induced by chimerics containing the Ad12 spacer along with flanking CR2/CR3 regions (similar to the cassettes we removed from Ad12) were moderately tumorigenic, but less so than those expressing the longer Ad12 N-terminal sequences. We also constructed Ad5-based E1A chimeric genes containing the Ad12 spacer region, and placed these into the Ad5 virus genome to produce chimeric viruses 5*ch*10 and 5*ch*17. These two chimeric viruses are wild-type for viral growth in Hela, A549, and 293 cells, for E1A expression, and for transformation of rat embryo cells in vitro. However, the transformants induced are nontumorigenic (as are Ad5 transformants) in syngeneic, immunocompetent rats, and neither of the chimeric viruses induce tumors in Hooded Lister rats, as is the case for Ad5 wild-type (Williams et al. 1995; J. Williams, unpublished results). Of course, these results are not really unexpected, as there is some evidence to suggest that adenoviral oncogenicity is determined by both E1 and other regions on the viral genome (Bernards et al. 1984; Sawada et al. 1988), and is likely, in the case of Ad5, to be influenced by viral growth and killing of target cells in the host (Williams 1973). In addition, the oncogenic potential of Ad5 may also be affected negatively by a determinant located in the C-terminus of the E1A protein (Boyd et al. 1993); this topic will be addressed in Sect. 5.

Thus, these investigations allow us to conclude that the spacer sequence plays a role in determining Ad12 oncogenicity and the tumorigenicity of Ad12 transformants. The next section describes the mechanism of MHC class I shut-off and addresses the question of whether the requirement of the spacer in tumorigenicity is related to MHC class I shut-off.

4 Reduction of MHC Class I Expression Is Necessary But Not Sufficient for Ad12 E1A-Mediated Tumorigenesis

The major histocompatibility complex (MHC) class I genes encode cell surface glycoproteins that mediate the cytotoxic T-cell lymphocyte (CTL) immune response. Cells become lysed when class I surface molecules present non-self antigens (e.g., viral antigens) as processed peptides to CTLs. However, cells that express very low surface levels of class I glycoproteins can escape being targeted by CTLs. As referred to above, such is the case with tumorigenic Ad12-transformed cells in which the surface levels of class I glycoproteins are greatly diminished, especially compared to non-tumorigenic Ad5 transformed cells (Eager et al. 1985; Schrier et al. 1983). This reduced class I surface expression applies to all of the MHC class I encoded alleles (e.g., H2K, D, and L in mouse). Consistent with their respective differential levels of class I surface expression, tumorigenic Ad12-transformed cells are much more resistant to lysis by syngeneic CTLs in vitro than nontumorigenic Ad5 transformed cells (Yewdell et al. 1988). The ability of Ad12 transformed cells to avoid CTL lysis was lost following treatment with interferon-γ (IFN-γ), which has the well-known effect of stimulating class I expression. In keeping with these in vitro results, rats bearing tumors generated by injection of Ad12-transformed cells have increased survival if the injected cells are first treated with IFN-γ (Hayashi et al. 1985; Tanaka et al. 1985). Significantly, low class I surface expression and its association with tumorigenesis are directly supported by the finding that cells from freshly excised tumors maintain reduced levels of class I glycoproteins (Eager et al. 1985).

4.1
MHC Class I Shut-Off in Ad12-Transformed Cells Is Due to Altered Binding of Transcription Factors NF-κB and COUP-TFII to the Class I Promoter

Collateral with low amounts of class I surface antigens on Ad12-transformed cells is the drastic reduction in the steady-state levels of class I mRNAs, as observed by Northern blot analyses (Eager et al. 1985; Schrier et al. 1983; Vasavada et al. 1986). Nuclear run-on assays demonstrated that a block in class I transcription accounts for the low levels of class I mRNAs (Ackrill and Blair 1986; Friedman et al. 1988). Extensive mutational analysis of the class I promoter further established that the class I enhancer (located between −205 bp to −159 bp upstream of the start site of transcription) is the target of 12-E1A (Ge et al. 1992). A transfected plasmid containing just the class I enhancer appended to class I basal TATA box promoter (−37 bp to +1) resulted in reporter gene activity that was strong in Ad5-transformed cells, but very weak in Ad12-transformed cells (Ge et al. 1992). These findings suggested that there might be differential binding activities of cellular proteins that recognize specific DNA sites within the class I enhancer.

The two DNA recognition sites within the class I enhancer are referred to as R1 and R2. The R1 site, which is closer to the promoter, has a recognition sequence for the nuclear transcription factor NF-κB. The R2 site has a recognition sequence for members of the nuclear hormone receptor family (Fig. 2) (reviewed in Ricciardi 1999). DNA band-shifts and super-shifts reveal that in nontumorigenic Ad5-transformed cells, there is strong binding of NF-κB to the R1 site (Liu et al. 1996). The fact that NF-κB is a strong transcriptional activator explains why there are substantial levels of class I mRNA and proteins expressed in Ad5-transformed cells. Significantly, as will be made apparent below, the binding activity to R2 site is minimal in A5-transformed cells. By direct contrast, tumorigenic Ad12-transformed cells exhibit negligible binding of NF-κB to the class I enhancer R1 site (Liu et al. 1996; Nielsch et al. 1991). Because NF-κB is the major transcriptional activator of class I transcription, its failure to bind the class I enhancer provides an explanation for the diminished class I mRNA and protein levels in Ad12 transformed cells. However, surprisingly, there is a dramatically strong DNA binding activity to the R2 site in Ad12-transformed cells. This strong R2 DNA binding activity is due to the nuclear hormone orphan receptor,

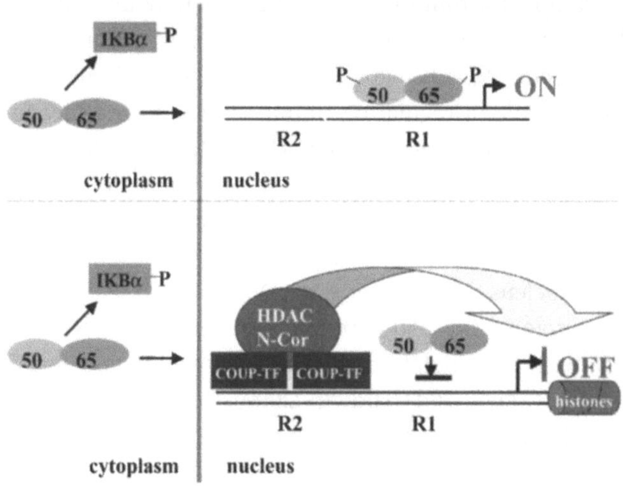

Fig. 2 The mechanism of MHC class I transcriptional repression in Ad12-transformed cells. *Upper-half* depicts normal activation of MHC class I transcription. In response to stimuli, IKB-α becomes phosphorylated and degraded. NF-κB, which is no longer retained in the cytoplasm by IKB-α, is translocated to the nucleus where it binds to the R1 site of the class I enhancer. Phosphorylation of NF-κB is critical to DNA binding (p50 subunit) and activation (p65 subunit). *Lower-half* depicts repression of MHC class I transcription in Ad12-transformed cells. First, NF-κB is translocated to the nucleus but fails to bind to the class I enhancer, because the p50 subunit is hypophosphorylated. Second, repressor COUP-TFII binds to the adjacent R2 site on the class I enhancer and forms a complex with HDAC and N-Cor that causes histone deacetylation and chromatin compaction as a means of preventing transcription

COUP-TFII, as revealed by super-shifting with a panel of antibodies specific for nuclear hormone receptors (Liu et al. 1994; Smirnov et al. 2001). Because COUP-TFII is known to function mainly as a transcriptional repressor (Zhou et al. 2000), its strong binding to the R2 site is consistent with decreased class I expression. This finding is fascinating because it is the only known example in which COUP-TFII binds to the well-studied MHC class I enhancer.

Thus, in Ad12-transformed cells, it would appear that the maintenance of reduced class I surface expression is assured by both loss of activator (NF-κB) binding and acquisition of repressor (COUP-TFII) binding to the transcriptional enhancer element. The mechanisms and consequences of these altered factor binding events and how they may coordinate in keeping class I transcription to a minimum are now discussed.

The transcription factor NF-κB, which comprises two subunits p50 and p65, plays a central role in regulating immune-responsive genes including the MHC class I alleles. Normally, NF-κB is retained in the cytoplasm in association with inhibitor, IκB. Following stimulation by cytokines, (e.g., TNF-α or IL-1β), IκB becomes phosphorylated and degraded by the ubiquitin pathway, permitting NF-κB to translocate to the nucleus. Once in the nucleus, NF-κB binds to DNA and stimulates transcription. A primary role of the p50 subunit is to bind DNA, whereas that of the p65 subunit is to stimulate transcription through its C-terminal activation domain (Mayo and Baldwin 2000). In both Ad5- and Ad12-transformed cells, there is an abundance of nuclear NF-κB. Apparently, a shared feature of their transformation mechanisms accounts for these nearly equivalent levels of nuclear NF-κB in both Ad5- and Ad12-transformed cells. Yet, as indicated above, only nuclear NF-κB in Ad12-transformed cells fails to bind the R1 site of the MHC class I enhancer (Fig. 2). The p50 subunit is largely responsible for this DNA binding deficiency of NF-κB. When the p50 and p65 subunits from Ad5- and Ad 12-transformed cells were isolated and exchanged in reconstitution experiments, the p50 subunits from Ad12-transformed cells failed to support DNA binding when recombined with the p65 subunit of from Ad5-transformed cells (Kushner and Ricciardi 1999). This block in NF-κB DNA binding is due to hypophosphorylation of the p50 subunit, as revealed by 2-D gel analysis of the different charged isoforms. The importance of phosphorylation of p50 is demonstrable by abolishing DNA binding activity through phosphatase treatment (Kushner and Ricciardi 1999) or by site-directed mutation of specific p50 phospho-residues (S. Hou and R. Ricciardi, unpublished results). Thus, even though NF-κB is copiously present in nuclei of Ad12-transformed cells, it fails to bind to the class I enhancer because the p50 subunit is underphosphorylated (Fig. 2) (Kushner and Ricciardi 1999).

Ad12-transformed cells contain elevated amounts of COUP-TFII mRNA and protein compared to Ad5-transformed cells (Smirnov et al. 2001). This correlates with the significantly greater DNA binding activity of COUP-TFII to the R2 site of the class I enhancer in Ad12-transformed cells. As indicated in Fig. 2, COUP-TFII binds as a dimer to the R2 site, and its C-terminal domain associates with a complex containing histone deacetylase (HDAC) and N-CoR (Smirnov et al. 2000, 2001). The association with HDAC, through protein:protein interaction, is consistent with COUP-TFII functioning as a repressor (Zhou et al. 2000) and serves to explain how transfected plasmid constructs with multiple R2 sites repress transcription in Ad12- (but not Ad5)-transformed cells (Kralli et al. 1992). In vivo analysis of the COUP-TFII complex using chromatin immunoprecipitation (ChIP) assays also indicates that COUP-TFII associates with HDAC and N-CoR, and that the surrounding chromatin of the class I enhancer (that contains the R2 site) is deacetylated (B. Zhou and R. Ricciardi, unpublished results). Thus, the net affect of HDAC activity, mediated through its association with COUP-TFII, is chromatin compaction that results in inhibition of class I transcription. Functional assays using NF-κB knockout cells confirmed that COUP-TFII can function as a repressor of MHC class I transcription (Smirnov et al. 2001). It is notable that this is the only known case where a COUP-TF factor regulates MHC class I repression.

4.2 Dynamics of MHC Class I Downregulation NF-κB and COUP-TFII

What might be the biological significance that tumorigenic Ad12-transformed cells possess both repressor COUP-TFII and disabled activator NF-κB? As mentioned above, even though NF-κB translocates to the nuclei of Ad12-transformed cells, it fails to bind DNA due to the hypophosphorylation of the p50 subunit (Kushner and Ricciardi 1999). Recent findings revealed that the cytokine inducers TNF-α and IL-1β are able to promote degradation of the NF-κB cytoplasmic inhibitor IκB-α, and permit the nuclear translocation of a phosphorylated form of NF-κB that is capable of binding DNA. Importantly, however, cytokine-induced NF-κB could not increase class I gene transcription unless repression by COUP-TFII was abolished by the inhibitor TSA or by mutating the R2 site. ChIP assays support the contention that these events occur in vivo (B. Zhou and R. Ricciardi, unpublished results). Thus, COUP-TFII repression overrides the ability of cytokine-induced NF-κB from binding DNA and stimulating class I transcription. It is interesting to speculate that there are reciprocal physiological conditions whereby COUP-TFII fails to occupy the R2 site. Under such conditions, the constitutive hypophosphorylated status of NF-κB would ensure that this activator fails to activate class I transcription. These dynamic measures of ensuring maintenance of class I transcriptional downregulation arguably contribute to the survival of Ad12-transformed cells in vivo, and their capacity to form tumors.

4.3
Ad12 E1A Alone Mediates MHC Class I Shut-Off

Of the two transforming genes (E1A and E1B) of adenovirus-12, E1A alone is responsible for shut-off of class I expression (Eager et al.1985; Schrier et al.,1983; Vasavada et al. 1986). This was definitively demonstrated by generating a cell line that produced Ad12 E1A as the only adenovirus gene product (Vasavada et al. 1986). Specifically, when Ad12 E1A was cotransfected into primary human embryonic kidney (HEK) cells, along with BKV TAg in order to achieve stable transformation, the Ad12-E1A/BK TAg cell lines expressed negligible class I HLA glycoproteins (HLA, A, B, and C) and mRNAs. By contrast, the HEK cells transformed with BK TAg alone expressed high levels of class I surface glycoproteins. Significantly, an independent study showed that every somatic hybrid cell line that was generated by fusing Ad5- and Ad12-transformed murine cells had a diminution of class I glycoproteins on the surface that was equivalent to that of the parental Ad12-transformed cell fusion partner (Ge et al. 1994). Taken together, these results indicate that Ad12 E1A actively mediates the shut-off of MHC class I glycoproteins.

4.4
The Ad12 E1A Spacer Encodes a Tumorigenic Function Distinct from MHC Class I Downregulation

Mapping studies indicated that the regions of Ad12 E1A responsible for class I repression are encoded within the first exon. It was logical to consider that nonconserved sequences in this first exon could mediate class I shut-off (Kushner et al. 1996; Telling and Williams 1994). The 20-aa spacer region was particularly intriguing because it is essential for tumorigenesis but is not essential for transformation, as described above. Interestingly, the Ad12 E1A spacer carries a structural feature, namely the contiguous run of alanine residues (Fig. 1), which is a characteristic of some other transcriptional repressors (Licht et al. 1990). Unlike Ad3 E1A (Larsen and Tibbetts 1987), no evidence for any autorepressive activity of Ad12 E1A (Lamberti and Williams 1988) was observed. However, it did seem possible that the Ad12 spacer could be involved in repres-

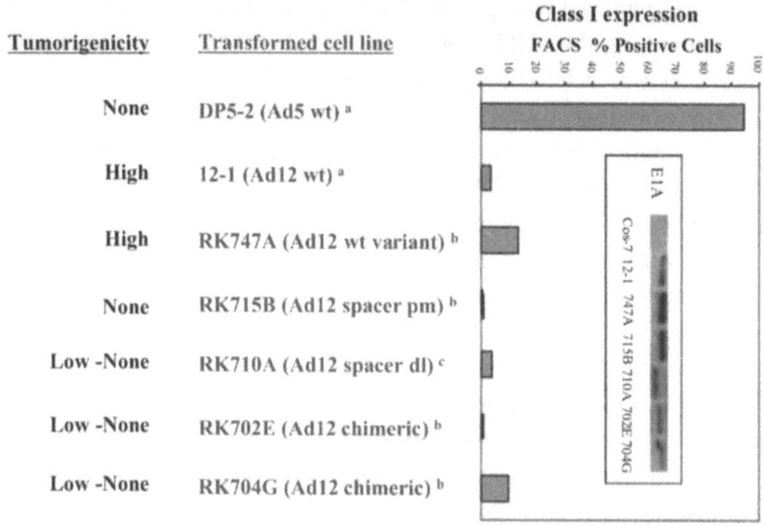

Fig. 3 Diminished MHC class I expression is not sufficient for tumorigenicity. The abilities of Ad5 wt, Ad12 wt, and Ad12 spacer mutant transformants (*center*) to generate tumors in syngeneic rats and repress MHC class I cell surface expression are presented in *left* and *right* panels, respectively. The tumorigenicity data are from Williams et al. (1995) and Table 3. FACS analysis was performed using anti-rat MHC class I MAb (clone OX-18, Cedarlane Laboratories) following established procedures (Smirnov et al. 2000). Western blot (*insert, right panel*) was performed using Ad12 E1A antibody 363 (Scott et al. 1984) to demonstrate the presence of Ad12 E1A in the spacer mutant cell lines; Cos-7 served as a negative control. *Superscripts* refer to the origins of the transformed lines: a, Jelinek et al. 1994; b, Williams et al. 1995; c, Table 3

sion of other viral or cellular enhancer-controlled cellular promoters, such as MHC class I. Significantly, spacer sequences do not occur in the EIA of Ad5 or other nontumorigenic strains of adenovirus.

We thus investigated whether the spacer is essential for reduction of class I synthesis by determining whether transformed cell lines expressing mutated spacer E1A proteins are no longer able to downregulate class I surface expression. As shown in Fig. 3, class I surface proteins were expressed on almost all of the Ad5 transformants (e.g. DP5–2 cells) but not on the wt A12 E1A transformants (12–1 cells) or the Ad12 wt variant transformants (747 cells). Surprisingly, class I downregulation still occurred in transformed cells lines expressing the point mutant Ad12 E1A (pm715) in which the fourth of the six contiguous alanines of the spacer was replaced with proline (see Fig. 1). Moreover ,transformed cell lines expressing the Ad12 E1A deletion mutant (dl710, in which just the 20R spacer was deleted), also failed to abolish class I repression. In addition, we examined transformants induced by Ad12-based chimeric viruses *ch*702 and *ch*704 (Telling and Williams 1994). The resulting chimeric Ad5/12 E1A products (ch704 and the splice variant ch702, Fig. 3) did not eliminate class I downregulation. These results solidly demonstrate that another region in the first exon of Ad12 E1A is responsible for class I repression. Most interestingly, this finding indicates that a function distinct from class I downregulation is encoded by the spacer and is necessary for tumorigenesis.

As alluded to above, one possible function of the spacer in tumorigenesis may be to enable avoidance of NK cell lysis. This is reasonable because it is well known that tumors lacking MHC class I antigens are more readily killed by NK cells (Lanier 2000). This is largely because certain receptors on NK cells (e.g., Ly49 in mouse and KIR in human) serve to inhibit attack on cells expressing class I antigens. It has been argued that Ad12 tumorigenic cells with their diminished MHC class I expression have also found a way to resist NK lysis (Kenyon and Raska 1986; Sawada et al. 1985). Should this function be provided by the spacer, then both the CTL and NK escape functions would be encoded by the Ad12 E1A gene. This concept is supported by an examination of MHCI antigen levels in Fischer (RT-1^l) BRK cells transformed by cotransfection with Ad2-based chimeric Ad2/Ad12 E1A and Ad12 E1B plasmids (Huvent et al. 1996). Although that study also concluded that the spacer does not appear to play a role in class I repression, it did not correlate low class I levels with an ability of the cell transformants to cause tumors in syngeneic rats, as tumorigenicity of transformants was not measured. In this context, it is of interest that Ad12 induces tumors at extremely low levels in Fischer rats, and Ad12 transformants do not form tumors in syngeneic, immunocompetent rats of that strain (Williams et al. 1995; J. Williams, unpublished results). Huvent et al. (1996) did suggest that NK resistance of the transformants may require both the spacer and another domain in the second exon of Ad12 E1A. However, tests carried out with a set of Hooded Lister BRK lines transformed by Ad5, Ad12, Ad12-based chimeric viruses and *pm* 715 indicate that loss or mutation of spacer does not result in increased susceptibility to NK cells (Williams et al. 1995; C. Lui and J. Williams, unpublished results). Therefore, the role of the spacer in NK nonsusceptibility of Ad12 transformants has not yet been resolved.

5
Reexamining the Hypothesis That Group C Adenovirus E1A Proteins Possess a Negative Modulator of Tumorigenicity

As mentioned above in Sect. 2.2, experiments using Ad2/5 plasmids expressing the E1A 243R protein have shown that the N-terminal portion of that protein, encoded by exon 1, is sufficient for complete transformation of primary BRK cells in cooperation with the *ras* oncogene (see references in Bayley and Mymryk 1994; Williams et al. 1995). Although the C-terminal region of E1A, encoded by exon 2, is not required for transformation with *ras*, it is required for plasmid-induced transformation of these cells in cooperation with E1B, and for transformation by the virus (Subramanian et al. 1991). In addition, the C-terminal region is required for immortalization of BRK cells, and deletion mapping was used to localize this activity to residues 225–238 (and 238–243) of the 243R protein (Quinlan and Douglas 1992; Boyd et al. 1993). Apparently, requirement for a second exon transforming function is nullified by activated *ras* but not by AdE1B. However, although this C-terminal region of E1A is not essential for cooperative transformation with *ras*, it does seem to influence this process negatively (Boyd et al., 1993; Douglas et al. 1991; Subramanian et al. 1989). Deletion of residues 225–238 (R271–284 in the 289R E1A protein) results in increased levels of transformation, greater degrees of tumorigenicity of transformants in newborn rats and athymic mice, and increased lung metastasis in the latter (Boyd et al. 1993). In addition to being a negative modulator of tumorigenesis and essential for immortalization, this short stretch of amino acids also binds the C-terminal binding protein (CtBP), a cellular phosphoprotein of 48 kDa. Mutational analysis within this region mapped the binding site to a highly conserved 5-aa element (R279–284) on the 289R E1A protein of Ad2/5 (Schaeper et al. 1995). This element in Ad12 (see Fig. 4 for location of that and the negative modulator) comprises residues 255–260 in the 266R protein, and also constitutes the core binding site for CtBP (Molloy et al. 1998; Schaeper et al. 1995). The nuclear localization signal (Lyons et al. 1987), which is also highly conserved among Ad serotypes, is C-terminally adjacent to the CtBP binding site.

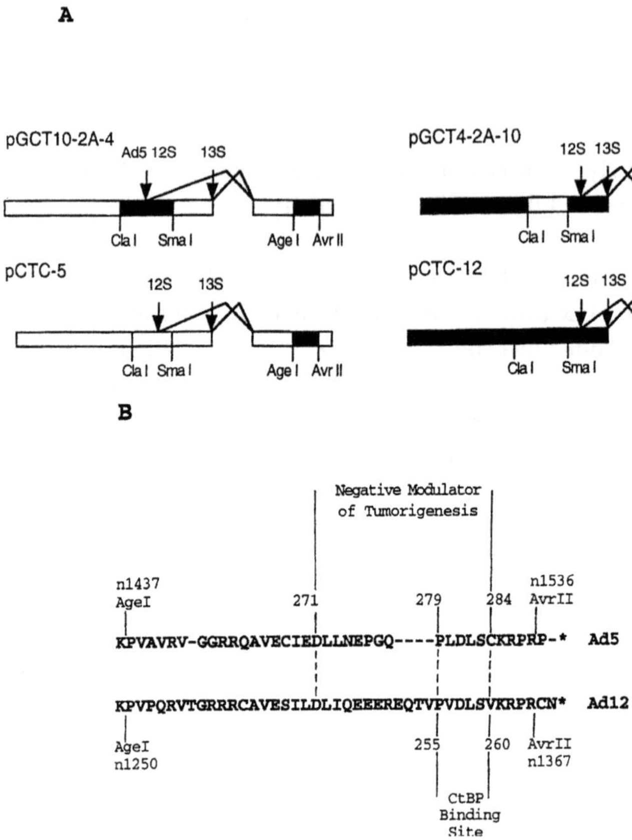

Fig. 4A, B Transfer of the Ad5 E1A C-terminal negative modulator sequence into the Ad12 E1A gene. **A** Structure of the Ad12/Ad5 chimeric E1A genes generated via reciprocal exchange of AgeI/AvrII (nts 1,250–1,367 in Ad12 and nts 1,437–1,536 in Ad5) cassettes. One of these (pPCTC-12) was used to create the Ad12-based chimeric virus Ch752 (see Sect. 5.1). Ad12 sequences are shown as *black blocks*, and Ad5 sequences are shown as *clear blocks*. **B** Comparative amino acid sequences in the Ad5 and Ad12 cassettes used to create these chimeric E1A genes, showing the locations of the negative modulator and the highly conserved PXDLS CtBP binding site. (Kimelman et al.1985; Schaeper et al.1995)

5.1
The Effects of Placing the Ad5 Negative Modulator in Ad12; Constructing and Analyzing Viruses Containing Chimeric E1A Genes

To determine whether the Ad5 C-terminal negative modulator of tumorigenesis acts within the context of virally transformed cells and isn't simply a product of the *ras* cooperative transformation system, we used a chimeric gene approach similar to the one used to show that the Ad12 spacer plays a role in oncogenesis and tumorigenesis (Telling and Williams 1995; see Sect. 3). This approach requires that small cassettes containing equivalent segments of Ad5 and Ad12 E1A genes can be reciprocally exchanged without reading frame disruption. To create C-terminal cassettes containing the sequence encoding the Ad5 E1A negative modulator and the equivalent sequence of the Ad12 gene, we used oligonucleotide mutagenesis (Kunkel 1985) to generate new *Age* I and *Avr* II restriction sites in sequences flanking these two regions. New *Age* I sites were generated in the same relative location in the second exons of Ad5 and Ad12 E1A by an A to G transition at n1437 and a T to G transversion at n1250, respectively. New *Avr* II sites were likewise generated at the same relative location in Ad5 and Ad12 second exons via a C to T transition at n1536 and an A to G transition at n 1367, respectively. The locations of these point mutations and new restriction sites are shown in Fig. 4. Methods for the transfer of doubly mutated segments from M13 constructs to E1A plasmids, and the construction of chimeric plasmids were similar to those described by Telling and Williams in 1994; to date two Ad12 and two Ad5-based chimeric E1A plasmids have been made (see Fig. 4). In pCTC-12, the Ad5 cassette was built into a wild-type Ad12 E1A plasmid, whereas in pGCT4–2A-10 it was built into an E1A chimeric plasmid from which the spacer cassette had been replaced by the equivalent Ad5 sequence. In pCTC-5, the Ad12 cassette was built into a wild-type Ad5 E1A plasmid, and in pGCT10–2A-4 it was inserted into a chimeric plasmid carrying the Ad12 spacer cassette. In all cases the M13 and plasmid constructs were sequenced to confirm that no undesirable alterations had occurred. The two Ad12-based chimeric plasmids were transferred into complete Ad12 genome by marker rescue (Lamberti and Williams 1990) of the Ad12 E1A deletion mutant *dl*701 (bp993–1,385 deleted), and correct cassette insertion was checked by sequencing viral DNA. The resultant viruses were found to grow and plaque on human A549 and HER3 cells as efficiently as Ad12 wild-type variant *var*747 (data not shown).

Transformation, tumor-induction, and analysis of transformant tumorgenicity in syngeneic, immunocompetent rats has to date been carried out only with *ch*752, the chimeric virus derived from pCTC-12. By comparison with Ad12 (*var*747) *ch*752 is strikingly defective for transformation of both primary BRK and primary rat embryo (RE) cells (Ta-

Table 4 Comparative transformation frequencies in primary cultures of BRK, Rat Embryo (RE), and BMK cells infected by Ad12 *var* 747, Ad12 E1A chimeric virus *ch* 752[a]

Cells	Virus	No. of foci/dish					
BRK	*var* 747	24	26	22	25	23	28
BRK	*ch* 752	0	1	2	2	1	0
BRK	none	0	0	0	0	0	0
RE	*var* 747	4	2	5	3	3	4
RE	*ch* 752	0	0	0	2	1	0
RE	None	0	0	0	0	0	0
BMK	*var* 747	28	30	27	27	25	32
BMK	*ch* 752	19	23	19	23	20	22
BMK	none	0	0	0	0	0	0

[a] The values correspond to the number of foci arising on individual replicate dishes. The procedures used for transformation were as described (Edbauer et al. 1988). In all cases the input multiplicity of infection was 10 pfu/cell. In the case of RE and BMK cells, immediately following infection the cells were moved by trypsinization, seeded on new dishes at 1×10^6 and 5×10^5 cells/dish, respectively, and grown at 36.5°C. The BRK cells were also grown at 36.5°C, but were not subcultured in this way. With BRK and BMK cultures, the final counts of foci were made at 31 days postinfection; in the case of RE cells final counts were made at 45 days.

ble 4). The reduction in transformation frequency is greatest on BRK cells, and is found at all multiplicities tested ranging from 1 to 100 pfu/cell (data not shown). The reductive effect is somewhat less on RE cells, and as we have consistently observed in the past, Ad12 and *var*747 transform them about fivefold less efficiently than BRK cells. Transformed foci induced by *ch*752 on cultures of these two cell types arose slightly later and grew more slowly than those generated by *var*747, although they were in general morphologically similar to the latter. When isolated and passaged to produce transformed cell lines, the chimeric cells initially grew somewhat more slowly than wild-type transformants, but once established (around 4–5 passages) their growth rates were similar to those of their Ad12 counterparts, and in general they grew to quite high densities. Somewhat surprisingly, although defective for transformation of rat cells, *ch*752 transforms BMK cells with frequencies very similar (about 20% lower) to those induced by *var*747 (see Table 4). In this case the *ch*752 foci arose at about the same time and grew just as vigorously as the wild-type foci. By comparison with *var*747, at the virus input levels used in these experiments, *ch*752 does not appear to induce

Table 5 Induction of tumors in Hooded Lister rats inoculated as newborns with Ad12 *ch* 752

Virus	Tumor-positive rats/rats inoculated[a]		
	6×10^8 pfu/ml	3×10^8 pfu/ml	1×10^8 pfu/ml
var 747[b]	N.D.	8/8	8/8
ch 752	3/19	2/17	3/22
Virus	Latency period (days)[c]		
	6×10^8 pfu/ml	3×10^8 pfu/ml	1×10^8 pfu/ml
var 747	N.D.	50±3	52±4
ch 752	94±15	147±18	165±18

[a] In all cases, 24–36-h-old newborn rats were inoculated intraperitoneally with 0·05 ml.

[b] The *var* 747-inoculated rats served as positive controls for both this experiment and the one shown in Table 3.

[c] Average time in days after inoculation at which palpable tumors were first detected ±standard deviation. Animals were killed when tumors reached a diameter of 3–4 cm, or 9 months after inoculation in animals in which no tumors developed.

N.D., not determined; pfu, plaque-forming unit.

an enhanced cytocidal, cytolytic, or cytopathic effect on any of these three cell types, more or less ruling out this as a possible reason for the reduced transformation frequencies observed in BRK and RE cultures. It is worth noting here that Ad5 also transforms BRK and BMK cells at frequencies some 20-fold lower than Ad12 (J. Williams, unpublished observations), and transforms RE cells at frequencies up to twofold higher than Ad12 (Williams et al. 1974; J. Williams, unpublished results). In strong contrast to Ad12, however, Ad5 induces a visibly cytocidal/cytolytic effect on BRK and BMK cells (somewhat less so on RE cells), which could perhaps explain, at least in part, its reduced capacity to transform these cells. As a rule, of course, at best only a very small proportion of the total number of cells in these primary cultures, which contain a varied mixture of cell types, are transformed, and it is probable that only a small subset of this mixture, most likely epithelial cells, are competent for transformation by adenoviruses. Although difficult to verify, a possible but unlikely explanation for the results obtained might be that infection by *ch*752 virus results in selective killing of BRK and RE, but not BMK target cells. An alternative explanation for the observed results, which could likely be substantiated, is that rat target cells infected by *ch*752 are incompletely or partially transformed, perhaps as a result of

Table 6 Tumorigenicity of transformed cell lines derived from BRK cells infected by the Ad12 E1A chimeric mutant *ch* 752

Cell line	Tumor incidence (no. of rats positive/total) by no. of cells inoculated[a]		
	1×10^7	3×10^6	1×10^6
RK747A	6/6 (8, 25)	6/6 (8, 25)[b]	6/6 (10, 31)
RK752A	6/6 (12, 60)	6/6 (17, 58)	N.D.
RK752C	6/6 (9, 36)	6/6 (12, 55)	5/6 (21, 52)
RK752D	6/6 (11, 41)	6/6 (15, 47)	3/6 (18, 60)
RK752E	3/6 (18, 82)	4/6 (26, 65)	0/6 (>90, 90)
RT752A	6/6 (8, 30)	6/6 (17, 36)	N.D.

[a] The conditions used for inoculation, examination of tumor growth and time of euthanization were identical to those used for BRK 710-derived cells (see Table 3). As in that case, some tumors appeared, grew slowly for some time and regressed (e.g., in rats inoculated with 3×10^6 RK 752D and with both 1×10^7 and 3×10^6 RK 752E); these animals were killed at 90 days postinoculation. In the case of rats inoculated with 1×10^6 RK 752E, no tumors showed up within the 90-day period (shown as >90). [b] Because these tests were carried out at or around the same time, the RK747A rats served as positive controls for both this experiment and the one shown in Table 3 (RK 710 cells).
N.D., not determined.

interference with or suppression of the normal progression of transformation events.

In view of these transformation results, it is perhaps not too surprising to find that the *ch*752 virus is also defective for tumor induction in Hooded Lister rats (Table 5). The few tumors which arose (8 out of 58 animals injected) did so with latency periods around three times as long as those induced by wild-type virus, suggesting that the *ch*752-transformed cells which gave rise to these tumors grew much more slowly in vivo than their wild-type counterparts. This is also true of cells transformed in vitro by *ch*752 (Table 6). In this case, four independently-derived 752 transformants, all of which are tumorigenic in syngeneic, immunocompetent adult rats, grew significantly more slowly than *var*747 transformed cell lines (Table 6 and see also Table 3). However, line RT752A, derived from a tumor induced in vivo by *ch*752 virus, behaved more like the 747 transformants, suggesting that a more tumorigenic line of cells is selected during the progressive development of the tumor in vivo. It should be noted that all of these 752 transformants show normal production of E1A and E1B proteins compared to 747 transformants (Y. Zhang and J. Williams, unpublished results). To date, MHC1 antigen levels and NK susceptibilities have not been determined in these cell lines. Collectively, these results, in which viruses and virally induced transformants were used, support the general conclusions obtained using cells transformed by wild-type and mutant 2/5 E1A and activated *ras* plasmids (Boyd et al. 1993). Those investigators also reported that in addition to negatively influencing transformation in vitro and transformant tumorigenicity in newborn syngeneic rats, the 2/5 E1A C-terminal region also modulates the capacity of transformants to form metastases in the lungs of athymic mice. Although it was reported that transplantable Ad5 tumors can metastasize to the lungs and liver of baby hamsters (Williams 1973), we have yet to find any clear evidence of metastasis to these sites in Hooded Lister rats injected with Ad12 virus or Ad12 transformants. Tumors formed by Ad12 *var*747 and *ch*752 transformants at the subcutaneous site of inoculation are solid sarcomatous masses, which do not spread locally or metastasize to internal organs such as lungs or liver. These differences can most likely be explained by the fact that adult immunocompetent rats were used as recipients in the Ad12 transformant studies, whereas the recipients of the Ad5 tumor cells and the Ad2/5-*ras* transformants were baby hamster and immunodeficient, athymic mice, respectively. The expression of *ras* in the latter cell lines undoubtedly also influences their capacity to grow aggressively in vivo. Tumors induced by Ad12 *var*747 and *ch*752 inoculated intraperitoneally into newborn rats were solid and sarcomatous, and some of them had soft, somewhat necrotic centers. In the case of *var*747, multiple tumors were often found in the peritoneal cavity, some of which grew more aggressively into the cavity wall and occasionally became visible as nodules just under the body surface, although lung or liver metastases were not observed. This aggressive growth was not displayed by tumors induced by *ch*752 virus, and they generally remained as single, solid masses.

5.2
How Does the C-Terminal Negative Modulator Influence Tumorigenesis?

The mechanism by which the C-terminal negative modulator influences tumorigenesis and metastasis of Ad2/5-*ras* transformants is not yet understood, but the prevailing model implicates the cellular C-terminal binding protein (CtBP). As mentioned above, this 48-kDa protein interacts with a 5-aa sequence (PXDLS) which is highly conserved in the E1A proteins of all adenoviruses, and comprises residues 279–283 within the negative modulator (R271–284) region in the Ad5 E1A 289R molecule (Boyd et al. 1993; Molloy et al. 1998; Schaeper et al. 1995). Although the role of CtBP in mammalian cells remains to be elucidated, the E1A CtBP binding domain has been shown to be involved in repressing CR1-dependent transactivation (Sollerbrant et al. 1996), suggesting that this cellular protein carries out a transcriptional regulatory function in adenovirus-infected and normal cells. Support for this view stems from the finding that CtBP interacts with another cellular protein, the C-terminal interacting protein (CtIP), and that this interaction can be disrupted by PLDLS-containing peptides (Schaeper et al. 1998). As with CtBP, the role of mammalian cell CtIP has still to be determined, but *Drosophila* homologs of this protein (e.g., Hairy) are known to be transcriptional repressors which interact with corepressors and are essential for embryonic development (Molloy et al. 2001; Nibu et al. 1998; Poortinga et al. 1998). Although it seems highly probable that mammalian CtBPs and their PXDLS-containing interacting partners are crucial players in the complex game of transcriptional regulation, it is by no means clear at present what, if any, role such processes might play in shaping the transformed cell phenotype, whether it be induced by E1A/*ras* plasmids or by infectious adenovirus. Furthermore, at least a few of the E1A functions necessary for complete transformation of rodent cells by the virus are likely to be different from those required for E1A/*ras* transformation, making direct comparison of events occurring in the respective systems somewhat tenuous at best. For example, in the virus system, E1A must cooperate with E1B to bring about complete transformation and tumorigenicity of BRK cells (see Sect. 2.2), and the integrity of the E1A second exon, while not required for plasmid-induced cooperative transformation with *ras*, is required for plasmid-induced cooperative transformation with E1B (see Sect. 5). Assuming for the moment that the plasmid and viral systems can be equated, and that the PXDLS domain, which is highly conserved in adenoviruses, is a critical component of negative modulation, it would appear that it does not function as such in Ad12, which is highly oncogenic. Nevertheless, our results with *ch*752-induced BRK transformants (and the tumors induced in vivo) suggest that the expression of the Ad5E1A negative modulator causes a significant reduction in the tumorigenicity of these otherwise fully transformed cells, suggesting that the phenomenon is not peculiar to E1A/*ras* transformants. The effect of the Ad5 C-terminal sequences on Ad12 transformation of BRK and RE cells in vitro, is, however, somewhat puzzling. Of course, the cassette transferred from Ad5 E1A to Ad12 E1A carries an additional 25-amino acid residues N-terminal to the PLDLS sequence (see Fig. 4), and that segment, or part of it, could contribute to the effect(s) observed. In this context it is worth noting again that deletion of residues 256–273 (R210 to 227 in the Ad5 243R protein) results in enhanced transformation with *ras* and failure of CtBP to bind to the mutant E1A protein (Douglas et al. 1991; Fischer and Quinlan 2000). This finding could be taken to mean that the upstream sequence acts independently of the PLDLS sequence to negatively modulate transformation. Alternatively, the deletion could bring about a spatial or conformational effect resulting in decreased affinity of the PLDLS sequence for CtBP, giving rise in turn to elevated transformation; this possibility seems more likely in view of the finding that peptides with identical PXDLS sequence flanked by divergent N- and C-terminal sequences display markedly different structures and binding affinities for CtBP (Molloy et al. 2001). Of course, the reductive effect of C-terminal cassette exchange on transformation of rat cells observed here might, instead, result from more extensive spatial and/or conformational perturbations of the E1A proteins. Potentially, this could lead to an alteration in the binding affinity of another E1A sequence for a cellular protein, thus interfering with a process which is required for the establishment of complete transformation. Whatever the reason for the negative effect on transformation, it occurs only in BRK and RE cells and not in BMK cells (Table 4); the host-range nature of this effect strongly suggests that a cellular factor is involved.

6 Concluding Remarks

The striking difference in the oncogenicity of the group A and group C adenoviruses and in the tumorigenicity of their respective transformants cannot be explained by any single process or factor acting alone. Although E1B certainly influences transformation and tumorigenicity, it is clear that the E1A products play a major role in developing the transformed cell phenotype, and as we have discussed above, multiple functions of the E1A genes likely act together to determine the outcome. The tactics used by adenoviruses to ensure transformation and tumorigenicity of rodent cells have undoubtedly been borrowed from the repertoire of tactics developed during the evolution of these viruses to ensure their replication and survival in their natural hosts. For example, the capacity of Ad12 E1A to repress transcription of MHCI genes, which is observed in Ad12 transformants, is considered to be essential to the establishment of persistent infections in humans by allowing infected cells to escape CTL-mediated immune surveillance (Pääbo et al. 1989; Ricciardi 1999). In contrast, during Ad5 infection of human target cells, the transport of class I molecules to the cell surface is impaired by action of an E3 gene product (reviewed in Wold and Tollefson 1999). In Ad5 transformants E3 is rarely, if ever, present and expressed, so these cells have normal levels of class I antigens and are susceptible to CTL action. These hypotheses and results help to explain the differential tumorigenicity of Ad12 and Ad5 transformants, and are consistent with the view that reduction of MHCI antigen levels is necessary for development of tumorigenicity in Ad12 transformants. In light of our results with the transformants induced by Ad12 spacer chimerics and mutants, however, we would argue that this event may be required, but is not sufficient for tumorigenesis. Evasion of NK cell surveillance has also been invoked as an important contributing element in Ad12 transformant tumorigenicity. As discussed above (Sects. 2.2 and 4.4), Ad12 transformants are nonsusceptible to these cells, whereas Ad5 transformants are generally highly susceptible. Again, however, results with the Ad12 spacer chimeric and mutant-induced transformants suggest that although NK nonsusceptibility may be required for tumorigenicity, it is not sufficient. Collectively, these results clearly show that the spacer region is not responsible for class I repression, and seems to have little or no influence on NK susceptibility, although it cannot be ruled out that the latter property is determined by both the spacer and an exon 2 domain.

At present, we do not know the mechanism by which the spacer determines Ad12 oncogenicity and the strong tumorigenicity of Ad12 transformants, but it almost certainly doesn't act alone. When the spacer sequence is placed in Ad5 E1A and rescued into Ad5 genomic background, the resultant chimeric viruses do not induce tumors in Hooded Lister rats, and transformants induced by these viruses are not tumorigenic in syngeneic rats. Thus, the spacer determinant may be required, but, it too, is not sufficient for oncogenesis and high tumorigenicity. Potentially, this result could stem from influence of the Ad5 C-terminal negative modulator (see Sect. 5). To test this possibility, we built the cassette containing the Ad12 C-terminal segment into an Ad5-based E1A plasmid containing the Ad12 spacer (see Fig. 4), and generated transformants by cooperative transfection with Ad12 E1B. None of these plasmid-induced transformants produced tumors in syngeneic, immunocompetent rats (J. Williams, unpublished results), suggesting that other regions of Ad12 are also required in addition to the spacer (and removal of the negative modulator), as we and others have proposed (Jelinek 1994; Williams et al. 1995). Of course, immunoevasion mechanisms must also be included in the overall scenario. Furthermore, whereas E1B is clearly required for complete transformation and important for tumor induction by virus, viral genes elsewhere on group A or group C genomes may also influence transformation in vitro, the phenotype of transformants, and/or viral oncogenicity (Bernard et al. 1984; Miller and Williams 1987; Rice et al. 1987; Sawada et al. 1988; Shiroki et al. 1984). Gradually, the combined viral and reductionist plasmid approaches are sketching out a picture of adenovirus oncogenic transformation as a complex and progressive series of events guided primarily by E1A gene functions, but influenced by the activities of E1B and other viral gene products. The picture is far from complete, but as further details are painted in and we learn more about the process of virally induced oncogenesis, we will assuredly also find out more about the factors that regulate virus-cell interactions in the natural host, *Homo sapiens*.

Acknowledgements. We are most grateful to Robin Rentka for valuable assistance with the preparation of this manuscript. Work from the authors' laboratories was supported by Public Health Service (NCI) grant CA32940 and a Winters Foundation grant to JFW, and Public Health Service (NCI) grant CA29797 to RPR.

References

Ackrill AM, Blair GE (1988) Regulation of major histocompatibility class I gene expression at the level of transcription in highly oncogenic adenovirus transformed rat cells. Oncogene 3:483–487

Bayley ST, Mymryk JS (1994) Adenovirus E1A proteins and transformation (Review). Int J Oncol 5:425–444

Bernards R, Houweling A, Schrier PI, Bos JL, van der Eb AJ (1982) Characterization of cells transformed by Ad5/Ad12 hybrid early region I plasmids. Virology 120:422–432

Bernards R, Schrier PI, Houweling A, Bos JL, van der Eb AJ, Zylstra M, Melief CJM (1983) Tumorigenicity of cells transformed by adenovirus type 12 by evasion of T-cell immunity. Nature (London) 350:776–779

Bernards R, de Leeuw M, Vaessen MJ, van der Eb AJ (1984) Oncogenicity by adenovirus is not determined by the transforming region only. J Virol 50:847–853

Boulanger PA, Blair GE (1991) Expression and interactions of human adenovirus oncoproteins. Biochem J 275:281–299

Boyd JM, Subramanian T, Schaeper U, La Regina M, Bayley S, Chinnadurai G (1993) A region in the C-terminus of adenovirus 2/5 E1A protein is required for association with a cellular phosphoprotein and important for the negative modulation of T24-*ras* mediated transformation, tumorigenesis and metastasis. EMBO J 12:469–478

Boyer TG, Martin ME, Lees E, Ricciardi RP, Berk AJ, (1999) Mammalian Srb/Mediator complex is targeted by adenovirus E1A protein. Nature 399:276–279.

Byrd PJ, Grand RJA, Breiding D, Williams J, Gallimore PH (1988) Host range mutants of adenovirus type 12 E1 defective for lytic infection, transformation and oncogenicity. Virology 163:155–165

Chadwick BS, Sambhara SR, Sasakura Y, Miller RG (1992) Effect of class I MHC binding peptides on recognition by natural killer cells. J Immunol 149:3150–3156

Chan V-L, Smith M (1984) In vitro generation of specific deletions in DNA cloned in M13 vectors using synthetic oligodeoxyribonucleotides: mutants in the 5'-flanking region of the yeast alcohol dehydrogenase II gene. Nucleic Acids Res.12:2407–2419.

Cook JL, Hibbs Jr JB, Lewis Jr. AM (1982) DNA virus-transformed hamster cell-host effector cell interactions: level of resistance to cytolysis correlated with tumorigenicity. Int J Cancer 30:795–803

Cook JL, May DL Wilson BA Holskin B Chen MJ Shalloway D, Walker TA (1989) Role of tumor necrosis factor-alpha in E1A oncogene-induced susceptibility of neoplastic cells to lysis by natural killer cells and activated macrophages. J Immunol 142:4527–4534

Douglas JL, Gopalakrishnan S, Quinlan MP (1991) Modulation of transformation of primary epithelial cells by the second exon of the Ad5 E1A 12S gene. Oncogene 6:2093–2103

Eager KB, Williams J, Breiding D, Pan S, Knowles B, Appela E, Ricciardi RP (1985) Expression of histocompatibility antigens H-2 K, D, and L is reduced in adenovirus-12-transformed mouse cells and is restored by interferon γ Proc Natl Acad Sci USA 82:5525–5529

Edbauer D, Lamberti C, Tong J, Williams J (1988) Adenovirus type 12 E1B 19-kilodalton protein is not required for oncogenic transformation in rats. J Virol 62:3265–3273

Fischer RS, Quinlan MP (1998) Expression of the Rb binding regions of E1A enables efficient transformation of primary epithelial cells by v-src. J. Virol 72:2815–2824

Fischer RS, Quinlan MP (2000) While E1A can facilitate epithelial cell transformation by several dominant oncogenes, the C-terminus seems only to regulate *rac* and *cdc* 42 function, but in both epithelial and fibroblastic cells. (2000) Virology 269:409–419

Friedman DJ, Ricciardi RP (1988) Adenovirus type 12 E1A gene represses accumulation of MHC class I mRNA at the level of transcription. Virology 165:303–305

Gallimore PH (1972) Tumor production in immunosuppressed rats with cells transformed in vitro by adenovirus type 2. J Gen Virol 16:99–102

Gallimore PH, Williams J (1982) An examination of adenovirus type 5 mutants for their ability to induce group C adenovirus tumor-specific antigenicity in rats. Virology 120:146–156

Gallimore PH, Byrd PJ, Grand RJA (1984a) Adenovirus genes involved in transformation. What determines the oncogenic phenotype?, p. 125–172. In Rigby PWJ, Wilkie NM (ed.), Symposium of the Society for General Microbiology. Viruses and Cancer. Cambridge University Press, Cambridge

Gallimore P, Byrd P, Grand R, Whittaker J, Breiding D, Williams J (1984b). An examination of the transforming and tumor-inducing capacity of a number of adenovirus type 12 early region 1, host-range mutants and cells transformed by subgenomic fragments of Ad12 E1 region. Cancer Cells 2:519–526

Gallimore PH, Williams J, Breiding D, Grand RJA, Rowe M., Byrd P (1986) Studies on adenovirus type-12 E1 region: gene expression, transformation of human and rodent cells, and malignancy. Cancer Cells 4:339–348

Ge R, Kralli A, Weinmann R, Ricciardi RP. (1992) Down-regulation of the major histocompatibility complex class I enhancer in adenovirus type 12-transformed cells is accompanied by an increase in factor binding. J.Virol. 66:6969–6978

Ge R, Liu X, Ricciardi RP (1994) E1A oncogene of adenovirus-12 mediates trans-repression of MHC class I transcription in Ad5/Ad12 somatic hybrid transformed cells. Virology. 203:389–392

Graham FL, Harrison T, Williams J (1978) Defective transforming capacity of adenovirus type 5 host-range mutants. Virology 86:10–21

Graham FL (1984) Transformation by and oncogenicity of human adenoviruses, p 339–398. In Ginsberg HS (ed.), The adenoviruses. Plenum Press, New York

Haddada H, Lewis Jr AM, Sogn JA, Coligan JE, Cook JL, Walker TA, Levine AS (1986) Tumorigenicity of hamster and mouse cells transformed by adenovirus types 2 and 5 is not influenced by the level of class I major histocompatibility antigens expressed on the cells. J Virol 83:9684–9688

Haddada H, Sogn JA, Coligan JE, Carbone M, Dixon K, Levine AS, Lewis, Jr, AM (1988) Viral gene inhibition of class I major histocompatibility antigen expression: not a general mechanism governing the tumorigenicity of adenovirus type 2-, adenovirus type 12-, and simian virus 40-transformed Syrian hamster cells. J Virol 2755–2761

Hayashi H, Tanaka K, Jay F, Khoury G, Jay G (1985) Modulation of the tumorigenicity of human adenovirus-12-transformed cells by interferon. Cell. 43:263–267

Houweling A, Van den Elsen PJ, van der Eb AJ (1980) Partial transformation of primary rat cells by the leftmost 4.5% fragment of adenovirus 5 DNA. Virology 105:537–550

Huebner RJ, Rowe WP, Lane WT (1962) Oncogenic effects in hamsters of human adenovirus types 12 and 18. Proc Natl Acad Sci USA 48:2051–2058

Huvent I, Cousin C, Kiss A, Bernard C, D'Halluin JC (1996) Susceptibility to natural killer cells and down regulation of MHC class I expression in adenovirus 12 transformed cells are regulated by different E1A domains. Virus Res 45:123–134

Javier RT, Raska K Jr, Shenk T (1992) Requirement for the adenovirus type 9 E4 region in production of mammary tumors. Science 257:1267–1271

Javier RT (1994) Adenovirus type 9 E4 open reading frame 1 encodes a transforming protein required for the production of mammary tumors in rats. J Virol 68:3917–3924

Jelinek T, Pereira DS, Graham FL (1994) Tumorigenicity of adenovirus-transformed rodent cells is influenced by at least two regions of adenovirus type 12 early region 1A. J Virol 68:888–896

Jones NC (1992) The multifunctional products of the adenovirus E1A gene, p. 87–113. In Doerfler W, Böhm P (ed), Malignant transformation by DNA viruses. VCH Publishers, Weinheim, Germany

Jonsson N, Ankerst J (1977) Studies on adenovirus type 9-induced mammary fibroadenomas in rats and their malignant transformation. Cancer 39:2513–2519

Kärre K, Ljunggren HG, Pointek G, Kiessling R (1986) Selective rejection of H-2 deficient lymphoma variants suggests alternative immune defense strategy. Nature 319:675–678

Kast WM, Offringa R, Peters PJ, Voordouw AC, Meloen RH, van der Eb AJ, Melief CJM (1989) Eradication of adenovirus E1-induced tumors by E1A-specific cytotoxic T lymphocytes. Cell 59:603–614

Kenyon DJ, Raska K Jr (1986) Region E1a of highly oncogenic adenovirus 12 in transformed cells protects against NK but not LAK cytolysis. Virology.155:644–654

Kimelmann D, Miller JS, Porter D, Roberts BE (1985) E1A regions of the human adenoviruses and of the highly oncogenic simian adenovirus 7 are closely related. J Virol 53:399–409

Kralli A, Ge R, Graeven U, Ricciardi RP (1992) Weinmann R. Negative regulation of the major histocompatibility complex class I enhancer in adenovirus type 12-transformed cells via a retinoic acid response element. J.Virol. 66:6979–6988

Krantz CK, Routes BA, Quinlan MP, Cook JL (1996) E1A second exon requirements for induction of target cell susceptibility to lysis by natural killer cells: implications for the mechanism of action. Virology 217:23–32

Kunkel TA (1985) Rapid and efficient site-specific mutagenesis without phenotypic selection. Proc Natl Acad Sci USA 82:488–492

Kushner DB, Pereira DS, Liu X, Graham FL, Ricciardi RP (1996) The first exon of Ad12 E1A excluding the transactivation domain mediates differential binding of COUP-TF and NF-κB to the MHC class I enhancer in transformed cells. Oncogene 12:143–151

Kushner DB, Ricciardi RP (1999) Reduced phosphorylation of p50 is responsible for diminished NF-κB binding to the major histocompatibility complex class I enhancer in adenovirus 12 transformed cells. Mol Cell Biol 99:2169–2179

Lamberti C, Williams J (1990) Differential requirement for adenovirus type 12 E1A gene products in oncogenic transformation. J Virol 64:4997–5007

Lanier LL (2000) The origin and functions of natural killer cells. Clin Immunol. 95:S14–18. Review

Larsen PL, Tibbetts C (1987) Adenovirus E1A gene autorepression: revertants of an E1A promoter mutation encode altered E1A proteins. Proc Natl Acad Sci USA 84:8185–8189

Lewis Jr, AM, Cook JL (1982) Spectrum of tumorigenic phenotypes among adenovirus 2, adenovirus 12 and simian virus 40 transformed Syrian hamster cells defined by host cellular immune-tumor cell interactions. Cancer Res 42:939–944

Licht JD, Grossel MU, Figge J, Hansen UM (1990) Drosophila Krüppel protein is a transcriptional repressor. Nature (London) 346:76–79

Liu X, Ge R, Ricciardi RP (1996) Evidence for the involvement of a nuclear NF-κB inhibitor in global down-regulation of the major histocompatibility complex class I enhancer in adenovirus type12-transformed cells. Mol Cell Biol 16:398–404

Lyons RH, Ferguson BQ, Rosenberg M (1987) pentapeptide nuclear localization signal in adenovirus E1A. Mol Cel Biol 7:2451–2456

Mayo MW, Baldwin AS (2000) The transcription factor NF-kappaB: control of oncogenesis and cancer therapy resistance. Biochim Biophys Acta1470:55–62. Review

Mautner V, Steinthorsdottir V, Bailey A (1995) Enteric adenoviruses. Curr Top Microbiol Immunol 199/II 229–282

McLorie W, McGlade CJ, Takayesu D, Branton PE (1991) Individual adenovirus E1B proteins induce transformation independently but by additive pathways. J Gen Virol 72:1467–1471

Mellow GH, Föhring B, Dougherty J, Gallimore PH, Raska K (1984) Tumorigenicity of adenovirus-transformed rat cells and expression of class I major histocompatibility antigen. Virology 134:951–961

Miller BW, Williams J (1987) Cellular transformation by adenovirus type 5 is influenced by the viral DNA polymerase. J Virol 61, 11:3630 3634

Molloy DP, Milner AE, Yakub IK, Chinnadurai G, Gallimore PH, Grand RJA (1998) Structural determinants present in the C terminal binding protein binding site of adenovirus early region 1A proteins. J Biol Chem 273:20867–20876

Molloy DP, Barral PM, Bremmner KH, Gallimore PH, Grand RJA (2001) Structural determinants outside the PXDLS sequence affect the interaction of adenovirus E1A, C-terminal interacting protein and Drosophila repressors with C-terminal binding protein. Biochim Biophys Acta 1546:55–70

Montell C, Courtois G, Eng C, Berk A (1984) Complete transformation by adenovirus 2 requires both E1A proteins. Cell 36:951–961

Moran E, Grodzicker T, Roberts RJ, Mathews MB, Zerler B. (1986) Lytic and transforming functions of individual products of the adenovirus E1A gene. J Virol 57:765–775

Nibu & Zhang H, Levine M (1998) Interaction of short-range repressors with *Drosophila* CtBP in the embryo. Science 280:101–104

Nielsch U, Zimmer SG, Babiss LE (1991) Changes in NF-kappa B and ISGF3 DNA binding activities are responsible for differences in MHC and beta-IFN gene expression in Ad5- versus Ad12-transformed cells. EMBO J 10:4169–4175

Pääbo S, Severinsson L., Andersson M, Martens I, Nilsson T, Peterson PA (1989) Adenovirus proteins and MHC expression. Adv Cancer Res 52:151–163

Poortinga G, Watanabe M, Parkhurst SM (1998) *Drosophila* CtBP: A hairy-interacting protein required for embryonic segmentation and hairy-mediated transcriptional repression. EMBO J 17:2067–2078

Pointek GE, Tanignchi K, Ljunggren HG, Grönberg A, Kiessling R, Klein G, Kärre K (1985) YAC-1 MHC class I variants reveal an association between decreased NK

sensitivity and increased H-2 expression after interferon treatment of in vivo passage. J Immunol 135:4281–4288
Quinlan MP, Douglas JL (1992) Immortalization of primary epithelial cells requires first and second exon functions of adenovirus type 5 12S. J. Virol 66:2020–2030
Raska K Jr, Gallimore PH (1982) An inverse relation of the oncogenic potential of adenovirus-transformed cells and their sensitivity to killing by syngeneic natural killer cells. Virology 123:8-18
Raska K Jr (1995) Functional domains of adenovirus E1A oncogenes which control interactions with effectors of cellular immunity. Curr Top Microbiol Immunol 199:131–148
Ricciardi RP. (1999) Adenovirus transformation and tumorigenicity In: Seth P (ed) Adenoviruses: Basic biology to gene therapy. Austin: RG Landes Co. 217–227.
Rice SA, Klessig DF, Williams J (1987) Multiple effects of the 72-kDa, adenovirus-specified DNA binding protein on the efficiency of cellular transformation. Virology 156:366–376
Roberts BE, Miller JS, Kimelmann D, Cepko CL, Lemischka IR, Mulligan RC (1985) Individual adenovirus type 5 early region 1A gene products elicit distinct alterations of cellular morphology and gene expression. J Virol 56:404–413
Routes JM, Bellgrau D, McGrory WJ, Bautista DS, Graham FL, Cook JL (1991) Anti-adenovirus type 5 cytotoxic T lymphocytes: immunodominant epitopes are encoded by the E1A gene. J Virol 65:1450–1457
Ruley HE (1983) Adenovirus early region 1A enables viral and cellular transforming genes to transform primary cells in culture. Nature (London) 304:602–606
Rumpf H, Esche H, Kirch H-C (1999) Two domains within the adenovirus type 12 E1A unique spacer have disparate effects on the interaction of E1A with p105Rb and the transformation of primary mouse cells. Virology 257:45–53
Sawada Y, Föhring B, Shenk T, Raska K Jr (1985) Tumorigenicity of adenovirus transformed cells: region E1A of adenovirus 12 confers resistance to natural killer cells Virology 147:413–421
Sawada Y, Urbanelli D, Raskova J, Shenk T, Raska K Jr (1986) Adenovirus tumor-specific transplantation antigen is a function of the E1A early region. J Exp Med 163:563–572
Sawada Y, Raska K Jr, Shenk T (1988) Adenovirus type 5 and type 12 recombinant viruses containing heterologous E1 genes are viable, transform rat cells, but are not tumorigenic in rats. Virology 166:281–284
Sawada Y, Raskova J, Fujinaga K, Raska K Jr (1994) Identification of functional domains of adenovirus tumor-specific transplantation antigen in types 5 and 12 by viable viruses carrying chimeric E1A genes. Int J. Cancer 57:598–603
Schaeper U, Boyd JM, Verma S, Uhlmann E, Subramanian T, Chinnadurai G (1995) Molecular cloning and characterization of a cellular phosphoprotein that interacts with a conserved C-terminal domain of adenovirus E1A involved in negative modulation of oncogenic transformation. Proc Natl Acad Sci 92:10467–10471
Schaeper U, Subramanian T, Lim L, Boyd JM, Chinnadurai G (1998) Interaction between a cellular protein that binds to the C-terminal region of adenovirus E1A (CtBP) and a novel cellular protein is disrupted by E1A through a conserved PLDLS motif. J Biol Chem 273:8549–8552

Schrier PI, Bernards R, Vaessen RTMJ, Houweling A, van der Eb AJ (1983) Expression of class I major histocompatibility antigens switched off by highly oncogenic adenovirus 12 in transformed rat cells. Nature (London) 305:771–775

Scott MO, Kimelman D, Norris D, Ricciardi RP (1984) Production of a monospecific antisera against the E1A protein of Ad12 and Ad5 by an Ad12 E1A/β-galactosidase fusion protein expressed in bacteria. J. Virol. 50:895–903

Shiroki K, Segawa K, Saito I, Shimojo H, Fujinaga K (1979) Products of the adenovirus 12 transforming genes and their functions. Cold Spring Harbor Symp Quant Biol 44:533–540

Shiroki K, Hashimoto S, Saito I, Fukui Y, Fukui Y, Hiroyuki K, Shimojo H (1984) Expression of the E4 gene is required for establishment of soft-agar colony-forming rat cell lines transformed by the adenovirus 12 E1 gene. J Virol 50, 3:854–863

Sjögren HO, Minowada J, Ankerst J (1968) Specific transplantation antigens of mouse sarcomas induced by adenovirus type 12. J Exp Med 125:689–701

Smirnov DA, Hou S, Ricciardi RP (2000) Association of histone deacetylase with COUP-TF in tumorigenic Ad12 transformed cells and its potential role in shut-off of MHC class I transcription. Virology 268:319–328

Smirnov DA, Hou S, Liu X, Claudio E, Siebenlist UK, Ricciardi RP (2001). COUP-TFII is up-regulated in adenovirus type 12 tumorigenic cells and is a repressor of MHC class I transcription. Virology 284:13–19

Soddu S, Lewis Jr AM (1992) Driving adenovirus type 12-transformed BALB/c mouse cells to express high levels of class I major histocompatibility complex proteins enhances, rather than abrogates, their tumorigenicity. J Virol 66:2875–2884

Sollerbrant K, Chinnadurai G, Svensson K (1996) The CtBP binding domain in the adenovirus E1A protein controls CR1-dependent transactivation. Nucleic Acids Res 24:2578–2584

Storkus WJ, Salter RD, Cresswell P, Dawson JR (1992) Peptide-induced modulation of target cell sensitivity to natural killing. J Immunol 149:1185–1190

Subramanian T, Malstrom SE, Chinnadurai G (1991) Requirement of the C-terminal region of adenovirus E1A for cell transformation in cooperation with E1B. Oncogene 6:1171–1173

Subramanian T, La Regina M, Chinnadurai G (1989) Enhanced *ras* oncogene mediated cell transformation and tumorigenesis by adenovirus 2 mutants lacking the C-terminal region of E1A protein. Oncogene 4:415–420

Tanaka K, Isselbacher KJ, Khoury G, Jay G (1985) Reversal of oncogenesis by the expression of a histocompatibility complex class I gene. Science 228:26–30

Telling GC, Williams J (1994) Constructing chimeric type 12/type 5 adenovirus E1A genes and using them to identify an oncogenic determinant of adenovirus type 12. J Virol 68, 2:877–887

Trentin JJ, Yabe Y, Taylor G (1962) The quest for human cancer viruses. Science 137:835–841

Trentin JJ, Bryan E (1966) Virus induced transplantation immunity to human adenovirus type 12 tumors of the hamster and mouse. Proc Soc Exp Biol Med 121:1216–1219

Urbanelli D, Sawada Y, Raskova J, Jones NC, Shenk T, Raska K (1989) C-terminal domain of the adenovirus E1A oncogene product is required for induction of cytotoxic T lymphocytes and tumor-specific transplantation immunity. Virology 173:607–614

van der Eb AJ, Mulder C, Graham FL, Houweling A (1977) Transformation with specific fragments of adenovirus DNAs. I. Isolation of specific fragments with transforming activity of adenovirus 2 and 5 DNA. Gene 2:115–132

van der Eb AJ, Zantema A (1992) Adenovirus oncogenesis. p 115–140. In Doerlfer W, Böhm P. (ed.) Malignant transformation by DNA viruses. VCH Publishers, Weinheim, Germany

van den Elsen PJ, de Pater S, Houweling A, van der Veer J, van der Eb AJ (1982) The relationship between region E1A and E1B of human adenoviruses in cell transformation. Gene 18:175–185

van Ormondt H, Galibert F (1984) Nucleotide sequences of adenovirus DNAs. Curr Top Microbiol Immunol 110:73–142

Vasavada R, Eager KB, Barbanti-Brodano G, Caputo A, Ricciardi RP. (1986) Adenovirus type 12 early region 1A proteins repress class I HLA expression in transformed human cells. Proc. Natl. Acad. Sci. USA 83:5257–61.

Williams JF, Young CSH, Austin PE (1974) Genetic analysis of human adenovirus type 5 in permissive and nonpermissive cells. Cold Spring Harbor Symp Quant Biol 39:427–437

Williams J (1986) Adenovirus genetics, p. 247–309. In Doerfler W (ed.), Adenovirus DNA: the viral genome and its expression. Martinus Nijhoff, The Hague, the Netherlands.

Williams J (1973) Oncogenic transformation of hamster embryo cells in vitro by adenovirus type 5. Nature 243:162–163

Williams J, Williams M, Liu C, Telling G (1995) Assessing the role of E1A in the differential oncogenicity of group A and group C human adenoviruses. Curr Top Microbiol Immunol 199/II 149–175

Winberg G, Shenk T (1984) Dissection of overlapping functions within the adenovirus type 5 E1A gene. EMBO J 3:1907–1912

Wold WS, Tollefson AE (1999) Adenovirus-host interactions to subvert host immune system. In: Seth P (ed) Adenoviruses: Basic biology to gene therapy. Austin: RG Landes Co. 245–252

Yewdell JW, Bennink JR, Eager KB, Ricciardi RP. (1988) CTL recognition of adenovirus-transformed cells infected with influenza virus: lysis by anti-influenza CTL parallels adenovirus-12-induced suppression of class I MHC molecules. Virol. 162:236–238

Zhou C, Tsai SY, Tsai M. (2000) >From apoptosis to angiogenesis: new insights into the roles of nuclear orphan receptors, chicken ovalbumin upstream promoter-transcription factors, during development. Biochim Biophys Acta 1470:M63–68. Review

Zinkernagel RM, Doherty PC (1979) MHC-restricted cytotoxic T cells: studies on the biological role of polymorphic major transplantation antigens determining T-cell restriction-specificity, function and responsiveness. Adv Immunol 27:51–177

3
Gene Therapy

Replicating Adenoviruses in Cancer Therapy

M. Dobbelstein

Institut für Virologie, Philipps-Universität Marburg, Robert Koch Str. 17,
35037 Marburg, Germany
E-mail: dobbelst@mailer.uni-marburg.de

Abstract The potential use of adenoviruses in therapy against cancer has evoked a rapidly moving field of research. Unlike conventional gene therapy vectors, oncolytic adenoviruses retain the ability to replicate. However, replication is restricted as much as possible to tumor cells, with the aim of eliminating these cells through viral cytotoxicity. The two key issues are to improve the efficiency of virus replication and cell killing while ensuring the specificity of these activities for tumor cells. Wild-type adenoviruses as such may already be usable for cancer therapy. Strategies to further improve efficiency and specificity include the partial or complete removal of viral genes. The idea is that functions carried out by the corresponding gene products are not required for replication in tumor cells, but are needed in normal cells. Accordingly, the removal of genes encoding E1B-55 kDa or E1B-19 kDa, or the mutation of E1A may improve the selective killing of tumor cells. On the other hand, the overexpression of the adenovirus death protein (ADP) may enhance viral spread and oncolytic efficiency. Other strategies to improve the specific oncolytic activity of replicating adenoviruses have been pursued. For instance, some promoters are active specifically in tumor cells, and these promoters were introduced into the viral genome, to regulate essential viral genes. Moreover, replicating viruses were engineered to express toxic proteins or drug converters. A number of these viruses have been tested successfully using tumor xenografts in nude mice as a model system. An oncolytic adenovirus lacking the E1B-55 kDa gene product, termed dl1520 or ONYX015, was injected into squamous cell carcinomas of head and neck in phase II clinical trials, and the results were encouraging when chemotherapy was applied in parallel. In the future, further progress might be achieved on the level of virus constructs, but also by refining and adjusting simultaneous conventional therapies, and by standardizing the assessment of the clinical outcome.

Recent progress has been made towards the use of replicating virus constructs in cancer therapy. The goal of these developments is to remove cancerous cells from patients with the help of viruses that selectively replicate in these cells. These viruses are generally termed oncolytic viruses. Some convenient properties of adenovirus make this virus particularly useful for this purpose. It infects a large number of human cell types, especially epithelial cells, which give rise to the vast majority of human malignancies. It can be grown easily and to high titers, and the creation of virus recombinants is well established. Finally, a large body of basic research has already been carried out on this virus, facilitating its manipulation. Various approaches to use adenovirus as a cancer drug have been reviewed (Alemany et al. 1999a, 2000; Curiel 2000; Galanis et al. 2001b; Gromeier 2001; Heise and Kirn 2000; Kirn 2000a; Kirn et al. 2001; Kirn and McCormick 1996; Smith and Chiocca 2000; Sunamura 2000; Wells 2000; Wodarz 2001). The aim of this chapter is to provide an integrated overview of these strategies.

1 Replicating Adenovirus As a Tool to Treat Cancer—A Question of Specificity and Efficiency

1.1 Adenovirus to Treat Cancer—A Brief Historical Overview

For almost a century, attempts have been made to use infectious agents, including viruses, in the therapy of cancer (Kirn 2000b). Wild-type adenovirus was employed for this purpose in a clinical trial in 1956, only 5 years after its discovery, and it was found to induce tumor necrosis when injected into cervical carcinomas (Smith et al. 1956). However, at that time, few other effective therapies such as chemotherapy were available, and clinical documentation did not follow today's standards. Therefore, the clinical benefit of the novel therapy is not obvious from that study. In the 1960s and 1970s, adenovirus was rarely used in cancer therapy (Yohn et al. 1968; Zielinski and Jordan 1969). Nonetheless, from the few studies it can be concluded that, in principle, it may be possible to eliminate tumor cells by the injection of wild-type adenovirus, without unacceptable side effects.

Beginning in the 1980s, adenoviruses were developed as gene therapy vectors, mostly by replacing the adenovirus E1 region with an expression cassette for the gene of interest. Deletion of the E1 region largely prevents the replication of virus. Such replication-deficient adenovirus vectors were used to treat various diseases, including cancer, e.g., by expressing the p53 tumor suppressor gene and/or the regulator of cell proliferation p16INK4/CDKN2 (Sandig et al. 1997). However, at least until now, the use of such viruses did not lead to a significant clinical benefit, presumably because the virus does not reach a sufficiently large proportion of tumor cells when applied to a patient.

Few attempts were made to improve the use of replication-competent adenoviruses in cancer therapy until 1996–1997, when two approaches were tested to make adenovirus replication selective for tumor cells: in one strategy, an adenovirus lacking the E1B-55 kDa gene product was used, thus abolishing an important p53-antagonism of adenovirus (Bischoff et al. 1996). This virus, dl1520 or ONYX015, has been tested in preclinical and clinical trials since then. In the other approach, the adenovirus E1 promoter was replaced with the promoter of the prostate-specific antigen, to achieve specific replication in the cells of prostate carcinoma (Rodriguez et al. 1997). Clinical studies are being pursued with this virus as well.

During the last 3 or 4 years, the published attempts to construct replicating adenoviruses with oncolytic potential became more and more numerous, and several of the newly created viruses are approaching clinical trials.

1.2 Replicating Adenovirus in Therapy—A Change of Paradigms

In the development of adenoviruses as gene therapy vectors, much effort and care has been taken to keep the amount of replication-competent viruses in a vector preparation as low as possible, e.g., by developing cell lines that express viral E1 and/or E4 proteins but do not contain overlapping sequences with the vector genome, to avoid homologous recombination (Brough et al. 1996; Gao et al. 2000; Zhou et al. 1996). The absence of replication-competent virus was widely regarded as essential for safety. Safety became even more of an issue after learning that the application of a genetically modified adenovirus (a first-generation vector) can have severe and even lethal side effects on patients (Lehrmann

1999; Stephenson 2001). Therefore, it seemed (and still seems) heretic to some gene therapists that replicating adenoviruses were recently considered for and employed in therapy. On the other hand, replicating adenoviruses (specifically ONYX015) have been the only recombinant adenoviruses that so far have shown a significant clinical benefit in cancer patients (Khuri et al. 2000; Lamont et al. 2000; Nemunaitis et al. 2000). The side effects were reported to be relatively mild, and certainly acceptable when treating an otherwise lethal disease. Thus, to our current knowledge, it appears that replication-competent viruses are, in principle, suitable to be used in patients and can potentially be developed into a serious option for cancer therapy.

1.3 Oncolytic Virotherapy Is Not Equivalent to Conventional Gene Therapy

The use of viruses in therapy is usually described as "gene therapy". If gene therapy means the introduction of genes into cells of a patient for therapeutic purposes, then the use of oncolytic viruses is still gene therapy. However, the term "gene therapy" is frequently associated with the aim of restoring or establishing a specific gene function in that cell. In many cases, this implies that the target cell needs to be kept alive as long as possible to carry out this function. In contrast, when using oncolytic viruses, the aim is simply to eliminate the target cells. This is not achieved by the expression of one particular gene, but rather by the cooperation of all viral genes and their products. Therefore, many of the common rules and aims of gene therapists—such as minimizing the expression of vector genes and avoidance of an immune response to transduced cells—do not apply or are turned into the opposite when using oncolytic viruses. Having said this, it certainly remains possible to destroy a tumor cell through a replicating virus, while expressing a transgene that helps to kill adjacent cells (see Sect. 2.4.2), thus combining classical gene therapy and viral oncolysis.

1.4 Specificity of Replication and Clinical Safety Versus Cytotoxic Efficiency and Clinical Efficacy

When trying to manipulate viruses to replicate selectively in cancer cells, the foremost goal was initially to avoid virus replication in normal tissue. In many cases, viral gene products and the corresponding functions were removed to achieve this. The result was frequently the construction of viruses that replicated poorly, if at all, in normal tissue. However, these changes also brought about impairments in the ability of the virus to replicate in and eliminate tumor cells. The best example of such an outcome is probably a herpesvirus that was devoid of two viral genes, one of them thymidine kinase. This rendered the virus virtually incapable of replicating in nontumorous neuronal tissue, whereas some replication competence was retained in cells derived from brain tumors. However, even in the tumor cells, the virus replicated much less efficiently than wild-type herpes virus. Correspondingly, the use of this virus in clinical trials was disappointing in that the administration was found to be safe but only weakly effective against the tumors (Markert et al. 2000; Mineta et al. 1995). The amount of virus particles that can be applied to a patient is limited by the titer and applicable volume, and furthermore, virus spread through a tumor may be crucial for therapeutic success. Hence, low replication efficiency cannot automatically be compensated for by increasing the dose of virus. The calculation of a therapeutic index based solely on the ratios of replication efficiency of wild-type versus mutant virus in normal and tumor cells (Alemany et al. 2000) can thus be misleading. A poorly replicating virus will not be useful, even if replication is highly specific for tumor cells. Hence, efficiency of virus replication—as a predictor of therapeutic efficacy—needs to be kept in focus as much as the specificity of replication for tumor cells. The following section describes approaches that have been used to reach both of these goals.

2
Strategies to Create Oncolytic Adenoviruses

2.1
"Natural" Oncolytic Activity of Wild-Type Adenovirus

A variety of attempts were made to manipulate the adenovirus genome, to render the virus more suitable for oncolytic therapy. However, it is currently unknown whether these recombinants are actually superior to wild-type adenovirus for this purpose. Correspondingly, it remains to be determined whether the specificity of replication or the efficiency of cell killing actually requires improvement compared to wild-type adenovirus, to allow successful application in clinical settings.

When adenoviruses (at least the widely studied adenovirus type 5) are propagated in tissue culture, it is easy to note that a large proportion of cell lines derived from human tumors (in particular malignant tumors of epithelial origin, i.e., carcinomas) allow the virus to replicate efficiently, whereas primary cells (e.g., human umbilical vein endothelial cells or human bronchial epithelial cells) generally do so only to a lesser extent (M. Dobbelstein, unpublished observations; Harada and Berk 1999; Koch et al. 2001). The reasons for this "natural" preference of virus replication for tumor cells are not entirely clear but may be related to the increased metabolic activity of tumor cells. Thereby, tumor cells can be expected to provide large amounts of the intermediates required for viral nucleic acid and protein synthesis. When virus is applied to a patient, tumor cells may also be supportive as viral hosts, because they have frequently evolved mechanisms to evade the immune response.

Further, many patients have neutralizing antibodies against adenovirus type 5. Among those who have not, virtually all develop neutralizing antibody titers after the first administration of an oncolytic adenovirus (Ganly et al. 2000; Mulvihill et al. 2001). Such antibodies can be expected to reduce the spread of virus from the injection site to distant sites of the body (Chen et al. 2000). Such neutralizing antibodies certainly pose a difficulty to the systemic application of adenovirus, but they are likely to increase the safety of local applications.

In any case, the natural selectivity of adenovirus for many cancerous cells appears to be an advantage for the use of adenovirus in cancer therapy. It is possible that this selectivity is sufficient to open a therapeutic window in some cases. Since the 1960s, no one has dared to administer wild-type adenovirus to a tumor patient. It is not known whether the administration of wild-type adenovirus in high amounts may be dangerous to a patient, especially if a large proportion of it reaches the bloodstream and, subsequently, the liver. A recent case of fatal liver failure after administration of a first-generation adenovirus vector (Lehrmann 1999; Stephenson 2001) raised safety concerns even further. However, it is possible that the injection of wild-type adenovirus, e.g., using repeatedly injected, escalating doses, may be of benefit to certain cancer patients. The realization of such an approach may depend on the readiness of clinicians and patients to accept risks, the efficacy of conventional therapeutic regimens, the availability of more specific viruses that nonetheless

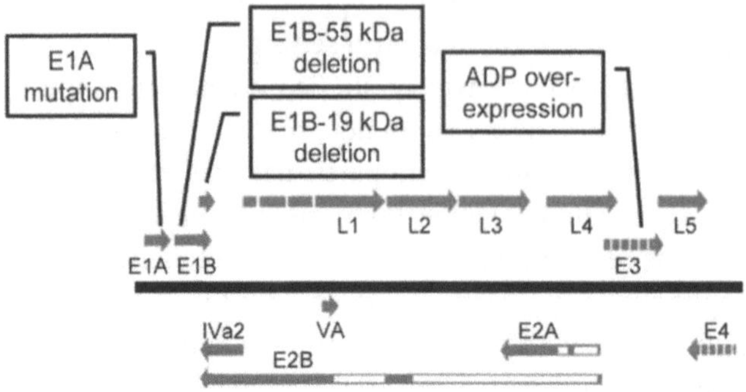

Fig. 1 Adenovirus genes removed or overexpressed to create oncolytic viruses. In an attempt to create replication-selective, oncolytic adenoviruses, the following changes were made to the adenovirus genome: mutations of E1A (Doronin et al. 2000; Fueyo et al. 2000; Heise et al. 2000b; Howe et al. 2000), deletion of E1B-55 kDa (Bischoff et al. 1996), deletion of E1B-19 kDa (Duque et al. 1999; Sauthoff et al. 2000), and overexpression of ADP (Doronin et al. 2000)

meet or exceed the efficiency of wild-type adenovirus, and possibly financial and legal considerations.

2.2 Partial or Complete Removal of Viral Genes

A widely used strategy to increase the selectivity of adenovirus replication for tumor cells above that of wild-type adenovirus is to remove one or several of the oncogenes from the adenovirus genome (Fig. 1). Even though little is known about the influence of these oncogenic activities on virus replication, it is assumed that viral oncogenes evolved because they confer some advantage to the virus when it replicates in normal cells. In tumor cells, however, the growth-controlling signaling pathways that would be targeted by viral oncogenes have frequently been modulated during tumorigenesis. Hence, the oncogenic activities of the viral gene products should no longer be required for virus replication in tumor cells. Eliminating these activities, e.g., by mutation or by deletion of the corresponding oncogene from the viral genome, should prevent virus replication in normal cells but should not impair replication in tumor cells. The idea was therefore that such virus mutants should be suitable for cancer therapy.

2.2.1 E1B-55 kDa—The Theoretical Doubts and Clinical Successes with ONYX015

It was reported that an adenovirus that does not express the E1B-55 kDa protein, termed dl1520 (Barker and Berk 1987) and renamed ONYX 015 (Bischoff et al. 1996) replicates poorly in cells that express wild-type p53, but efficiently in cells with mutant or absent p53 (Bischoff et al. 1996). These observations led to the use of dl1520/ONYX 015 for cancer therapy. The virus was injected into tumors with the aim of destroying them. This concept had preliminary success in animals and patients (Ganly et al. 2000; Heise et al. 1999a,b; Kirn et al. 1998; Mulvihill et al. 2001; Rogulski et al. 2000a; You et al. 2000). Moreover, the application of dl1520/ONYX 015 in combination with cytostatic drugs was found superior to chemotherapy alone in phase II clinical trials to treat head and neck cancer (Khuri et al. 2000; Lamont et al. 2000; Nemunaitis et al. 2000, 2001b). However, the theoretical basis of this approach has been challenged by a number of groups who all found that replication of this virus does not correlate with the p53 status of infected cells (Goodrum and Ornelles 1998; Hall et al. 1998; Harada and Berk 1999; Hay et al. 1999; Rothmann et al. 1998; Steegenga et al. 1999; Turnell et al. 1999). More recently, loss of p14ARF, a regulator of mdm-2 and p53 (Pomerantz et al. 1998; Stott et al. 1998; Zhang et al. 1998) in tumor cells, was reported to facilitate the replication of dl1520/ONYX 015 in a cell line (HCT-116) with wild-type *p53* (Ries et al. 2000). However, it remains to be determined whether p53 and p14ARF can generally cooperate to antagonize virus growth. Thus, the role of p53 in adenovirus replication remains unclear from these studies.

To address this role of p53, several approaches can be taken. The most obvious approach is to compare the replication efficiency of wild-type adenovirus with a virus that lacks E1B-55 kDa and therefore cannot inactivate p53. However, such a mutant virus is also defective in all the additional activities that involve E1B-55 kDa. It is known that E1B-55 kDa is not only necessary for p53 inactivation, but also for the modulation of mRNA export during adenovirus infection (Babiss et al. 1985; Dobbelstein et al. 1997; Gabler et al. 1998; Horridge and Leppard 1998; Leppard and Shenk 1989; Pilder et al. 1986; Weigel and Dobbelstein 2000). In addition, there might be unknown activities of E1B-55 kDa that would also be missing in a virus that does not express the protein. As an alternative approach, it would be desirable to create a mutant of E1B-55 kDa that is not capable of binding p53 but potentially able to perform all other functions of E1B-55 kDa. This mutant could then be tested in a recombinant adenovirus to assess the role of p53 binding during virus replication. However, most mutations in E1B-55 kDa have led to pleiotropic effects on its functions (Kao et al. 1990; Rubenwolf et al. 1997; Yew and Berk 1992; Yew et al. 1990), possibly due to a shift in protein conformation. Nonetheless, a mutant of E1B-55 kDa that selectively lacks p53 binding became available recently (Shen et al. 2001). Further studies will be required to fully characterize this virus regarding its replication ability in various cell types. However, this mutation of E1B-55 kDa may affect unknown functions of the protein in addition to the inhibition of p53. Therefore, a third approach was taken to assess the role of p53 in virus replication. Cells of various origin were infected with wild-type adenovirus, and simultaneously, an adenoviral vector was used to express a mutant version of p53, termed p53mt24–28, that is resistant to the inhibitory effect of E1B-55 kDa (Roth et al. 1998). p53mt24–28 remained stable and active in adenovirus-infected cells. However, even in the presence of high levels of active p53, adenovirus replication was not detectably impaired (Koch et al. 2001).

These results suggest that dl1520/ONYX015 might merely represent an attenuated adenovirus with little additional specificity for tumor cells compared to wild-type adenovirus. The attenuation seems to be attributable to missing functions of E1B-55 kDa other than p53 binding, such as the modulation of mRNA export. Because the virus lacks some of the E3 gene products, this may also limit virus spread in vivo. The lack of these functions can be expected to attenuate virus replication in cells of tumors *and* normal tissue. On the other hand, virus spread was observed in tumor biopsies but not in the adjacent normal tissue from patients who had been treated with dl1520/ONYX015 (Khuri et al. 2000). This may, however, be due to the "natural" selectivity of adenovirus replication for tumor cells (cf. Sect. 2.1). In any case, the clinical successes of dl1520/ONYX015 are encouraging and suggest that more could be achieved by further developed oncolytic adenoviruses.

2.2.2
E1A

The two most thoroughly characterized pathways that control cell proliferation are the p14ARF-mdm2-p53 pathway and the cdk-pRb-E2F pathway. In the latter case, growth-regulatory signals are transmitted from cyclin-dependent kinases (cdks) to the retinoblastoma protein (pRb) family through protein phosphorylation. pRb, in turn, regulates the activity of the E2F transcription factors. Like the p53 pathway, growth-suppressing signaling through pRb is disrupted at some point in the vast majority of all malignancies. On the other hand, adenovirus interferes with the Rb pathway through its E1A gene products that directly bind and inactivate pRb and its relatives p107 and p130. If one supposes a positive effect of pRb inactivation on adenovirus replication, it is logical to use an adenovirus that lacks the ability to inactivate pRb in cancer therapy. Such a virus should be impaired in normal cells but not in tumor cells, because the latter do not have a functional pRb pathway. Indeed, an adenovirus with a deletion of the pRb binding region was reported to replicate well in glioma cells but not in cells with a functional Rb pathway (Fueyo et al. 2000). This principle was also tested using adenoviruses with an E1A mutation, termed dl1101/1107 (Howe et al. 1990), that abolishes the interaction of E1A proteins with pRb (Doronin et al. 2000, 2001). In addition, the E1A mutations dl1101/1107 eliminate the interaction of E1A with a transcriptional cofactor, p300. The missing ability of E1A to bind pRb as well as to bind p300 may therefore affect virus replication in these experiments. In addition, these viruses contained mutations that increase the expression of the adenovirus death protein (see Sect. 2.3.1). Selectivity for tumor cells was observed with these viruses. However, the mutation of E1A reduced the ability of adenovirus to grow in normal cells but also in tumor cells. This effect may be due to residual activity of pRb in many tumor cells, or related to other activities of E1A that are mediated by the same region that binds pRb. In any case, this kind of E1A mutation appears to result in a too-strong attenuation of the virus to be useful in clinical settings.

In many anogenital cancers, especially cervical carcinomas, the genes E6 and E7 of oncogenic human papilloma viruses (HPVs) are expressed. E7 binds and inactivates pRb, and therefore, it was tested whether adenovirus mutants with deletions in the pRb binding region of E1A might selectively replicate in HPV-transformed keratinocytes. This was not found to be the case in an organotypic model system. However, more extensive deletions within E1A (in conserved regions I and II) led to some preference of virus replication for cells that express HPV oncoproteins (Balague et al. 2001).

A different E1A mutant was recently tested for tumor-specific growth, and the results seemed more encouraging. An adenovirus mutant termed dl922–947 (Whyte et al. 1989) grew at least as well as wild-type adenovirus in a number of tumor-derived cell lines, but only poorly in normal cells (Heise et al. 2000b). This mutant was identified by screening a large panel of previously constructed adenovirus mutants. The E1A mutation is located in the second conserved region (CR2) of E1A, which binds to pRb. However, not all E1A mutants lacking pRb binding had a replication phenotype similar to dl922–947. Thus, some biochemical properties of this mutant, other than the lack of binding to pRb, may be important for the observed tumor-selective replication.

2.2.3 E1B-19 kDa

Other deletion mutations of adenovirus might improve the destruction of tumor cells. The adenovirus E1B-19 kDa protein is encoded by the second open reading frame of the E1B gene. It interacts with the cellular bax (Chen et al. 1996; Han et al. 1996; White et al. 1992) and bak (Farrow et al. 1995) proteins at intracellular membranes and functions as a potent inhibitor of apoptosis. Deleting the E1B-19 kDa coding region results in enhanced viral cytotoxicity in a variety of tumor cells (Sauthoff et al. 2000) and in cells of chronic lymphatic leukemia (Medina et al. 1999). However, adenovirus lacking E1B-19 kDa grows in an attenuated fashion, e.g., in HeLa cells, presumably because it induces these cells to undergo apoptosis before virus replication is complete (Subramanian et al. 1984). On the other hand, many tumor cells overexpress inhibitors of apoptosis, such as the bcl-2 protein, or otherwise interfere with apoptotic signaling pathways (Evan and Vousden 2001). It remains to be determined whether these tumor-specific inhibitors of apoptosis may allow an adenovirus lacking E1B-19 kDa to replicate with reasonable efficiency, and how this compares to replication and cytotoxicity in normal cells.

2.2.4 Deletion of Other Adenovirus Genes—A Perspective for Further Improvements?

The deletion of adenovirus genes that do not belong to the E1 gene cluster may also help to create tumor-selective adenoviruses. The virus-associated (VA) RNA species I and II bind and inhibit the cellular enzyme PKR, which otherwise becomes activated by virus infections (Mathews and Shenk 1991). Because active PKR can shut down protein translation, and thereby block the synthesis of virus particles, the expression of VA RNA contributes to the efficiency of virus replication. On the other hand, some tumor cells attenuate the PKR signaling pathway. For instance, the expression of mutated, oncogenic *ras* was reported to diminish PKR activity through an unknown mechanism, and this apparently enhances the replication of a PKR-sensitive reovirus (Coffey et al. 1998; Strong et al. 1998). Thus, it might also be possible to achieve the selective replication of adenoviruses in certain tumor cell species by deleting one or both genes that encode VA RNA. We did not observe a correlation between the growth of a virus lacking VAI and the presence of *ras* mutations in host cells (M. Schümann and M. Dobbelstein, unpublished observations), but this might be due to downstream alterations of the ras signalling pathway.

Some proteins encoded by the adenovirus E4 region also interact with cellular growth-control factors. E4 open reading frame (orf) 1 (at least when derived form adenovirus type 9) binds and inactivates the DLG tumor suppressor (Lee et al. 1997), E4orf3 associates with the promyelocytic leukemia (PML) protein (Carvalho et al. 1995; Doucas et al. 1996), E4orf6 destabilizes p53 in cooperation with E1B-55 kDa (Moore et al. 1996; Querido et al. 1997; Roth et al. 1998), and E4orf6/7 associates with E2F proteins (Huang and Hearing 1989b). All of these cellular targets are functionally altered in at least some tumor cells, and therefore, the deletion of any of the corresponding E4 reading frames may theoretically contribute to selective virus replication in tumors. Among the gene products mentioned above, only the E4orf6 protein was needed for efficient growth of adenovirus in transformed cells. All the other E4 open reading frames, when deleted individually, did not alter the efficiency of virus replication in these cells (Huang and Hearing 1989a). It seems worthwhile to test the growth of these virus mutants in a set of primary cells, or in cotton rats (cf. Sect. 3.3), to see whether the deletion of E4 open reading frames may reduce virus replication in normal cells while sustaining replication in tumor cells.

2.3
Overexpression of Viral Genes

Instead of eliminating genes, it was also attempted to overexpress a viral gene to construct oncolytic adenoviruses (Fig. 1). This might increase the efficiency of cell lysis and/or virus replication.

2.3.1
The Adenovirus Death Protein

The adenovirus death protein (ADP) is expressed by the E3 region, predominantly during the late phase of infection (Burgert and Blusch 2000). It encodes a palmitoylated (Hausmann et al. 1998) transmembrane protein that localizes to the endoplasmic reticulum, the Golgi apparatus, and the nuclear membranes. Recombinant adenoviruses lacking the ADP coding region form smaller plaques than wild-type virus. Apparently, ADP increases the release of virus from the infected cells (Tollefson et al. 1996a,b). Based on these findings, adenoviruses were engineered to overexpress ADP, and this was combined with the dl1101/1107 mutation of the E1A gene (cf. Sect. 2.2.2) (Doronin et al. 2000). These viruses were further developed to contain a tumor-specific promoter driving the expression of the E4 region (Doronin et al. 2001). The viruses were tested for their ability to eliminate tumor cells, and indeed, specific tumor cell lysis was observed in cell culture and a nude mouse model. Although the effect of ADP overexpression was not tested separately in these experiments, it appears likely that this strategy may increase the efficiency of adenovirus-mediated oncolysis.

2.3.2
Overexpression of Other Adenovirus Genes—A Perspective for Further Improvements?

In addition to ADP, the E4orf4 protein has also been reported to induce cell death through the modulation of the cellular phosphatase PP2A (Marcellus et al. 1998; Shtrichman and Kleinberger 1998), and this effect was found to be more pronounced in a number of tumor cells, compared

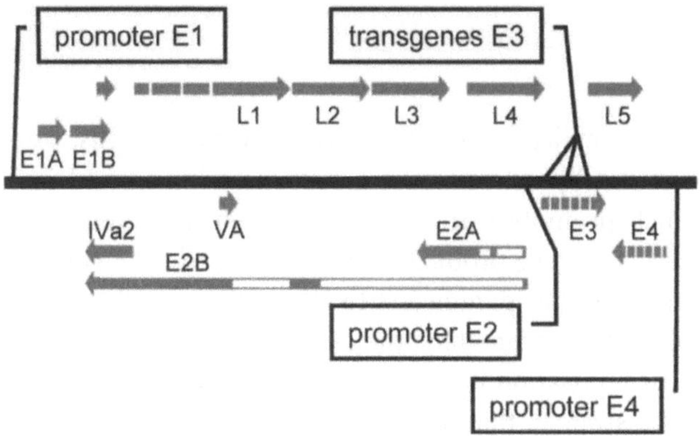

Fig. 2 Promoters and transgenes inserted to create oncolytic viruses. In an attempt to create replication-selective, oncolytic adenoviruses, tumor-specific promoters were introduced into the adenovirus genome, to regulate the expression of E1A (Alemany et al. 1999b; Hallenbeck et al. 1999; Hernandez-Alcoceba et al. 2000; Kurihara et al. 2000; Ohashi et al. 2001; Rodriguez et al. 1997), E2 (Brunori et al. 2001), and E4 (Doronin et al. 2001; Hernandez-Alcoceba et al. 2000). Transgenes were introduced into the E3 region (Hawkins and Hermiston 2001a,b; Hawkins et al. 2001; Lambright et al. 2001; Rogulski et al. 2000b; Wildner 1999; Wildner et al. 1999a,b)

to normal cells (Shtrichman et al. 1999). Hence, the overexpression of E4orf4 may also increase the efficiency of cell lysis and virus spread. This remains to be tested in future experiments.

2.4
Introduction of Nonviral DNA Sequences Into the Adenovirus Genome

Besides the removal or overexpression of viral genes, important strategies to create oncolytic adenoviruses involve the insertion of nonviral genetic material into the adenovirus genome (Fig. 2). Either tumor-selective promoters are inserted to specifically control the expression of viral genes, or alternatively, transgenic coding regions are placed into the genome of adenoviruses, thereby creating a replication-competent expression vector.

2.4.1
Promoters

To achieve the selective replication of DNA viruses in tumor cells, promoter(s) that are specifically active in these cells were introduced into the viral genome. Such promoters are frequently taken from genes encoding tumor markers, i.e., gene products that are released to the bloodstream, and that can be used as parameters to follow the growth of a tumor in a patient. Using these promoters, the expression of essential viral genes, and hence virus replication itself, can be restricted to tumor cells.

The example of this strategy that has been most advanced in clinical settings is represented by recombinant adenoviruses called CN706 (Rodriguez et al. 1997) and CV787 (Yu et al. 1999), which have the E1 promoter replaced by the cellular promoter for prostate-specific antigen (PSA). Expression of PSA is well-restricted to prostate carcinoma cells, and the determination of PSA in the serum serves as a marker for the progression of the disease. CN706 has been shown to express the E1A protein selectively in prostate carcinoma cells, and virus replication was found to be restricted accordingly, in vitro and in vivo (Rodriguez et al. 1997).

Other promoters have been used to create replication-selective adenoviruses. Details are given in Table 1. These promoters were inserted into the viral genome to control the early expression of essential viral genes, namely E1, E2, and E4 (Fig. 2). Different degrees of selectivity and efficiency of virus replication in tumor cells have been observed. Some of these viruses had additional changes made to enhance their oncolytic potential. These changes might be useful for therapeutic applications, but in many cases, the different manipulations to the viral genome have not been tested in an isolated fashion, making it difficult to judge the contribution of every single change. Still other promoters have not yet been introduced into replication-competent viruses, but instead were used to express a toxic transgene from nonreplicative first-generation adenovirus vectors, with the aim of destroying tumor cells specifically. In addition to the promoters listed in Table 1, the promoters used for transgene expression were the upstream sequence of the carcinoembryonic antigen that is expressed in many colorectal carcinomas and other malignant tumors of the gastrointestinal tract (Brand et al. 1998; Kijima et al. 1999; Lan et al. 1996; Tanaka et al. 1996), a promoter derived from the enhancer region of the glial fibrillary acidic protein gene which is ac-

Table 1 Tumor-specific promoters employed to create replication-selective adenoviruses

Promoter/gene	Tumor specificity of the promoter	Viral gene(s) controlled	Reference(s)	Special features
Prostate-specific antigen (PSA)	Prostate carcinoma	E1A	Rodriguez et al. 1997; Yu et al. 1999	Constructs with (CV787) or without (CN706) E3 region
Alpha-fetoprotein	Hepatocellular carcinoma	E1A	Hallenbeck et al. 1999	Otherwise wild-type adenovirus
Alpha-fetoprotein	Hepatocellular carcinoma	E1A	Ohashi et al. 2001	Virus lacking E1B-55 kDa
Alpha-fetoprotein	Hepatocellular carcinoma	E1A	Alemany et al. 1999b	Two complementing viruses
Synthetic estrogen responsive promoter	Breast carcinoma	E1A and E4	Hernandez-Alcoceba et al. 2000	–
DF3/MUC1	Breast carcinoma	E1A	Kurihara et al. 2000	Expression of tumor necrosis factor-alpha
Surfactant protein B (SPB)	Alveolar and bronchial cancer cells, but also adult type II alveolar epithelial cells and bronchial epithelial cells	E4	Doronin et al. 2001	E1A mutation that abolishes pRb and p300 binding; ADP overexpression
Synthetic promoter responsive to TCF and beta-catenin	Colorectal carcinoma	E1B, E2	Brunori et al. 2001	–

tive in glial cells and glioblastoma (Chen et al. 1998), a hypoxia-inducible promoter (Binley et al. 1999), and many others (Dachs et al. 1997). These promoters, at least in principle, could also be used to regulate adenoviral genes with the aim of making new replication-selective adenoviruses. Further, to achieve tumor-specific replication, it is plausible to use promoters that are activated by known oncoproteins, e.g., c-myc or E2F, or repressed by tumor suppressor proteins, e.g., p53 or pRb.

At least in principle, the reverse strategy may also be followed, although no practical example for it is available yet: A promoter that is active only in normal tissue, not in tumor cells (e.g., a p53-responsive promoter) could be used to regulate the expression of a gene product that prevents virus replication (e.g., a dominant negative E1A or E2F mutant). It remains to be determined whether such an approach is feasible.

The major determinants that decide the success of these approaches are the selectivity and strength of the promoter(s) used. In the case of CN706, E1A expression can be made specific for prostate cancer cells using the PSA promoter. However, it is obvious that even in prostate carcinoma cells, E1A expression is much lower after infection with CN706, compared with wild-type adenovirus (Rodriguez et al. 1997), and this may limit the efficiency of virus replication at least in vivo. Moreover, the activity of a promoter can generally be affected by the context of an adenovirus genome. This has become clear through a large number of attempts to create tissue-specific adenovirus vectors for gene therapy. For instance, the packaging signal within the genomic DNA of adenovirus, immediately upstream of the E1 promoter, has a strong enhancer activity, and this can partially or entirely destroy the tissue selectivity of a promoter that is used to replace the viral E1 promoter. The enhancer activity of the packaging signal might be circumvented by the use of insulination sequences (Steinwaerder and Lieber 2000; Vassaux et al. 1999), through transcriptional repressor binding sites or by moving the packaging signal to the opposite end of the viral genome. However, these strategies again may impair the overall expression of E1A even in tumor tissue. Given these particular difficulties in regulating E1 expression, it may be easier to make the expression of E2 and/or E4 tumor-specific. For instance, the E2-promoter has successfully been rendered dependent on the activity of beta-catenin, thereby restricting virus replication largely to cells with constitutively active TCF, such as colon carcinoma cells (Brunori et al. 2001). More examples are presented in Table 1.

2.4.2
Transgenic Coding Regions

In addition to destroying tumor cells by the cytotoxic effect of replicating adenovirus, it is possible to express transgenes from such a virus that might accelerate the killing of the infected cell, or might help to kill cells in the neighborhood. To express a transgene from a replicating virus, it has most commonly been inserted into the E3 region (Fig. 2). This region is not essential, and its partial deletion does not impair virus replication, at least in cell culture. However, the disruption of the ADP coding region can significantly diminish the efficiency of virus spread (Tollefson et al. 1996a,b), as discussed in Sect. 2.3.1. Therefore, it seems desirable to leave ADP expression intact when engineering an oncolytic adenovirus.

The effect of inserting a transgene into different sites within the E3 region on the expression of this transgene has been carefully analyzed (Hawkins and Hermiston 2001a,b; Hawkins et al. 2001). It was found that the expression levels and time characteristics of a transgene during infection closely parallel the characteristics of the endogenous adenovirus gene that was replaced by it. Thus, by choosing the appropriate insertion site, it is possible to fine-tune the expression pattern of a transgene.

The E4 region may also be suitable to accommodate transgenes, as long as the essential E4orf6 protein is expressed, but this has not been thoroughly tested yet.

What genes can be expressed to enhance the oncolytic potential? The candidates should confer a significant bystander effect to noninfected cells in the neighborhood of an infected tumor cell. Thus, secreted proteins can be used that mediate a cytotoxic effect on surrounding cells and/or enhance the immune response to the tumor. Alternatively, drug-converting enzymes can be expressed that activate a cytostatic drug within the infected cells, followed by the diffusion of the active metabolite to the environment.

As an example of the latter possibility, the thymidine kinase gene of herpes simplex virus (HSVtk), which converts acyclothymidine to its active form, was overexpressed by oncolytic adenoviruses that either lacked or contained the E1B-55 kDa gene product. Some reports come to the conclusion that HSVtk did increase oncolytic activity (Morris and

Wildner 2000; Wildner and Morris 2000a), whereas others did not observe a beneficial effect (Lambright et al. 2001).

Cytosine desaminase (CD) is an enzyme that metabolizes and thereby activates the drug 5-fluorocytosine. The active metabolite helps to kill not only the cells that express CD, but also spreads to cells in the neighborhood. Thereby, CD can be expected to increase the cytotoxicity of an oncolytic virus when 5-fluorocytosine is administered together with the virus. This principle was successfully tested with replication-competent adenovirus vectors, in the presence and in the absence of E1B-55 kDa (Freytag et al. 1998; Rogulski et al. 2000b).

Drug converters can not only be used to enhance the oncolytic potential of an adenovirus, but also to trace the spread of virus in vivo. To this end, enzymes ("marker transgenes") are used to change the conformational structure of contrast agents. Then, the presence and distribution of virus can be assessed by nuclear magnetic resonance imaging (Bell and Taylor-Robinson 2000).

The cytokines interleukin-2 (IL-2) and interleukin-12 (IL-12) were also expressed from replication-competent adenoviruses lacking E1B-55 kDa (Motoi et al. 2000), but the contribution of the cytokines to tumor cell elimination in a patient is difficult to predict preclinically, due to the lack of an immunocompetent animal model (cf. Sect. 3.1).

2.5 More Possibilities to Target Tumor Cells Selectively

The fiber protein of adenovirus interacts with receptors. Modifications of this protein can render the virus infective for cells that are difficult to infect with wild-type adenovirus, e.g., due to weak expression of the coxsackievirus and adenovirus receptor (CAR) (Bergelson et al. 1997). Modifications in the fiber protein have been reported to enhance the infection of glioma cells with an E1B-55 kDa-deleted adenovirus (Shinoura et al. 1999), and also to broaden the infectability of tumor cells by an E1A-modified virus (Suzuki et al. 2001).

As a different strategy to achieve tumor-selective adenovirus replication, the use of two complementing viruses in cancer therapy has been suggested. Each virus is unable to replicate on its own, but the two viruses complement the replication potential of each other (Alemany et al. 1999b). This can be achieved by deleting the E1 region in one virus, and all the other regions except E1 in the other. This will presumably increase the confinement of virus replication to the injection site. Near this site, more than one virus particle can be expected to hit the cells, and this would result in complementation of the two viruses. If, however, the virus particles reach distant sites, e.g., through the bloodstream, they will be more diluted, making it unlikely that two virus particles hit the same cell. One of the two viruses alone, however, is unable to replicate, limiting the damage to tissue at distant sites. Further, the use of two complementing viruses allows ample space for the insertion of transgenes, to increase the oncolytic potential. However, it remains to be determined whether splitting the virus genome will compromise virus spread within the tumor and therefore reduce the oncolytic efficiency toward the injected tumor mass.

Another system for the conditional expression of genes—and therefore potentially the conditional replication of viruses—employs the fact that the homologous recombination of DNA in a cell depends on the DNA replication activity (Steinwaerder et al. 2001). The following strategy was therefore developed: a promoterless transgene was introduced into an adenovirus vector, flanked by homologous recombination sites. When this vector was introduced into the liver of an animal with liver metastases, the transgene was rearranged by homologous recombination, and brought into conjunction with a promoter. As a result, the transgene was expressed only in the cancerous cells. In contrast, transgene expression was not detected in the parenchymal cells of the liver, presumably because they rarely divide. In principle, this strategy could be employed to make the expression of an essential viral gene dependent on the presence of ongoing DNA replication in a cell, thereby targeting the virus to cancerous cells in an environment of quiescent cells. The feasibility and efficiency of such an approach are currently unknown.

Finally, adenoviruses may be optimized for oncolytic purposes by randomization of the viral genome, either through treatment with chemicals, such as nitrous acid (Williams et al. 1971), or by "sloppy" polymerases to amplify portions of the genome. If such a randomization procedure is followed by selection of those mutants that grow in tumor cells (in culture or animal models), then oncolysis may be enhanced by a priori unknown mechanisms. This strategy is currently being developed, with encouraging results (Y. Shen, personal communication).

3 Animal Models

3.1 Lack of an Ideal Animal Model

The replication of human adenoviruses is largely restricted to human host cells. Therefore, animal studies to evaluate oncolytic adenoviruses provide only limited predictive value for the outcome in patients. Nonetheless, some information can be gathered from animal models, and this can at least provide preliminary evidence for the usefulness of such viruses in therapy.

3.2 Human Tumor Xenografts in Nude Mice

Because human adenoviruses are restricted to human cells, only tumors of human origin can be successfully treated by these viruses. Human cells are immediately rejected by the immune system of healthy animals, and therefore, tumors of human cells can only be established in immunocompromised animals, mostly nude mice. These mice develop tumors upon injection of malignant cells from a different species. Different injection sites (e.g., subcutaneous, intravenous, intraglandular) serve as model systems for different tumor locations. Oncolytic viruses can then be injected by any route (cf. Sect. 4.2) to examine their effect on the tumor. In this way, animal models have been created to study the treatment of a number of tumors with oncolytic viruses. Details are given in Table 2.

Although this system is probably the most useful among the currently available models, it has severe shortcomings: First, these mice have strong defects in their immune response. The immune system, however, is known to play an important role in tumor growth or rejection (Rosenberg 2001), and these effects are likely to be modified by adenovirus infection. Much of the immune response and its modulation by oncolytic adenoviruses cannot be studied in this system. Secondly, murine cells replicate human adenoviruses poorly or not at all. Therefore, the system does not allow the accurate assessment of the specificity and safety of an oncolytic adenovirus. Based on this model, it cannot be predicted

Table 2 Xenograft models for different tumor species, treated with oncolytic adenoviruses

Tumor species	Recombinant adenovirus	Reference(s)
Gastrointestinal tract		
Hepatocellular carcinoma	*Alpha-fetoprotein* promoter to regulate E1A	Hallenbeck et al. 1999
Hepatocellular carcinoma	*Alpha-fetoprotein* promoter to regulate E1A; deletion of E1B-55 kDa	Ohashi et al. 2001
Hepatocellular carcinoma	*Alpha-fetoprotein* promoter to regulate E1A; two complementing viruses	Alemany et al. 1999b
Hepatocellular carcinoma (Hep2B and HepG3 cells)	dl1520/ONYX015; deletion of E1B-55 kDa	Vollmer et al. 1999
Hepatic metastases	Replication-dependent homologous recombination	Steinwaerder et al. 2001
Colon carcinoma (HT29 cells)	Deletion of E1B-55 kDa and E1B-19 kDa	Duque et al. 1999
Colon carcinoma (HT29 cells)	Deletion of E1B-55 kDa; expression of herpes simplex-1 thymidine kinase (HSVtk)	Wildner et al. 1999a
Peritoneal carcinomatosis from colon carcinoma (HT29 cells)	Deletion of E1B-55 kDa; expression of herpes simplex-1 thymidine kinase (HSVtk)	Wildner and Morris 2000b
Genitourinary tract		
Cervical carcinoma (C33A cells)	dl1520/ONYX015; deletion of E1B-55 kDa	Bischoff et al. 1996
Cervical carcinoma (C33A cells)	Expression of a cytosine deaminase (CD)/herpes simplex type 1 thymidine kinase (HSV-1 TK) fusion gene	Rogulski et al. 2000b
Cervical carcinoma (HeLa cells)	Deletion of E1B-55 kDa and E1B-19 kDa	Duque et al. 1999
Cervical carcinoma (ME180 cells)	Deletion of E1B-55 kDa; expression of herpes simplex-1 thymidine kinase (HSVtk)	Wildner et al. 1999b
Ovarian carcinoma (MDAH 2774 cells)	Deletion of E1B-55 kDa; expression of herpes simplex-1 thymidine kinase (HSVtk)	Wildner and Morris 2000a
Breast carcinoma	dl922–947, mutation of E1A	Heise et al. 2000b

Table 2 (continued)

Tumor species	Recombinant adenovirus	Reference(s)
Breast carcinoma	*DF3/MUC1* promoter to regulate E1A; expression of tumor necrosis factor-alpha	Kurihara et al. 2000
Prostate cancer	Promoter of *prostate-specific antigen (PSA)* to regulate E1A expression	Chen et al. 2001; Rodriguez et al. 1997; Yu et al. 1999, 2001
Respiratory tract		
Lung carcinoma with properties of type II alveolar epithelial cells (A549 cells)	Deletion of E1B-55 kDa; expression of herpes simplex-1 thymidine kinase (HSVtk)	Wildner and Morris 2000a
Lung carcinoma with properties of type II alveolar epithelial cells (A549 cells)	Mutation of E1A; overexpression of ADP	Doronin et al. 2000
Pulmonary adenocarcinoma (H441 ells)	Mutation of E1A; overexpression of ADP; promoter of *surfactant protein B (SPB)* to regulate E4 expression	Doronin et al. 2001
Central nervous system		
Glioma	Mutation of E1A	Fueyo et al. 2000
Glioma	Deletion of E1B-55 kDa; fiber mutation	Shinoura et al. 1999
Others		
Head and neck squamous cell carcinoma (HNSCC)	Expression of herpes simplex-1 hymidine kinase (HSVtk)	Morris and Wildner 2000
Epidermoid carcinoma (A431 cells)	Deletion of E1B-55 kDa and E1B-19 kDa	Duque et al. 1999
Malignant melanoma (A375 cells)	Deletion of E1B-55 kDa; expression of herpes simplex-1 thymidine kinase (HSVtk)	Wildner et al. 1999b

whether or not a virus construct will spread to the normal tissue of a patient.

It will be difficult to improve the mouse system. However, it might be possible to use murine adenoviruses to treat murine tumors in immunocompetent mice. Genetic modifications of these viruses may then serve as models for their human counterparts. The problem with this strategy lies mainly in the extensive differences between human and murine adenoviruses. Nonetheless, this approach might be useful to study the immune response to a tumor that was injected with adenoviruses.

3.3 Cotton Rats

The cotton rat *Sigmodon hispidus* is one of the few animals where human adenovirus infection will lead to a disease, i.e., pneumonia (Ginsberg et al. 1990). Therefore, cotton rats can serve as a model system for adenovirus pathogenesis. However, it is difficult to establish a tumor model in these animals, and human adenoviruses replicate poorly in a cotton rat-derived tumor cell line (M. Dobbelstein, unpublished observations). Thus, cotton rats can only serve to evaluate the attenuation of an adenovirus in normal cells, and this system is therefore valuable to assess safety. However, it will probably not be suitable as a model for adenoviral oncolysis.

4 From Bench to Bedside: Clinical Trials

Most of the currently available clinical data were derived from the use of dl1520/ONYX015 in phase I and phase II trials (Ganly et al. 2000; Heise et al. 2000c; Kirn et al. 1998; Lamont et al. 2000; Nemunaitis et al. 2001a; Von Hoff et al. 1998). These experiences were recently summarized (Kirn 2001).

4.1 Tumors to Be Treated

Several tumor species have been chosen for treatment with oncolytic adenoviruses. Three obvious conditions should lead the choice of target tumors:

1. No other successful therapy is available, making it necessary to use experimental approaches.
2. The tumor can be reached for local treatment, e.g., with an injection needle or catheter.
3. Cells derived from this tumor species can be infected in vitro with reasonable efficiency.

These criteria are met by head and neck cancer (squamous cell carcinomas) in patients who previously underwent chemotherapy and/or irradiation, but now present with refractory or recurrent tumor masses. Patients in this situation were the first to be treated with dl1520/ONYX015 (Ganly et al. 2000; Kirn et al. 1998). In addition, attempts have been made to treat patients with the following tumors with oncolytic adenoviruses: pancreatic carcinoma (Von Hoff et al. 1998), primary and secondary hepatic tumors (Habib et al. 2001), and prostate cancer (Yu et al. 2001). Xenograft animal models for many more tumors have successfully been tested with respect to the effect of oncolytic adenoviruses (cf. Sect. 3.2 and Table 2), raising hopes that some of these approaches may be taken to clinics in the near future.

4.2 How to Deliver the Virus

Three major routes of application can be distinguished for the delivery of oncolytic adenoviruses to tumors:

1. Intratumoral injection. This is the only possibility that has been extensively tested to date. In the case of head and neck cancer, many tumors can be reached percutaneously with a regular injection needle. In the documented protocols, tumors were injected from multiple injection sites into multiple directions, and it was attempted to “fill” the entire injection channel with virus suspension (Khuri et al. 2000). This protocol ensures that the distance between the instillation sites of virus and any tumor cell is kept minimal. If one assumes that the virus spreads only a certain distance through the tumor, and that intratumoral septation with fibrotic material may further limit virus spread, it seems important to infiltrate a tumor as thoroughly as possible. In the future, this may be further improved by adding proteolytic enzymes to the injected virus, with the aim of destroying intra-

tumoral connective tissue (Kuriyama et al. 2000, 2001). A more sophisticated technique has been used to instill adenovirus into pancreatic carcinomas. The tumors were localized with the help of computer tomography, and a working channel was introduced by endoscopy, allowing intratumoral injection through the wall of the duodenum (Von Hoff et al. 1998). The advantage of intratumoral injection is the very high local density of virus that can be achieved. The obvious disadvantages are that not all tumors can easily be reached with a simple injection device, and that distant metastases, including undetectable micrometastases, are very unlikely to be targeted.

2. Locally restricted injection. Examples of this include the injection into an artery that supplies the organ under treatment, e.g., the perfusion of the hepatic artery (Habib et al. 2001) to reach the liver and thereby any primary or secondary liver tumor. Another possibility is the injection into a body cavity, such as the peritoneum (Heise et al. 2000a; Wildner and Morris 2000b) or the urinary bladder (thus far only tested with nonreplicative adenovirus vectors; Watanabe et al. 2000), to treat peritoneal carcinomatosis or bladder cancer, respectively. This strategy can be expected to target (micro)metastases within the region under therapy, but it is unlikely to reach distant metastases.
3. Systemic application. In this approach, adenovirus is injected intravenously and spreads through the bloodstream. The major obstacle to this strategy appears to be the antibody response. Many patients have neutralizing antibodies against adenovirus types 2 and 5 (the only subtypes that have been developed into oncolytic agents to date) even before therapy, and virtually all patients develop such antibodies after the first injection of virus. These antibodies are likely to keep the vast majority of intravenously injected virus particles from infecting any cell, and antibody complexes may cause damage to organs, e.g., the kidney. Nonetheless, at least a small proportion of intravenously injected adenovirus did reach tumors in animal models and patients with metastases in the liver (Habib et al. 2001) and the lung (Nemunaitis et al. 2001a). Thus, intravenous application appears as an option that may become feasible in the future. Possibilities for improvement include the coating of adenovirus by synthetic polymers, to block the interaction between virus and neutralizing antibodies (Croyle et al. 2001 and references therein).

4.3 Safety Considerations

In general, the side effects of treatment with oncolytic adenoviruses were observed to be mild. Even the systemic application of dl1520/ONYX015 in doses up to 2×10^{13} particles induced acceptable side effects, such as mild to moderate fever, rigors, and a dose-dependent transient elevation in serum transaminases (a sign of transient damage to liver cells) (Nemunaitis et al. 2001a). Thus, side effects do not seem the primary concern in the further improvement of oncolytic adenoviruses. Nonetheless, some individuals appear to be more susceptible than others to damage induced by high amounts of adenovirus particles. One case of a fatal outcome has been reported and was directly attributed to the application of an adenovirus vector. In this case, a young patient with ornithine transcarbamylase deficiency was treated by injection of a high dose (roughly 10^{13} particles) of a nonreplicative first-generation adenovirus vector into the liver artery. Although other patients did tolerate comparable amounts of virus, this patient responded with fatal liver failure, for reasons that are not fully understood (Lehrmann 1999; Stephenson 2001). At least in mice, the majority of a systemically (i.e., intravenously) applied adenovirus is retained in the liver (Huard et al. 1995), and hence, liver toxicity seems to be one of the major potential hazards that should be kept in mind when therapeutically using adenoviruses. Additional concerns may consist in anaphylactic reactions. Such allergic hyperreactions are occasionally observed in patients who were vaccinated with a preparation containing animal proteins, e.g., from chicken (James et al. 1995). Although no such incident has been reported in connection with the use of therapeutic adenoviruses to date, similar reactions may be encountered with adenovirus preparations. Further, replicating adenoviruses pose at least a theoretical risk to individuals who are in contact with a patient undergoing therapy. Although the risk of spread seems low for the same reasons that are likely to limit the side effects in patients, it has been proposed with good reason to further investigate this potential hazard (Kirn et al. 2001). In summary, although the reported side effects of replication-selective adenoviruses in tumor therapy all seem acceptable, the possibility of rare but more severe complications must be taken into consideration and weighed against the potential benefits.

4.4
Tradition and Innovation—Combining Conventional Therapy with Oncolytic Viruses

The application of dl1520/ONYX015 alone to head and neck tumors had limited success (Nemunaitis et al. 2001b), but the combination of the virus with chemotherapy (cisplatin and 5-fluorouracil) lead to a partial or complete clinical response in one-third of the treated patients each (Khuri et al. 2000; Lamont et al. 2000), in a population of patients who had been unsuccessfully treated by conventional means before entering the trials. The increased efficacy of treatment with chemotherapy and virus in combination was also observed in animal studies. Interestingly, the sequence of treatment did matter, and the application of virus simultaneously with or before chemotherapy worked better than the reverse order (Heise et al. 2000c). Clearly, the interplay of chemotherapy and virotherapy is only beginning to be elucidated, and therapeutic improvements can be expected from more detailed studies on this matter.

4.5
Currently Available Clinical Data—Efficiency Matters

Oncolytic adenoviruses, in particular dl1520/ONYX015, underwent a number of phase I studies to assess their safety. Local injection into head and neck tumors was reported to elicit mild side effects that did not exceed flu-like symptoms (Ganly et al. 2000). Another phase I study aiming at the treatment of liver tumors showed that dl1520 was well tolerated when administered directly intratumorally, intra-arterially, or intravenously up to a dose of 3×10^{11} infectious units (Habib et al. 2001). Injection of the dl1520/ONYX015 virus into pancreatic carcinomas resulted in mild and transient pancreatitis in one out of 22 patients and was well tolerated by the others (Mulvihill et al. 2001). Even upon systemic application of up to 2×10^{13} virus particles of dl1520/ONYX015, no dose-limiting toxicity was identified. Mild to moderate fever, rigors, and a dose-dependent transient elevation of serum transaminases were the most common adverse events (Nemunaitis et al. 2001a). More phase I studies are underway to treat ovarian carcinomatosis (Heise et al. 2000a) with dl1520/ONYX015 and prostate cancer with CN706 (Yu et al. 2001), but the results have not been published yet.

Thus, it seems that the application of replication-competent adenovirus by various routes is reasonably safe, especially considering the severity of a malignant disease. The big question is how effective this treatment will turn out to be for various tumor species. This can only be evaluated by phase II and phase III trials, some of which are currently underway. The only available data from phase II studies report successes when treating patients with head and neck cancer with a combination of dl1520/ONYX015 and chemotherapy, the complete response rate being 33% in the first 6 months (Khuri et al. 2000; Lamont et al. 2000). Virus injection was reported to be safe but ineffective in a phase II study aiming at the treatment of liver tumors (Habib et al. 2001). Thus, the currently available clinical data are encouraging, but demanding further improvements.

5 Future Perspectives: Much Room for Improvement

The encouraging but limited clinical successes of replication-selective, oncolytic adenoviruses clearly imply that more work needs to be done to improve their clinical benefit.

5.1 What Determines the Efficiency of Virus Replication in Various Cell Types?

One of the big advantages of developing adenoviruses as therapeutic tools is the fact that with these viruses, extensive studies have been performed for decades to dissect the mechanisms of viral replication and host-cell interaction (Shenk 1996). Nonetheless, it is still difficult to determine the cell-specific limiting factors of adenovirus replication and cytotoxicity. In other words, we currently do not know which virus-expressed factors determine how well or how poorly a virus replicates in and kills a given cell type. This type of knowledge determines what factors should be regulated by tumor-specific promoters, or should be mutated or removed because they are not required for replication in tumor cells. To fully clarify this issue, it will be necessary to determine the replication efficiency and cytotoxicity of many virus recombinants in many cell types systematically. Before that, it will be important to precisely define the criteria to assess replication and cell killing, and to apply these criteria consistently throughout a large study.

5.2 Immune Response and Pathogenesis

Studies of the replication of adenoviruses in cell culture are routinely done in tissue culture dishes or flasks where the cells adhere as a monolayer, or grow in suspension. This is only remotely similar to the situation in a malignant tumor, which is a compact, three-dimensional array of malignant cells that is frequently interspersed with blood vessels, connective tissue, and other noncancerous cells. It will be important to establish cell culture models that come closer to the spatial organization of a malignant tumor. In particular, the cell-to-cell spread of an oncolytic virus can probably be much more accurately studied in such systems (Balague et al. 2001).

At the moment, tumor xenografts in nude mice are still the only practical and widely used model system to study virus-mediated oncolysis in vivo. Unfortunately, these mice lack major parts of the immune response and therefore do not reflect the situation in a patient, where tumor growth is known to be strongly influenced by the immune system (Rosenberg 2001). It will thus be important to identify suitable animal models, using immunocompetent animals and matched tumor models and viruses (cf. Sect. 3.2), to study the pathogenesis of oncolytic viruses, and their interplay with the immune system. At present, only patients and biopsy material can be used to study these issues. However, an experimental system will be required to conveniently study the influence of parameters such as virus load, injection site, chemo- and irradiation therapy, expression of cytokines, tumor location, etc.

5.3 Beyond Gene Regulation: More Ways of Training Adenovirus to Eliminate Tumor Cells

Thus far, the specificity of oncolytic adenoviruses was largely achieved through the modulation of early genes and their expression. Less work has been done to achieve tumor cell specificity of oncolytic adenoviruses through the modification of structural proteins. One report suggested that a modification of the fiber knob improved the ability of an oncolytic adenovirus to enter glioma cells (Shinoura et al. 1999), and more extensive modifications of fiber proteins may ensure specific tumor cell tropism in the future. Moreover, in conventional gene therapy with nonreplicating adenovirus vectors, immunologic retargeting has been achieved via conjugates comprised of an anti-fiber knob Fab fragment and a targeting moiety consisting of a ligand or antireceptor antibody (Curiel 1999). Gene delivery was then mediated by receptors for folate, FGF, and EGF. In addition to cell-specific gene delivery, this strategy has allowed enhanced gene delivery to target cells lacking the native adenoviral receptor, CAR. These strategies can be expected to improve the targeting of replication-competent adenoviruses as well.

Adenovirus type 2 and type 5, or chimeras of both, are the only adenovirus types that have been used in therapy to date, because they are by far the most widely studied virus types. However, there are many more types of adenovirus, and these have different biological properties, including tissue specificity. A systematic screen of other adenovirus subtypes for their oncolytic potential is likely to contribute to the improvement of oncolysis by this virus.

The spread of adenovirus through a tumor is probably a crucial determinant of its clinical efficacy. To enhance virus spread, the overexpression of ADP (Doronin et al. 2000) is one possibility. Different ways can be envisioned to increase cell lysis by apoptotic or nonapoptotic mechanisms. However, this may counteract the efficiency of virus particle synthesis during the late phase of infection. Alternatively, it seems an attractive possibility to "teach" adenovirus to induce cell fusion. Recombinant first-generation adenovirus vectors have been developed to express fusogenic membrane glycoproteins (FMGs), and these nonreplicative viruses did induce tumor regression in an animal model of glioma (Galanis et al. 2001a). It remains to be determined whether cell fusion through FMG expression can also be achieved by replicating adenoviruses, and how this would affect virus replication, cytotoxicity, and oncolytic activity.

On a larger scale, virus spread might be impaired by connective tissue within tumors. To overcome this barrier, proteases were co-injected with virus preparations in experimental systems, and the results seemed encouraging (Kuriyama et al. 2000, 2001). To further optimize this approach, it might be possible to express proteases as transgenes from adenoviruses.

5.4 Systemic Treatment: Finding the Metastases

One of the biggest questions in the development of oncolytic viruses is whether it might be possible to inject the virus systemically, and make the virus find its target cells, rather than having to inject it locally. Recent advances do suggest that such an approach might be feasible (Habib et al. 2001; Heise et al. 1999b, 2000b; Nemunaitis et al. 2001a). However, to make this a potent therapy, further modifications are likely to be required. The goals are to reduce virus neutralization by antibodies, decrease virus elimination by cells other than the tumor cells, increase infectivity for tumor cells, enhance intratumoral spread despite the low amount of virus particles that can be expected to reach the tumor, and minimize any potential toxic effects to the body. A combination of modifications to the viral surface, and increased specificity and efficiency of virus replication may help to achieve these goals.

5.5 Clinical Research Directions

Oncolytic virotherapy can be expected to achieve major improvements not just by modifications of the virus itself. Rather, the administration of virus must be flanked by conventional therapy, and only clinical research can help to find and optimize these procedures. The clinical regimens that might help to improve oncolysis include the following:

1. Conventional chemotherapy. To optimize the interplay between chemotherapy and virotherapy, the influence of chemotherapeutics on virus replication (Heise et al. 1997) will need to be further evaluated.
2. Tumor irradiation. Again, the interplay with virus replication awaits evaluation.
3. Removal of virus-neutralizing antibodies by plasmapheresis (a procedure that removes plasma proteins but returns blood cells to the patient).
4. Immunotherapy that aims at increasing the antitumor immune response (Rosenberg 2001), including the application of cytokines, or the addition / stimulation of cytotoxic T cells directed against adenovirus-infected cells (e.g., by a recombinant vaccine or by T-cell stimulation in vitro).

Besides the optimization of oncolytic therapy as such, it will also be of great importance to precisely define criteria and parameters in accord with definitions of clinical efficacy and safety. Virotherapy of cancer is still in the pioneer phase, and there are few specific standards to which all involved clinical centers adhere. In the future, such standards should include the definition of virus amounts (particles? plaque forming units? fluorescence forming units?), the determination of intratumoral virus spread, the assessment of biopsy specimens by conventional and molecular pathology, the definitions of partial and complete responses, the determination of neutralizing antibodies, and the assessment of toxic effects. Recommendations by appropriate committees will be required to unify the assessment procedures, and to allow accurate comparison of different viruses in different studies.

Both basic and clinical research on oncolytic adenoviruses are still in their infancies. There is some reason to expect substantial progress in the near future. There is a chance, but still no certainty, that virotherapy may become a standard in cancer therapy and, like chemotherapy, radiation, and surgery, contribute to the successful treatment of cancer patients.

Acknowledgments. I thank Lynda Hawkins, Terry Hermiston, and David Kirn for helpful discussions, Hendrik Dobbelstein, Urs Hobom, Kristina Löhr, and Judith Roth for critically reading the manuscript, Hans-Dieter Klenk for continuous support, and the German Research Foundation, the Wilhelm Sander Foundation, the Mildred Scheel Foundation, the P. E. Kempkes Foundation, and the Fazit Foundation for funding of our work.

References

Alemany, R., Balague, C. and Curiel, D. T. (2000). Replicative adenoviruses for cancer therapy. Nat Biotechnol 18(7), 723–7

Alemany, R., Gomez-Manzano, C., Balague, C., Yung, W. K., Curiel, D. T., Kyritsis, A. P. and Fueyo, J. (1999a). Gene therapy for gliomas: molecular targets, adenoviral vectors, and oncolytic adenoviruses. Exp Cell Res 252(1), 1–12

Alemany, R., Lai, S., Lou, Y. C., Jan, H. Y., Fang, X. and Zhang, W. W. (1999b). Complementary adenoviral vectors for oncolysis. Cancer Gene Ther 6(1), 21–5

Babiss, L. E., Ginsberg, H. S. and Darnell Jr., J. E. (1985). Adenovirus E1B proteins are required for accumulation of late viral mRNA and for effects on cellular mRNA translation and transport. Mol Cell Biol 5, 2552–2558

Balague, C., Noya, F., Alemany, R., Chow, L. T. and Curiel, D. T. (2001). Human papillomavirus e6e7-mediated adenovirus cell killing: selectivity of mutant adenovirus replication in organotypic cultures of human keratinocytes. J Virol 75(16), 7602–11

Barker, D. D. and Berk, A. J. (1987). Adenovirus proteins from both E1B reading frames are required for transformation of rodent cells by viral infection and DNA transfection. Virology 156(1), 107–21

Bell, J. D. and Taylor-Robinson, S. D. (2000). Assessing gene expression in vivo: magnetic resonance imaging and spectroscopy. Gene Ther 7(15), 1259–1264

Bergelson, J. M., Cunningham, J. A., Droguett, G., Kurt-Jones, E. A., Krithivas, A., Hong, J. S., Horwitz, M. S., Crowell, R. L. and Finberg, R. W. (1997). Isolation of a common receptor for Coxsackie B viruses and adenoviruses 2 and 5. Science 275(5304), 1320–3

Binley, K., Iqball, S., Kingsman, A., Kingsman, S. and Naylor, S. (1999). An adenoviral vector regulated by hypoxia for the treatment of ischaemic disease and cancer. Gene Ther 6(10), 1721–7

Bischoff, J. R., Kirn, D. H., Williams, A., Heise, C., Horn, S., Muna, M., Ng, L., Nye, J. A., Sampson-Johannes, A., Fattaey, A. and McCormick, F. (1996). An adenovirus mutant that replicates selectively in p53-deficient human tumor cells. Science 274(5286), 373–6

Brand, K., Loser, P., Arnold, W., Bartels, T. and Strauss, M. (1998). Tumor cell-specific transgene expression prevents liver toxicity of the adeno-HSVtk/GCV approach. Gene Ther 5(10), 1363–71

Brough, D. E., Lizonova, A., Hsu, C., Kulesa, V. A. and Kovesdi, I. (1996). A gene transfer vector-cell line system for complete functional complementation of adenovirus early regions E1 and E4. J Virol 70(9), 6497–501

Brunori, M., Malerba, M., Kashiwazaki, H. and Iggo, R. (2001). Replicating adenoviruses that target tumors with constitutive activation of the wnt signaling pathway. J Virol 75(6), 2857–65

Burgert, H. G. and Blusch, J. H. (2000). Immunomodulatory functions encoded by the E3 transcription unit of adenoviruses. Virus Genes 21(1–2), 13–25

Carvalho, T., Seeler, J. S., Ohman, K., Jordan, P., Petterson, U., Akusjarvi, G., Carmo-Fonseca, M. and Dejean, A. (1995). Targeting of adenovirus E1A and E4-ORF3 proteins to nuclear matrix-associated PML bodies. J Cell Biol 131, 45–56

Chen, G., Branton, P. E., Yang, E., Korsmeyer, S. J. and Shore, G. C. (1996). Adenovirus E1B 19-kDa death suppressor protein interacts with Bax but not with Bad. J Biol Chem 271(39), 24221–5

Chen, J., Bezdek, T., Chang, J., Kherzai, A. W., Willingham, T., Azzara, M. and Nisen, P. D. (1998). A glial-specific, repressible, adenovirus vector for brain tumor gene therapy. Cancer Res 58(16), 3504–7

Chen, Y., DeWeese, T., Dilley, J., Zhang, Y., Li, Y., Ramesh, N., Lee, J., Pennathur-Das, R., Radzyminski, J., Wypych, J., Brignetti, D., Scott, S., Stephens, J., Karpf, D. B., Henderson, D. R. and Yu, D. C. (2001). CV706, a prostate cancer-specific adenovirus variant, in combination with radiotherapy produces synergistic antitumor efficacy without increasing toxicity. Cancer Res 61(14), 5453–60

Chen, Y., Yu, D. C., Charlton, D. and Henderson, D. R. (2000). Pre-existent adenovirus antibody inhibits systemic toxicity and antitumor activity of CN706 in the nude mouse LNCaP xenograft model: implications and proposals for human therapy. Hum Gene Ther 11(11), 1553–67

Coffey, M. C., Strong, J. E., Forsyth, P. A. and Lee, P. W. (1998). Reovirus therapy of tumors with activated Ras pathway. Science 282(5392), 1332–4

Croyle, M. A., Chirmule, N., Zhang, Y. and Wilson, J. M. (2001). "Stealth" adenoviruses blunt cell-mediated and humoral immune responses against the virus and allow for significant gene expression upon readministration in the lung. J Virol 75(10), 4792–801

Curiel, D. T. (1999). Strategies to adapt adenoviral vectors for targeted delivery. Ann N Y Acad Sci 886, 158–71

Curiel, D. T. (2000). The development of conditionally replicative adenoviruses for cancer therapy. Clin Cancer Res 6(9), 3395–9

Dachs, G. U., Dougherty, G. J., Stratford, I. J. and Chaplin, D. J. (1997). Targeting gene therapy to cancer: a review. Oncol Res 9(6–7), 313–25

Dobbelstein, M., Roth, J., Kimberly, W. T., Levine, A. J. and Shenk, T. (1997). Nuclear export of the E1B 55-kDa and E4 34-kDa adenoviral oncoproteins mediated by a rev-like signal sequence. Embo J 16(14), 4276–4284

Doronin, K., Kuppuswamy, M., Toth, K., Tollefson, A. E., Krajcsi, P., Krougliak, V. and Wold, W. S. (2001). Tissue-specific, tumor-selective, replication-competent adenovirus vector for cancer gene therapy. J Virol 75(7), 3314–24

Doronin, K., Toth, K., Kuppuswamy, M., Ward, P., Tollefson, A. E. and Wold, W. S. (2000). Tumor-specific, replication-competent adenovirus vectors overexpressing the adenovirus death protein. J Virol 74(13), 6147–55

Doucas, V., Ishov, A. M., Romo, A., Juguilon, H., Weitzman, M. D., Evans, R. M. and Maul, G. G. (1996). Adenovirus replication is coupled with the dynamic properties of the PML nuclear structure. Genes Dev 10(2), 196–207

Duque, P. M., Alonso, C., Sanchez-Prieto, R., Lleonart, M., Martinez, C., de Buitrago, G. G., Cano, A., Quintanilla, M. and Ramon y Cajal, S. (1999). Adenovirus lacking the 19-kDa and 55-kDa E1B genes exerts a marked cytotoxic effect in human malignant cells. Cancer Gene Ther 6(6), 554–63

Evan, G. I. and Vousden, K. H. (2001). Proliferation, cell cycle and apoptosis in cancer. Nature 411(6835), 342–8

Farrow, S. N., White, J. H., Martinou, I., Raven, T., Pun, K. T., Grinham, C. J., Martinou, J. C. and Brown, R. (1995). Cloning of a bcl-2 homologue by interaction with adenovirus E1B 19 K. Nature 374(6524), 731–3

Freytag, S. O., Rogulski, K. R., Paielli, D. L., Gilbert, J. D. and Kim, J. H. (1998). A novel three-pronged approach to kill cancer cells selectively: concomitant viral, double suicide gene, and radiotherapy. Hum Gene Ther 9(9), 1323–33

Fueyo, J., Gomez-Manzano, C., Alemany, R., Lee, P. S., McDonnell, T. J., Mitlianga, P., Shi, Y. X., Levin, V. A., Yung, W. K. and Kyritsis, A. P. (2000). A mutant oncolytic adenovirus targeting the Rb pathway produces anti- glioma effect in vivo. Oncogene 19(1), 2–12

Gabler, S., Schutt, H., Groitl, P., Wolf, H., Shenk, T. and Dobner, T. (1998). E1B 55-kilodalton-associated protein: a cellular protein with RNA- binding activity implicated in nucleocytoplasmic transport of adenovirus and cellular mRNAs. J Virol 72(10), 7960–71

Galanis, E., Bateman, A., Johnson, K., Diaz, R. M., James, C. D., Vile, R. and Russell, S. J. (2001a). Use of viral fusogenic membrane glycoproteins as novel therapeutic transgenes in gliomas. Hum Gene Ther 12(7), 811–21

Galanis, E., Vile, R. and Russell, S. J. (2001b). Delivery systems intended for in vivo gene therapy of cancer: targeting and replication competent viral vectors. Crit Rev Oncol Hematol 38(3), 177–92

Ganly, I., Kirn, D., Eckhardt, S. G., Rodriguez, G. I., Soutar, D. S., Otto, R., Robertson, A. G., Park, O., Gulley, M. L., Heise, C., Von Hoff, D. D. and Kaye, S. B. (2000). A phase I study of Onyx-015, an E1B attenuated adenovirus, administered intratumorally to patients with recurrent head and neck cancer. Clin Cancer Res 6(3), 798–806

Gao, G. P., Engdahl, R. K. and Wilson, J. M. (2000). A cell line for high-yield production of E1-deleted adenovirus vectors without the emergence of replication-competent virus. Hum Gene Ther 11(1), 213–9

Ginsberg, H. S., Horswood, R. L., Chanock, R. M. and Prince, G. A. (1990). Role of early genes in pathogenesis of adenovirus pneumonia. Proc Natl Acad Sci U S A 87(16), 6191–5

Goodrum, F. D. and Ornelles, D. A. (1998). p53 status does not determine outcome of E1B 55-kilodalton mutant adenovirus lytic infection. J Virol 72(12), 9479–90

Gromeier, M. (2001). Viruses as therapeutic agents against malignant disease of the central nervous system. J Natl Cancer Inst 93(12), 889–90

Habib, N. A., Sarraf, C. E., Mitry, R. R., Havlik, R., Nicholls, J., Kelly, M., Vernon, C. C., Gueret-Wardle, D., El-Masry, R., Salama, H., Ahmed, R., Michail, N., Edward, E. and Jensen, S. L. (2001). E1B-deleted adenovirus (dl1520) gene therapy for patients with primary and secondary liver tumors. Hum Gene Ther 12(3), 219–26

Hall, A. R., Dix, B. R., O'Carroll, S. J. and Braithwaite, A. W. (1998). p53-dependent cell death/apoptosis is required for a productive adenovirus infection. Nat Med 4(9), 1068–72

Hallenbeck, P. L., Chang, Y. N., Hay, C., Golightly, D., Stewart, D., Lin, J., Phipps, S. and Chiang, Y. L. (1999). A novel tumor-specific replication-restricted adenoviral vector for gene therapy of hepatocellular carcinoma. Hum Gene Ther 10(10), 1721–33

Han, J., Sabbatini, P., Perez, D., Rao, L., Modha, D. and White, E. (1996). The E1B 19 K protein blocks apoptosis by interacting with and inhibiting the p53-inducible and death-promoting Bax protein. Genes Dev 10(4), 461–77

Harada, J. N. and Berk, A. J. (1999). p53-Independent and -dependent requirements for E1B 55 K in adenovirus type 5 replication. J Virol 73(7), 5333–44

Hausmann, J., Ortmann, D., Witt, E., Veit, M. and Seidel, W. (1998). Adenovirus death protein, a transmembrane protein encoded in the E3 region, is palmitoylated at the cytoplasmic tail. Virology 244(2), 343–51

Hawkins, L. K. and Hermiston, T. W. (2001a). Gene delivery from the E3 region of replicating human adenovirus: evaluation of the ADP region. Gene therapy 8(15), 1132–1141

Hawkins, L. K. and Hermiston, T. W. (2001b). Gene delivery from the E3 region of replicating human adenovirus: evaluation of the E3B region. Gene therapy 8(15), 1142–1148

Hawkins, L. K., Johnson, L., Bauzon, M., Nye, J. A., Castro, D., Kitzes, G. A., Young, M. D., Holt, J. K., Trown, P. and Hermiston, T. W. (2001). Gene delivery from the E3 region of replicating human adenovirus: evaluation of the 6.7 K/gp19 K region. Gene Therapy 8(15), 1123–1131

Hay, J. G., Shapiro, N., Sauthoff, H., Heitner, S., Phupakdi, W. and Rom, W. N. (1999). Targeting the replication of adenoviral gene therapy vectors to lung cancer cells: the importance of the adenoviral E1b-55kD gene. Hum Gene Ther 10(4), 579–90

Heise, C., Ganly, I., Kim, Y. T., Sampson-Johannes, A., Brown, R. and Kirn, D. (2000a). Efficacy of a replication-selective adenovirus against ovarian carcinomatosis is dependent on tumor burden, viral replication and p53 status. Gene Ther 7(22), 1925–9

Heise, C., Hermiston, T., Johnson, L., Brooks, G., Sampson-Johannes, A., Williams, A., Hawkins, L. and Kirn, D. (2000b). An adenovirus E1A mutant that demonstrates potent and selective systemic anti-tumoral efficacy. Nat Med 6(10), 1134–9

Heise, C. and Kirn, D. H. (2000). Replication-selective adenoviruses as oncolytic agents. J Clin Invest 105(7), 847–51

Heise, C., Lemmon, M. and Kirn, D. (2000c). Efficacy with a replication-selective adenovirus plus cisplatin-based chemotherapy: dependence on sequencing but not p53 functional status or route of administration. Clin Cancer Res 6(12), 4908–14

Heise, C., Sampson-Johannes, A., Williams, A., McCormick, F., Von Hoff, D. D. and Kirn, D. H. (1997). ONYX-015, an E1B gene-attenuated adenovirus, causes tumor-specific cytolysis and antitumoral efficacy that can be augmented by standard chemotherapeutic agents. Nat Med 3(6), 639–45

Heise, C. C., Williams, A., Olesch, J. and Kirn, D. H. (1999a). Efficacy of a replication-competent adenovirus (ONYX-015) following intratumoral injection: intratumoral spread and distribution effects. Cancer Gene Ther 6(6), 499–504

Heise, C. C., Williams, A. M., Xue, S., Propst, M. and Kirn, D. H. (1999b). Intravenous administration of ONYX-015, a selectively replicating adenovirus, induces antitumoral efficacy. Cancer Res 59(11), 2623–8

Hernandez-Alcoceba, R., Pihalja, M., Wicha, M. S. and Clarke, M. F. (2000). A novel, conditionally replicative adenovirus for the treatment of breast cancer that allows controlled replication of E1a-deleted adenoviral vectors. Hum Gene Ther 11(14), 2009–24

Horridge, J. J. and Leppard, K. N. (1998). RNA-binding activity of the E1B 55-kilodalton protein from human adenovirus type 5. J Virol 72(11), 9374–9

Howe, J. A., Demers, G. W., Johnson, D. E., Neugebauer, S. E., Perry, S. T., Vaillancourt, M. T. and Faha, B. (2000). Evaluation of E1-mutant adenoviruses as conditionally replicating agents for cancer therapy. Mol Ther 2(5), 485–95

Howe, J. A., Mymryk, J. S., Egan, C., Branton, P. E. and Bayley, S. T. (1990). Retinoblastoma growth suppressor and a 300-kDa protein appear to regulate cellular DNA synthesis. Proc Natl Acad Sci U S A 87(15), 5883–7

Huang, M. M. and Hearing, P. (1989a). Adenovirus early region 4 encodes two gene products with redundant effects in lytic infection. J Virol 63(6), 2605–15

Huang, M. M. and Hearing, P. (1989b). The adenovirus early region 4 open reading frame 6/7 protein regulates the DNA binding activity of the cellular transcription factor, E2F, through a direct complex. Genes Dev 3(11), 1699–710

Huard, J., Lochmuller, H., Acsadi, G., Jani, A., Massie, B. and Karpati, G. (1995). The route of administration is a major determinant of the transduction efficiency of rat tissues by adenoviral recombinants. Gene Ther 2(2), 107–15

James, J. M., Burks, A. W., Roberson, P. K. and Sampson, H. A. (1995). Safe administration of the measles vaccine to children allergic to eggs. N Engl J Med 332(19), 1262–6

Kao, C. C., Yew, P. R. and Berk, A. J. (1990). Domains required for in vitro association between the cellular p53 and the adenovirus 2 E1B 55 K proteins. Virology 179(2), 806–14

Khuri, F. R., Nemunaitis, J., Ganly, I., Arseneau, J., Tannock, I. F., Romel, L., Gore, M., Ironside, J., MacDougall, R. H., Heise, C., Randlev, B., Gillenwater, A. M., Bruso, P., Kaye, S. B., Hong, W. K. and Kirn, D. H. (2000). A controlled trial of intratumoral ONYX-015, a selectively-replicating adenovirus, in combination with cisplatin and 5-fluorouracil in patients with recurrent head and neck cancer. Nat Med 6(8), 879–85

Kijima, T., Osaki, T., Nishino, K., Kumagai, T., Funakoshi, T., Goto, H., Tachibana, I., Tanio, Y. and Kishimoto, T. (1999). Application of the Cre recombinase/loxP system further enhances antitumor effects in cell type-specific gene therapy against carcinoembryonic antigen-producing cancer. Cancer Res 59(19), 4906–11

Kirn, D. (2000a). Replication-selective oncolytic adenoviruses: virotherapy aimed at genetic targets in cancer. Oncogene 19(56), 6660–9

Kirn, D. (2001). Clinical research results with dl1520 (Onyx-015), a replication- selective adenovirus for the treatment of cancer: what have we learned? Gene Ther 8(2), 89–98

Kirn, D., Hermiston, T. and McCormick, F. (1998). ONYX-015: clinical data are encouraging. Nat Med 4(12), 1341–2

Kirn, D., Martuza, R. L. and Zwiebel, J. (2001). Replication-selective virotherapy for cancer: Biological principles, risk management and future directions. Nat Med 7(7), 781–7

Kirn, D. H. (2000b). Replication-selective microbiological agents: fighting cancer with targeted germ warfare. J Clin Invest 105(7), 837–9

Kirn, D. H. and McCormick, F. (1996). Replicating viruses as selective cancer therapeutics. Mol Med Today 2(12), 519–27

Koch, P., Gatfield, J., Lober, C., Hobom, U., Lenz-Stoppler, C., Roth, J. and Dobbelstein, M. (2001). Efficient Replication of Adenovirus Despite the Overexpression of Active and Nondegradable p53. Cancer Res 61(15), 5941–7

Kurihara, T., Brough, D. E., Kovesdi, I. and Kufe, D. W. (2000). Selectivity of a replication-competent adenovirus for human breast carcinoma cells expressing the MUC1 antigen. J Clin Invest 106(6), 763–71

Kuriyama, N., Kuriyama, H., Julin, C. M., Lamborn, K. and Israel, M. A. (2000). Pretreatment with protease is a useful experimental strategy for enhancing adenovirus-mediated cancer gene therapy. Hum Gene Ther 11(16), 2219–30

Kuriyama, N., Kuriyama, H., Julin, C. M., Lamborn, K. R. and Israel, M. A. (2001). Protease pretreatment increases the efficacy of adenovirus-mediated gene therapy for the treatment of an experimental glioblastoma model. Cancer Res 61(5), 1805–9

Lambright, E. S., Amin, K., Wiewrodt, R., Force, S. D., Lanuti, M., Propert, K. J., Litzky, L., Kaiser, L. R. and Albelda, S. M. (2001). Inclusion of the herpes simplex thymidine kinase gene in a replicating adenovirus does not augment antitumor efficacy. Gene Ther 8(12), 946–53

Lamont, J. P., Nemunaitis, J., Kuhn, J. A., Landers, S. A. and McCarty, T. M. (2000). A prospective phase II trial of ONYX-015 adenovirus and chemotherapy in recurrent squamous cell carcinoma of the head and neck (the Baylor experience). Ann Surg Oncol 7(8), 588–92

Lan, K. H., Kanai, F., Shiratori, Y., Okabe, S., Yoshida, Y., Wakimoto, H., Hamada, H., Tanaka, T., Ohashi, M. and Omata, M. (1996). Tumor-specific gene expression in carcinoembryonic antigen–producing gastric cancer cells using adenovirus vectors. Gastroenterology 111(5), 1241–51

Lee, S. S., Weiss, R. S. and Javier, R. T. (1997). Binding of human virus oncoproteins to hDlg/SAP97, a mammalian homolog of the Drosophila discs large tumor suppressor protein. Proc Natl Acad Sci U S A 94(13), 6670–5

Lehrmann, S. (1999). Virus treatment questioned after gene therapy death. Nature 401, 517–518

Leppard, K. N. and Shenk, T. (1989). The adenovirus E1B 55 kd protein influences mRNA transport via an intranuclear effect on RNA metabolism. Embo J 8(8), 2329–36

Marcellus, R. C., Lavoie, J. N., Boivin, D., Shore, G. C., Ketner, G. and Branton, P. E. (1998). The early region 4 orf4 protein of human adenovirus type 5 induces p53-independent cell death by apoptosis. J Virol 72(9), 7144–53

Markert, J. M., Medlock, M. D., Rabkin, S. D., Gillespie, G. Y., Todo, T., Hunter, W. D., Palmer, C. A., Feigenbaum, F., Tornatore, C., Tufaro, F. and Martuza, R. L. (2000). Conditionally replicating herpes simplex virus mutant, G207 for the treatment of malignant glioma: results of a phase I trial. Gene Ther 7(10), 867–74

Mathews, M. B. and Shenk, T. (1991). Adenovirus virus-associated RNA and translation control. J Virol 65(11), 5657–62

Medina, D. J., Sheay, W., Goodell, L., Kidd, P., White, E., Rabson, A. B. and Strair, R. K. (1999). Adenovirus-mediated cytotoxicity of chronic lymphocytic leukemia cells. Blood 94(10), 3499–508

Mineta, T., Rabkin, S. D., Yazaki, T., Hunter, W. D. and Martuza, R. L. (1995). Attenuated multi-mutated herpes simplex virus-1 for the treatment of malignant gliomas. Nat Med 1(9), 938–43

Moore, M., Horikoshi, N. and Shenk, T. (1996). Oncogenic potential of the adenovirus E4orf6 protein. Proc Natl Acad Sci U S A 93(21), 11295–301

Morris, J. C. and Wildner, O. (2000). Therapy of head and neck squamous cell carcinoma with an oncolytic adenovirus expressing HSV-tk. Mol Ther 1(1), 56–62

Motoi, F., Sunamura, M., Ding, L., Duda, D. G., Yoshida, Y., Zhang, W., Matsuno, S. and Hamada, H. (2000). Effective gene therapy for pancreatic cancer by cytokines mediated by restricted replication-competent adenovirus. Hum Gene Ther 11(2), 223–35

Mulvihill, S., Warren, R., Venook, A., Adler, A., Randlev, B., Heise, C. and Kirn, D. (2001). Safety and feasibility of injection with an E1B-55 kDa gene-deleted, replication-selective adenovirus (ONYX-015) into primary carcinomas of the pancreas: a phase I trial. Gene Ther 8(4), 308–15

Nemunaitis, J., Cunningham, C., Buchanan, A., Blackburn, A., Edelman, G., Maples, P., Netto, G., Tong, A., Randlev, B., Olson, S. and Kirn, D. (2001a). Intravenous infusion of a replication-selective adenovirus (ONYX-015) in cancer patients: safety, feasibility and biological activity. Gene Ther 8(10), 746–59

Nemunaitis, J., Ganly, I., Khuri, F., Arseneau, J., Kuhn, J., McCarty, T., Landers, S., Maples, P., Romel, L., Randlev, B., Reid, T., Kaye, S. and Kirn, D. (2000). Selective replication and oncolysis in p53 mutant tumors with ONYX-015, an E1B-55kD gene-deleted adenovirus, in patients with advanced head and neck cancer: a phase II trial. Cancer Res 60(22), 6359–6366

Nemunaitis, J., Khuri, F., Ganly, I., Arseneau, J., Posner, M., Vokes, E., Kuhn, J., McCarty, T., Landers, S., Blackburn, A., Romel, L., Randlev, B., Kaye, S. and Kirn, D. (2001b). Phase II trial of intratumoral administration of ONYX-015, a replication-selective adenovirus, in patients with refractory head and neck cancer. J Clin Oncol 19(2), 289–98

Ohashi, M., Kanai, F., Tateishi, K., Taniguchi, H., Marignani, P. A., Yoshida, Y., Shiratori, Y., Hamada, H. and Omata, M. (2001). Target gene therapy for alpha-fetoprotein-producing hepatocellular carcinoma by E1B55k-attenuated adenovirus. Biochem Biophys Res Commun 282(2), 529–35

Pilder, S., Moore, M., Logan, J. and Shenk, T. (1986). The adenovirus E1B-55 K transforming polypeptide modulates transport or cytoplasmic stabilization of viral and host cell mRNAs. Mol Cell Biol 6, 470–476

Pomerantz, J., Schreiber-Agus, N., Liegeois, N. J., Silverman, A., Alland, L., Chin, L., Potes, J., Chen, K., Orlow, I., Lee, H. W., Cordon-Cardo, C. and DePinho, R. A. (1998). The Ink4a tumor suppressor gene product, p19Arf, interacts with MDM 2 and neutralizes MDM 2's inhibition of p53. Cell 92(6), 713–23

Querido, E., Marcellus, R. C., Lai, A., Charbonneau, R., Teodoro, J. G., Ketner, G. and Branton, P. E. (1997). Regulation of p53 levels by the E1B 55-kilodalton protein and E4orf6 in adenovirus-infected cells. J Virol 71(5), 3788–98

Ries, S. J., Brandts, C. H., Chung, A. S., Biederer, C. H., Hann, B. C., Lipner, E. M., McCormick, F. and Michael Korn, W. (2000). Loss of p14ARF in tumor cells facilitates replication of the adenovirus mutant dl1520 (ONYX-015). Nat Med 6(10), 1128–33

Rodriguez, R., Schuur, E. R., Lim, H. Y., Henderson, G. A., Simons, J. W. and Henderson, D. R. (1997). Prostate attenuated replication competent adenovirus (ARCA) CN706: a selective cytotoxic for prostate-specific antigen-positive prostate cancer cells. Cancer Res 57(13), 2559–63

Rogulski, K. R., Freytag, S. O., Zhang, K., Gilbert, J. D., Paielli, D. L., Kim, J. H., Heise, C. C. and Kirn, D. H. (2000a). In vivo antitumor activity of ONYX-015 is influenced by p53 status and is augmented by radiotherapy. Cancer Res 60(5), 1193–6

Rogulski, K. R., Wing, M. S., Paielli, D. L., Gilbert, J. D., Kim, J. H. and Freytag, S. O. (2000b). Double suicide gene therapy augments the antitumor activity of a replication-competent lytic adenovirus through enhanced cytotoxicity and radiosensitization. Hum Gene Ther 11(1), 67–76

Rosenberg, S. A. (2001). Progress in human tumour immunology and immunotherapy. Nature 411(6835), 380–4

Roth, J., König, C., Wienzek, S., Weigel, S., Ristea, S. and Dobbelstein, M. (1998). Inactivation of p53 but not p73 by adenovirus type 5 E1B 55-kilodalton and E4 34-kilodalton oncoproteins. J Virol 72(11), 8510–8516

Rothmann, T., Hengstermann, A., Whitaker, N. J., Scheffner, M. and zur Hausen, H. (1998). Replication of ONYX-015, a potential anticancer adenovirus, is independent of p53 status in tumor cells. J Virol 72(12), 9470–8
Rubenwolf, S., Schutt, H., Nevels, M., Wolf, H. and Dobner, T. (1997). Structural analysis of the adenovirus type 5 E1B 55-kilodalton-E4orf6 protein complex. J Virol 71(2), 1115–23
Sandig, V., Brand, K., Herwig, S., Lukas, J., Bartek, J. and Strauss, M. (1997). Adenovirally transferred p16INK4/CDKN2 and p53 genes cooperate to induce apoptotic tumor cell death. Nat Med 3(3), 313–9
Sauthoff, H., Heitner, S., Rom, W. N. and Hay, J. G. (2000). Deletion of the adenoviral E1b-19kD gene enhances tumor cell killing of a replicating adenoviral vector. Hum Gene Ther 11(3), 379–88
Shen, Y., Kitzes, G., Nye, J. A., Fattaey, A. and Hermiston, T. (2001). Analyses of single-amino-acid substitution mutants of adenovirus type 5 e1b-55 k protein. J Virol 75(9), 4297–307
Shenk, T. (1996). Adenoviridae: The viruses and their replication. In Virology 3 edit. (Fields, B. N., Knipe, D. M. and Howley, P. M., eds.), Vol. 2, pp. 2111–2148. 2 vols. Lippincott-Raven, Philadelphia
Shinoura, N., Yoshida, Y., Tsunoda, R., Ohashi, M., Zhang, W., Asai, A., Kirino, T. and Hamada, H. (1999). Highly augmented cytopathic effect of a fiber-mutant E1B-defective adenovirus for gene therapy of gliomas. Cancer Res 59(14), 3411–6
Shtrichman, R. and Kleinberger, T. (1998). Adenovirus type 5 E4 open reading frame 4 protein induces apoptosis in transformed cells. J Virol 72(4), 2975–82
Shtrichman, R., Sharf, R., Barr, H., Dobner, T. and Kleinberger, T. (1999). Induction of apoptosis by adenovirus E4orf4 protein is specific to transformed cells and requires an interaction with protein phosphatase 2A. Proc Natl Acad Sci U S A 96(18), 10080–5
Smith, E. R. and Chiocca, E. A. (2000). Oncolytic viruses as novel anticancer agents: turning one scourge against another. Expert Opin Investig Drugs 9(2), 311–27
Smith, R. R., Huebner, R. J., Rowe, W. P., Schatten, W. E. and Thomas, L. B. (1956). Studies on the use of viruses in the treatment of carcinoma of the cervix. Cancer 9(6), 1211–1218
Steegenga, W. T., Riteco, N. and Bos, J. L. (1999). Infectivity and expression of the early adenovirus proteins are important regulators of wild-type and DeltaE1B adenovirus replication in human cells. Oncogene 18(36), 5032–43
Steinwaerder, D. S., Carlson, C. A., Otto, D. L., Li, Z. Y., Ni, S. and Lieber, A. (2001). Tumor-specific gene expression in hepatic metastases by a replication- activated adenovirus vector. Nat Med 7(2), 240–3
Steinwaerder, D. S. and Lieber, A. (2000). Insulation from viral transcriptional regulatory elements improves inducible transgene expression from adenovirus vectors in vitro and in vivo. Gene Ther 7(7), 556–67
Stephenson, J. (2001). Studies illuminate cause of fatal reaction in gene-therapy trial. Jama 285(20), 2570
Stott, F. J., Bates, S., James, M. C., McConnell, B. B., Starborg, M., Brookes, S., Palmero, I., Ryan, K., Hara, E., Vousden, K. H. and Peters, G. (1998). The alternative product from the human CDKN2A locus, p14(ARF), participates in a regulatory feedback loop with p53 and MDM 2. Embo J 17(17), 5001–14

Strong, J. E., Coffey, M. C., Tang, D., Sabinin, P. and Lee, P. W. (1998). The molecular basis of viral oncolysis: usurpation of the Ras signaling pathway by reovirus. Embo J 17(12), 3351–62

Subramanian, T., Kuppuswamy, M., Gysbers, J., Mak, S. and Chinnadurai, G. (1984). 19-kDa tumor antigen coded by early region E1b of adenovirus 2 is required for efficient synthesis and for protection of viral DNA. J Biol Chem 259(19), 11777–83

Sunamura, M. (2000). Mutant adenoviruses selectively replication-competent in tumor cells. Adv Exp Med Biol 465, 65–71

Suzuki, K., Fueyo, J., Krasnykh, V., Reynolds, P. N., Curiel, D. T. and Alemany, R. (2001). A conditionally replicative adenovirus with enhanced infectivity shows improved oncolytic potency. Clin Cancer Res 7(1), 120–6

Tanaka, T., Kanai, F., Okabe, S., Yoshida, Y., Wakimoto, H., Hamada, H., Shiratori, Y., Lan, K., Ishitobi, M. and Omata, M. (1996). Adenovirus-mediated prodrug gene therapy for carcinoembryonic antigen- producing human gastric carcinoma cells in vitro. Cancer Res 56(6), 1341–5

Tollefson, A. E., Ryerse, J. S., Scaria, A., Hermiston, T. W. and Wold, W. S. (1996a). The E3–11.6-kDa adenovirus death protein (ADP) is required for efficient cell death: characterization of cells infected with adp mutants. Virology 220(1), 152–62

Tollefson, A. E., Scaria, A., Hermiston, T. W., Ryerse, J. S., Wold, L. J. and Wold, W. S. (1996b). The adenovirus death protein (E3–11.6 K) is required at very late stages of infection for efficient cell lysis and release of adenovirus from infected cells. J Virol 70(4), 2296–306

Turnell, A. S., Grand, R. J. and Gallimore, P. H. (1999). The replicative capacities of large E1B-null group A and group C adenoviruses are independent of host cell p53 status. J Virol 73(3), 2074–83

Vassaux, G., Hurst, H. C. and Lemoine, N. R. (1999). Insulation of a conditionally expressed transgene in an adenoviral vector. Gene Ther 6(6), 1192–7

Vollmer, C. M., Ribas, A., Butterfield, L. H., Dissette, V. B., Andrews, K. J., Eilber, F. C., Montejo, L. D., Chen, A. Y., Hu, B., Glaspy, J. A., McBride, W. H. and Economou, J. S. (1999). p53 selective and nonselective replication of an E1B-deleted adenovirus in hepatocellular carcinoma. Cancer Res 59(17), 4369–74

Von Hoff, D. D., Goodwin, A. L. and Garcia, L. (1998). Advances in the treatment of patients with pancreatic cancer: improvement in symptoms and survival time. The San Antonio Drug Development Team. Br J Cancer 78(Suppl 3), 9–13

Watanabe, T., Shinohara, N., Sazawa, A., Harabayashi, T., Ogiso, Y., Koyanagi, T., Takiguchi, M., Hashimoto, A., Kuzumaki, N., Yamashita, M., Tanaka, M., Grossman, H. B. and Benedict, W. F. (2000). An improved intravesical model using human bladder cancer cell lines to optimize gene and other therapies. Cancer Gene Ther 7(12), 1575–80

Weigel, S. and Dobbelstein, M. (2000). The nuclear export signal within the e4orf6 protein of adenovirus type 5 supports virus replication and cytoplasmic accumulation of viral mRNA. J Virol 74(2), 764–72

Wells, W. A. (2000). Smarter viruses. Onyx pharmaceuticals, Inc. Chem Biol 7(12), R223–4

White, E., Sabbatini, P., Debbas, M., Wold, W. S., Kusher, D. I. and Gooding, L. R. (1992). The 19-kilodalton adenovirus E1B transforming protein inhibits programmed cell death and prevents cytolysis by tumor necrosis factor alpha. Mol Cell Biol 12(6), 2570–80

Whyte, P., Williamson, N. M. and Harlow, E. (1989). Cellular targets for transformation by the adenovirus E1A proteins. Cell 56(1), 67–75

Wildner, O. (1999). In situ use of suicide genes for therapy of brain tumours. Ann Med 31(6), 421–9

Wildner, O., Blaese, R. M. and Morris, J. C. (1999a). Therapy of colon cancer with oncolytic adenovirus is enhanced by the addition of herpes simplex virus-thymidine kinase. Cancer Res 59(2), 410–3

Wildner, O. and Morris, J. C. (2000a). The role of the E1B 55 kDa gene product in oncolytic adenoviral vectors expressing herpes simplex virus-tk: assessment of antitumor efficacy and toxicity. Cancer Res 60(15), 4167–74

Wildner, O. and Morris, J. C. (2000b). Therapy of peritoneal carcinomatosis from colon cancer with oncolytic adenoviruses. J Gene Med 2(5), 353–60

Wildner, O., Morris, J. C., Vahanian, N. N., Ford, H., Jr., Ramsey, W. J. and Blaese, R. M. (1999b). Adenoviral vectors capable of replication improve the efficacy of HSVtk/GCV suicide gene therapy of cancer. Gene Ther 6(1), 57–62

Williams, J. F., Gharpure, M., Ustacelebi, S. and McDonald, S. (1971). Isolation of temperature-sensitive mutants of adenovirus type 5. J Gen Virol 11(2), 95–101

Wodarz, D. (2001). Viruses as antitumor weapons: defining conditions for tumor remission. Cancer Res 61(8), 3501–7

Yew, P. R. and Berk, A. J. (1992). Inhibition of p53 transactivation required for transformation by adenovirus early 1B protein. Nature 357(6373), 82–5

Yew, P. R., Kao, C. C. and Berk, A. J. (1990). Dissection of functional domains in the adenovirus 2 early 1B 55 K polypeptide by suppressor-linker insertional mutagenesis. Virology 179(2), 795–805

Yohn, D. S., Hammon, W. M., Atchison, R. W. and Casto, B. C. (1968). Oncolytic potentials of nonhuman viruses for human cancer. II. Effects of five viruses on heterotransplantable human tumors. J Natl Cancer Inst 41(2), 523–9

You, L., Yang, C. T. and Jablons, D. M. (2000). ONYX-015 works synergistically with chemotherapy in lung cancer cell lines and primary cultures freshly made from lung cancer patients. Cancer Res 60(4), 1009–13

Yu, D. C., Chen, Y., Dilley, J., Li, Y., Embry, M., Zhang, H., Nguyen, N., Amin, P., Oh, J. and Henderson, D. R. (2001). Antitumor synergy of CV787, a prostate cancer-specific adenovirus, and paclitaxel and docetaxel. Cancer Res 61(2), 517–25

Yu, D. C., Chen, Y., Seng, M., Dilley, J. and Henderson, D. R. (1999). The addition of adenovirus type 5 region E3 enables calydon virus 787 to eliminate distant prostate tumor xenografts. Cancer Res 59(17), 4200–3

Zhang, Y., Xiong, Y. and Yarbrough, W. G. (1998). ARF promotes MDM 2 degradation and stabilizes p53: ARF-INK4a locus deletion impairs both the Rb and p53 tumor suppression pathways. Cell 92(6), 725–34

Zhou, H., O'Neal, W., Morral, N. and Beaudet, A. L. (1996). Development of a complementing cell line and a system for construction of adenovirus vectors with E1 and E2a deleted. J Virol 70(10), 7030–8

Zielinski, T. and Jordan, E. (1969). [Remote results of clinical observation of the oncolytic action of adenoviruses on cervix cancer]. Nowotwory 19(3), 217–21

Adenovirus Vectors: Biology, Design, and Production

M. J. Imperiale[1] · S. Kochanek[2]

[1] Department of Microbiology and Immunology, Center for Gene Therapy, University of Michigan Medical School, 1500 E. Medical Center Drive, 6304 Cancer Center, Ann Arbor, MI 48109-0942, USA

[2] Center for Molecular Medicine (ZMMK), University of Cologne, Kerpener Str. 34, 50931 Cologne, Germany

E-mail: stefan.kochanek@medizin.uni-koeln.de

Abstract The use of adenovirus as a gene transfer vehicle arose from early reports of recombinant viruses carrying heterologous DNA fragments. Adenovirus vectors offer many advantages for gene delivery: they are easy to propagate to high titers, they can infect most cell types regardless of their growth state, and in their most recent embodiments they can accommodate large DNA inserts. In this chapter, the development of adenovirus vectors is reviewed, from the use of so-called first-generation, E1-deleted viruses to the latest generation high-capacity, helper-dependent vectors. Examples of their use in the clinic are described, as are the current areas in which improvements to these vectors are being explored.

1 Introduction

Somatic gene therapy has the potential to provide novel treatments for many inherited and acquired diseases. Central to a successful implementation of somatic gene therapy is the development of gene transfer technologies that allow efficient delivery of genes into target cells. Adenovirus vectors have many features that make them very promising reagents to achieve this goal. With relatively low toxicity, they can efficiently transduce a wide variety of different primary and cultured cell types in vitro and in vivo. In addition, at least in a more recent vector type, they can deliver DNA fragments with sizes of up to 35 kb, which allows the simultaneous expression of several genes or the tight control of gene expression by including regulatory control elements. This chapter focuses on several aspects of development, design, and production of adenovirus vectors, together with a discussion of some of their more promising applications in vivo, both in animal models and human clinical trials.

2 Historical Perspective

In principle, the stage for the development of adenovirus vectors was set in the late 1950s when formalin-inactivated adenovirus vaccines were developed to prevent the frequent occurrence of adenovirus-mediated respiratory illnesses. Since the use of human cell lines for vaccine production was not allowed, it was attempted to adapt human adenoviruses for growth in nonhuman cell lines (Hartley et al. 1956). Vaccines prepared in rhesus monkey cells effectively prevented respiratory disease in military recruits (Pierce et al. 1968). However, vaccination of volunteers was terminated after the discovery that these vaccines were contaminated with Simian Virus 40 (SV40). Surprisingly, removal of SV40 from the adenovirus vaccines proved to be impossible, the reason for which became clear when it was discovered that the vaccines frequently contained hybrid viruses carrying SV40 DNA that was covalently linked to adenovirus DNA. Usually, because of a block in the expression of late viral proteins, wild-type human adenovirus cannot be propagated in monkey cells. This block can be relieved by helper functions of the large T antigen of SV40 (reviewed in Klessig 1984). Two different types of adenovirus populations were isolated, defective and nondefective. Defective viruses required the presence of a helper virus; nondefective viruses could be grown in human cells in the absence of a helper virus (Lewis et al. 1966). The latter were found to contain relatively small parts of SV40 DNA that did not disrupt essential adenovirus genes, whereas the defective viruses had disruptions and, in general, deletion of essential adenovirus genes due to the insertion of SV40 DNA. In monkey cells, these defective hybrid viruses could only be propagated with the help of a wild-type adenovirus (Lewis and Rowe 1970). In virus stocks, the defective hybrid viruses always constituted less than 30% of the total virus mixture. Frequently they contained several complete copies of the SV40 genome arranged in tandem arrays. As a consequence, frequent rearrangements were generated by either intra- or intermolecular recombination. One of the Ad-SV40 hybrids contained 5.5 copies of the SV40 genome within the largely deleted adenovirus genome (Gluzman and van Doren 1983). The SV40 portion was arranged in two head-to-tail tandem copies, each approximately 2.7 genomes in length and with an axis of symmetry at the center of the genome. Thus, the entire molecule was organized as a giant inverted repeat with a total size of 35 kb. At both ends of the molecule, 3.5 kb of identical adenovirus termini were present that included the E1 gene and the packaging signal. Two years earlier, an adenovirus-human DNA hybrid had been isolated following repeated passaging of adenovirus type 12 (Ad12) on human KB cells at a high multiplicity of infection (Deuring et al. 1981). These hybrids contained head-to-head repeats of human chromosomal DNA that were located between two identical DNA fragments of 1 kb size that were derived from the left Ad12 terminus (essentially containing the left ITR and the packaging signal). Due to the peculiar structure of their DNAs, these recombinants were called symmetrical recombinants (SYRECs). SYRECs could be easily maintained over years together with wild-type Ad12 virus. Partial purification of SYRECs by CsCl density equilibrium centrifugation was possible because of their lower density compared to wild-type Ad12. These SYRECs were an important conceptual foundation for the later development of the high-capacity adenovirus vectors described below.

Defective adenoviruses were initially used in several research areas (Klessig 1984; Berkner 1988): (1) For a functional analysis of the major late promoter (MLP) and the adenovirus tripartite leader that are critical for transcription and translation of late adenovirus genes. (2) For the overexpression and functional analysis of the SV40 large T antigen. These experiments also took advantage of the helper function of T antigen to remove the translational block of late adenovirus transcripts present in simian cells. Thus, SV40 T antigen served as both the gene of interest and the selective force. (3) For the expression of nonselected genes. In these experiments, the defective virus contained two foreign expression units: the gene of interest (e.g., mouse polyomavirus T antigens, HSV1-TK, or the alpha subunit of the human chorionic gonadotropin gene; Yamada et al. 1985; Mansour et al. 1985) and, in addition, the SV40 T antigen that was required for propagation of the defective adenovirus in simian cells.

In other studies, the adenovirus *cis*-acting sequences that mediate encapsidation of the viral genome into virions at the late stage of productive infection were defined. The existence of a packaging signal was suggested by both the structure of the hybrid viruses described above and the structure of viral genomic fragments isolated from incomplete particles (Daniell 1976; Tibbetts 1977). The location of an encapsidation signal was mapped more precisely by studies involving Ad16 incomplete particles that contained subgenomic viral DNA fragments (Hammarskjold and Winberg 1980). Finally, the packaging signal of Ad5 was characterized at the molecular level in a series of elegant studies that involved the generation and analysis of a large number of packaging-deficient adenovirus mutants (Hearing et al. 1987; Gräble and Hearing 1990, 1992).

3
E1-Deleted Vectors

The most common vectors in use for gene transfer studies are deleted in the E1 (E1A+E1B) region, and are sometimes referred to as first-generation adenovirus vectors (Fig. 1). This deletion allows for two important features. First, it reduces the replication potential of the vector in most cell types. Second, it provides space for insertion of the transgene of interest. Cloning of genes was first accomplished by an overlap recombination method (Chinnadurai et al. 1979), but today there are dozens of different ways to build these vectors (reviewed in Danthinne and Imperiale 2000). Viruses with deletions in E1 can be propagated easily in the 293 cell line (Graham et al. 1977) or newer cells such as 911 or PER.C6 (see below), so preparing large volumes of high-titer stocks of these viruses is not a problem. Many E1-deleted vectors also carry partial deletions in the E3 region, as they were derived from the Ad5 mutant, *dl*309 (Jones

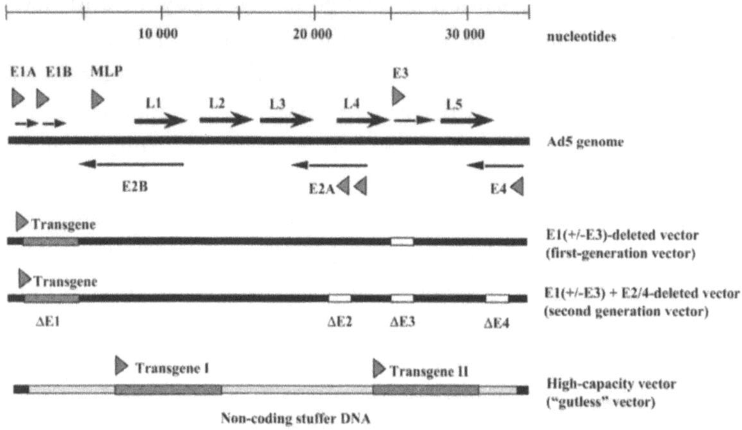

Fig. 1 Organization of the adenovirus genome and the different adenovirus vector types as discussed in the text. Promoters are indicated by *arrowheads*, transcribed genes by *arrows*. The genes that are transcribed early during infection are the E1A, E1B, E2, E3, and E4 genes. The main gene products that are generated late during infection are transcribed from the major late promoter (*MLP*), which directs a very long pre-mRNA (*MLTU*, major late transcription unit). Different RNA species (L1–L5) that code for structural proteins are generated by alternative splicing and differential polyadenylation. (For clarity, not all adenoviral genes and gene products are indicated). First-generation adenovirus vectors are characterized by deletion of the E1 genes, and optionally by a deletion in the E3 region. Second-generation adenovirus vectors, in addition to the deletion of the E1 function, are characterized by the additional deletion of the E2 and/or E4 genes. High-capacity adenovirus vectors have most of the viral genome removed and retain only the noncoding viral termini (ITRs and packaging signal). In high-capacity adenoviral vectors, stuffer DNA is included on the vector genome for stability and expression reasons

and Shenk 1978). The E3 region plays a role in the pathogenesis of adenovirus infections, mainly by interfering with the immune response against the virus, and therefore is dispensable for growth in culture.

The first attempts to introduce either reporter or therapeutic genes into animals were performed using E1-deleted viruses, under the assumption that such viruses would not replicate, and therefore would not express any viral genes that could stimulate an immune response. Investigators soon learned, however, that these viruses did express some of their genes, and expression of the transgenes was short-lived due to cellular immune responses that apparently eliminated the transduced cells. In retrospect, this result could have been predicted based on a number of studies showing that E1-deleted viruses, although defective, are not completely blocked for replication (Shenk et al. 1980; Gaynor and Berk 1983; Imperiale et al. 1984). The ability of these vectors to replicate their DNA and express late genes is likely correlated with levels of the cellular transcription factor E2F, which drives expression of the adenovirus E2 genes, which encode the viral DNA replication proteins (Kovesdi et al. 1986).

Despite their limitations for applications that require long-term gene expression, E1-deleted viruses are an effective delivery vehicle for applications that do not require persistent transgene expression, such as killing of cancer cells or vaccination. In fact, in these instances the immune response provides added benefits. There are numerous examples (far too many to catalog in their entirety here) in which these first-generation adenoviruses have shown promise as therapeutic agents, most of which are in oncology. One such approach is to use adenovirus to deliver a wild-type p53 gene into tumors that have lost p53 function through mutation or deletion, which is the case in well over half of all cancers. This vector has proven safe and efficacious in Phase I and II clinical trials for a number of tumor sites, including non-small cell lung cancer and head and neck squamous cell carcinomas (Clayman et al. 1998; Roth et al. 1998; Swisher et al. 1999), and Phase III trials are currently commencing. Interestingly, it also seems to kill tumor cells harboring wild-type p53. Another use of recombinant adenovirus in cancer has been to deliver cytotoxic genes to bone marrow preparations prior to autologous transplantation, in an effort to purge the bone marrow of any contaminating tumor cells. An adenovirus expressing Bcl-X_S, a pro-apoptotic factor, has been shown in animal models to be effective along these lines (Clarke et al. 1995). A large number of other adenovirus vectors expressing genes whose products convert prodrugs into active entities, such as thymidine kinase or cytosine deaminase, have also been tested in a variety of tumor models as well as in clinical trials (Morris et al. 2000; Rogulski et al. 2000; Trask et al. 2002). Cancer is not the only disease in which adenovirus has been used with some success, however. Investigators have also used adenoviruses expressing factors that promote blood vessel growth, such as vascular endothelial growth factor (VEGF), to attempt to bypass clogged arteries. In these trials, the recombinant virus is injected directly into the heart muscle of patients with coronary disease or into the extremities of patients with peripheral vascular disease, in order to recruit a new blood supply (Rosengart et al. 1999a,b; Rajagopalan et al. 2001).

A variation on the E1-deleted class of vectors has been the development of conditionally replication-competent vectors for cancer treatment. E1-deleted viruses carrying toxic transgenes are limited in that they can kill the cell they initially infect, but cannot spread efficiently. Thus, investigators looked for a way to allow the virus to complete its life cycle during the initial infection of tumor cells but not normal cells, and then spread to neighboring tumor cells. Perhaps the best known, and most widely studied, of these are viruses containing mutations in the E1B region. The rationale behind this group of viruses is the observation that a major role of the E1B region is to inactivate p53 function. This is required for efficient viral replication because E1A induces a p53 response, leading to apoptosis of the cell and drastically diminished viral yields unless E1B is present (White et al. 1986; Rao et al. 1992). Therefore, it was argued that E1B-viruses would be able to replicate in p53-negative but not p53-positive cells, and that because many tumor cells contain mutations in p53, such a virus would selectively kill tumors. In vitro studies bore this prediction out (Bischoff et al. 1996), but in vivo it has been demonstrated that the ability of these viruses to kill cancer cells is not correlated with the p53 status of the cell: rather, the E1B-viruses appear to have a broader ability to kill cancer cells in general while sparing surrounding normal tissue. Indeed, this approach has proven efficacious against locoregional tumors such as those of the head and neck, and trials have been initiated to attack other sites including pancreatic and ovarian cancer (Khuri et al. 2000; Ganly et al. 2000; Nemunaitis et al. 2000, 2001). Another variation on this theme has been to mutate the E1A gene with the same ultimate goal, namely to allow the virus to replicate only in tumor cells. Here, the domains of the E1A proteins that bind to and inactivate the pRb tumor suppressor protein have been mutated. Such viruses can only replicate in *RB1*-cells, in which E2F activity is elevated independent of E1A (Heise et al. 2000). Some groups have recently taken both the E1A- and E1B-viruses and added therapeutic genes in the E3 region, resulting in so-called "armed" conditionally replicating viruses (Nanda et al. 2001). Additionally, Wold's group has shown that overexpression of the adenovirus death protein (ADP) from the E3 region results in even more efficient spread of these viruses (Doronin et al. 2000). At the time this chapter was written, these types of viruses were undergoing initial evaluations in clinical trials for various cancers.

A second type of conditionally replicating viruses are those in which a viral promoter has been replaced with a tissue-specific promoter with the hope of conferring cell type-specific replication on the virus. This has been accomplished by changing the E1A promoter, allowing its expression and subsequent activation of the other viral genes (Rodriguez et al. 1997; Hallenbeck et al. 1999; Hernandez-Alcoceba et al. 2000; Kurihara et al. 2000; Adachi et al. 2001; Benjamin et al. 2001; Matsubada et al. 2001), or the E4 promoter (Doronin et al. 2001), resulting in controlled late gene expression. In each case these manipulations result in complete viral replication and cell killing only in the appropriate cellular environment. A more recent variation on this theme is the use of a two-component system, in which one virus expresses a transcription factor under the control of a target-specific promoter, and the second virus expresses the therapeutic gene only in the presence of that transcription factor (Ueda et al. 2001; Qiao et al. 2002).

4 Vectors With Mutations in Other Viral Genes

The inability of E1-deleted viruses to effect long-term expression has led to attempts to disable other viral genes that are involved in replication. One transcription unit that has been targeted is E2 (Fig. 1). Initially, researchers tried taking advantage of a temperature-sensitive mutant in the 72 K DNA-binding protein gene, E2A. Unfortunately, this mutant is leaky at 37°C, and gene expression from such viruses was still relatively short-lived (Engelhardt et al. 1994a,b; Yang et al. 1994). Subsequently, investigators have deleted the DNA polymerase and preterminal protein genes (E2B), resulting in drastically reduced replication and a concomitant decrease in the immune response. Expression from these viruses has been reported to last at least six months in animal studies, a clear improvement over E1-deleted vectors (Amalfitano et al. 1998; Ding et al. 2001). The second gene that has been removed from the virus is the E4 gene. E4 encodes a number of proteins that are required for viral replication and, more specifically, for late gene expression. Thus, E4-viruses do not express major viral antigens. The results with these viruses have been mixed, however. First, it is difficult to grow such viruses because some E4 products are toxic to the cell, making the derivation of helper cell lines difficult. Second, it appears that these viruses work, in terms of prolonged transgene expression, in some tissues but not others (Brough et al. 1997; Armentano et al. 1997, 1999; Dedieu et al. 1997; Wang et al. 1997; Lusky et al. 1998). Therefore, although the E4 viruses hold promise, more work is required to understand their biology. The E3 region, as discussed earlier, is often deleted from vectors because it is not required for growth in culture, and its removal provides additional space for the transgene. Investigators have been interested in restoring the E3 region to the vector under the control of a strong constitutive promoter, however, so as to take advantage of its anti-immune response activities and thereby allow for longer-term gene expression (Lee et al. 1995; Bruder et al. 1997).

5
High-Capacity Adenovirus Vectors

High-capacity adenovirus (HC-Ad) vectors [also called pseudoadenovirus (PAV), helper-dependent (HD-Ad), gutted, or gutless adenovirus vectors] have been developed to address both the capacity and the immunogenicity problems of first- and second-generation adenovirus vectors (Mitani et al. 1995; Kochanek et al. 1996; Fisher et al. 1996; Clemens et al. 1996; Kumar-Singh and Chamberlain 1996; Parks et al. 1996). The genome of these vectors contains only two viral elements: the inverted terminal repeats (ITRs), which are essential for replication of the viral DNA, and the packaging signal that is located close to the left terminus and that is required for encapsidation of the DNA into capsids late during productive infection (Fig. 1). Since the size of the ITRs and the packaging signal adds up to a total of less than 0.6 kb, the potential cargo of these vectors is approximately 37 kb, being restricted only by the overall encapsidation limit of 38 kb for Ad5. Both the ITRs and the packaging signal are noncoding, resulting in the lack of expression of any viral protein (both early and late) or RNA (VAI and VAII) functions from the vector following transduction of target cells, an important safety feature for gene therapy in humans. It is important to note that our current experience with the biology of first- and second-generation adenovirus vectors is based mainly on the results of experiments performed in rodents. One should not forget, then, that the natural course of an infection of rodent cells with a human adenovirus is abortive or semi-abortive. Thus, toxicity (or the lack thereof) following adenovirus-mediated gene transfer in rodents cannot be expected to be fully predictive for the situation in humans, a lesson that was unfortunately learned through the death of a young man in a gene transfer trial (Raper et al. 2002). Furthermore, in general the experiments in rodents are performed in genetically inbred strains, whereas in humans interindividual differences in immune and nonimmune responses to Ad5 vector-mediated gene transfer are a possibility that cannot be considered to be remote (Crystal et al. 2002; Harvey et al. 2002).

An important early question concerned the performance of this vector type in vivo. As described above, helper-dependent vectors had been produced inadvertently in cell culture at a time even before helper-independent vectors had been developed. Intuitively one probably would consider it likely that this vector would transduce cells as efficiently as had been observed in the case of E1-deleted vectors. However, experimental observations by Lieber and coworkers (1996, 1997) indicated that gene transfer with HC-Ad vectors of a small (9-kb) size leads to a transient (at a maximum days) and low-level expression of the transgene both in vitro and in vivo. Currently, it is unclear what step during infection (e.g., cellular entry, transport to the nucleus, DNA stability in the nucleus) is defective. Using HC-Ad vectors with genome sizes above 27 kb, a large number of different genes have been introduced successfully into target cells following in vivo gene transfer. The first vector that was used in vivo expressed the 14-kb dystrophin cDNA from a 6.5-kb muscle-specific promoter. Dystrophin is deficient in patients with Duchenne muscular dystrophy (DMD), an X-chromosomal, fatal, degenerative muscle disease. Dystrophin is an essential structural component of the muscle cell membrane and links intracellular actin filaments with the dystrophin-associated proteins (DAPs) in the sarcolemma. In the absence of dystrophin, the membrane is fragile and degeneration of muscle fibers ensues. The vector was delivered by intramuscular injection into the hind-limb of the *mdx* mouse, a genetic model for DMD. Following gene transfer, stable expression of the 400-kDa full-length dystrophin protein was observed. Immunofluorescence analysis demonstrated correct localization of the transgenic dystrophin to the sarcolemma and restoration of the DAPs to the muscle membrane, and the histological phenotype was significantly improved (Clemens et al. 1996). Since this first proof-of-concept experiment, HC-Ad vectors have been delivered to several tissues including liver, CNS, and eye. An example that firmly established advantageous properties of HC-Ad vectors compared to first-generation vectors was a vector that carried the 19-kb human alpha1-antitrypsin (hAAT) gene locus (Schiedner et al. 1998). AAT is abundantly expressed in hepatocytes and to a lower degree in macrophages and inhibits elastase produced by neutrophilic granulocytes. Hepatic gene transfer in immunocompetent C57Bl/6J mice resulted in tissue-specific and durable gene expression for longer than one year. The injection of high vector doses resulted in very high (5 mg/ml) and stable hAAT serum levels (Morral et al. 1998). Mice that were injected with an identical dose of a first-generation vector expressing hAAT as a cDNA from the murine PGK promoter suffered severe liver damage, documented by strong elevation of liver enzymes detected in the serum and by significant liver abnormalities found on histological analysis.

5.1 Stuffer DNA in HC-Ad Vectors

For practical and/or feasibility reasons, most HC-Ad vectors will carry genes or expression cassettes that are smaller than 37 kb, the capacity limit of HC-Ad vectors. The smallest genomes that were observed with the Ad5-SV40 hybrids mentioned above were about 25 kb in size (Hassell et al. 1978). These findings suggested a selective advantage during rescue or propagation of larger over smaller virus genomes. In experiments designed to study the size dependence of successful HC-Ad vector amplification by the Cre-*loxP* production system (see Sect. 6), only vectors with genome sizes of at least 27 kb allowed efficient and stable amplification (Parks and Graham 1997). Using smaller DNAs as starting material, rearrangements and amplifications led to larger genomes with sizes closer to the wild-type size that finally dominated the vector stock. The regular occurrence of vector rearrangements and amplifications has also been the experience when HC-Ad vectors carrying small genomes were used as starting material. The structures that were observed were either symmetric dimers (Morsy et al. 1998) or mixtures of head-to-head, head-to-tail, or tail-to-tail concatamers (Fisher et al. 1996; Haecker et al. 1996; Kumar-Singh and Farber 1998). Additional evidence for the impact of vector genome size on functional competence of adenovirus vectors mentioned above comes from the performance of adenovirus vector particles that contained expression constructs with sizes of less than 10 kb (Lieber et al. 1996). Following in vitro and in vivo gene transfer, vector DNA levels in target cells were found to be low and transient. Consequently, transgene expression was detectable only for a few hours. Together these observations indicate that genome size and potentially also the composition of the vector genome play important roles during vector production and for gene expression following gene transfer. Therefore, in many HC-Ad vectors, additional "stuffer" DNA must be incorporated into the vector genome if the expression construct has a smaller size than 27 kb. With respect to the expression cassette itself, in several cases expression has been demonstrated to be considerably stronger from vectors that carried a genomic locus compared to vectors that carried small cDNA expression cassettes (Schiedner et al. 1998, Kim et al. 2001). Although the orientation and the location of the transgene within the vector genome do not significantly influence levels of gene expression, the overall choice of stuffer DNA does have an impact on gene expression from HC-Ad vectors. For example, vectors carrying stuffer DNA derived from the bacteriophage lambda genome have an inferior in vivo performance compared to vectors that carry human DNA as stuffer (Parks et al. 1999; Schiedner et al. 2002). The presence of a matrix-associated region (MAR) element on the vector genome may further improve expression under certain conditions (Schiedner et al. 2002). Taken together, these findings indicate that functions of both the particular expression construct and of the stuffer DNA should be considered to achieve optimal in vivo performance of HC-Ad vectors.

6
Vector Production

It is clear from the foregoing discussion that the production of adenovirus vectors differs between vectors that have mutations in one or two viral genes and those that are devoid of all viral genes. As mentioned above, vectors with mutations in the E1 genes can be produced in cell lines that complement the E1 defect. Traditionally, this has been the 293 cell line that was established by transfection of human embryonic kidney cells with sheared Ad5 DNA. In fact, the history of adenovirus vector development is a good example of how certain discoveries can determine the direction of the research for years and even decades. The availability of the 293 cell line led to the development of gene transfer vectors that were Ad2- or Ad5-based. However, it might be asked whether adenovirus subgroup C-based vectors will be viable therapeutics for a number of currently favored gene therapy applications, because the prevalence of serum antibodies directed against Ad2 and Ad5 is very high in the human population. For this reason it is unclear whether intravenous injection of Ad5-based vectors will result in efficient and predictable liver gene transfer in humans, for example, unless the presence of anti-Ad5 serum antibodies has been excluded in a particular patient prior to enrollment in the gene transfer procedure. An additional reason why the development of adenovirus vectors based on other serotypes was delayed has been the notorious difficulty of transforming primary human cell lines with adenovirus E1 functions. For two decades, the 293 cell line was the only adenovirus-transformed human cell line that was suitable for production of E1-deleted vectors. Since the E1 functions of Ad5 may not fully complement E1 functions from other serotypes (Abrahamsen et al. 1997), it would have been difficult to develop vectors from serotypes that would face a very low antibody prevalence in the population. The 293 cell line has been extremely valuable for many developments in molecular biology, and it is the cell line that is being used for vector production in most of the ongoing clinical trials. A problem that has been observed during vector production using 293 cells, however, is the generation of replication-competent adenovirus (RCA). In this cell line, a large DNA fragment derived from the left terminus of Ad5, which has significant sequence overlap with the genomes of first- and second-generation vectors and also with helper viruses that are used in the production process of HC-Ad vectors, is integrated into the cellular chromosome. RCA is generated regularly due to homologous recombination between the integrated viral DNA and the infecting virus genome (Lochmuller et al. 1994). Cell lines have been developed more recently that preclude the generation of RCA by homologous recombination. These cells are derived from either human embryonic retina (Fallaux et al. 1998) or human amniocytes (Schiedner et al. 2000), currently the only two known cell types known to be reproducibly transformable by adenovirus E1 functions. The design of the new complementing cell lines avoids sequence overlap between the integrated transforming functions and the vector and, therefore, prevents the generation of RCA during production. Although it has been more difficult to produce helper cell lines for the support of E2B- and E4-deleted vectors, such lines do exist.

HC-Ad vectors cannot be produced similar to helper-independent vectors, in which most viral functions are provided from the vector. Adenovirus is a relatively large DNA virus that expresses some 20 to 30 different protein and RNA functions, many of which are either cytotoxic or are expressed at different stages and/or different levels during a productive infectious cycle. Thus, the chances of generating complementing cell lines that provide appropriate levels of all viral functions in *trans* are very slim, meaning that a helper virus must be used and subsequently eliminated from the end product. The first system that was used to produce HC-Ad vectors was based on the use of an E1-deleted helper virus with a mutated packaging signal that was derived from a previously described packaging adenovirus mutant (Gräble and Hearing 1990; Kochenek et al. 1996). Although the recombinant vector DNA with the wild-type packaging signal was packaged efficiently into viral capsids, the packaging of the helper virus was impaired, resulting in preferential production of the vector. The purity of the vector could be improved somewhat by CsCl equilibrium centrifugation due to slight differences in the densities of the two populations of particles. By this strategy it was possible to reduce the helper virus contamination to around 1% following two rounds of CsCl centrifugation. Although this system has been used successfully by different researchers to produce HC-Ad vectors, it is less suitable for clinical application in humans, because scale-up is difficult and reversion or genome instability of the helper virus mutant is likely.

The currently preferred production system is based on the excision of the packaging signal of the helper virus by a prokaryotic or yeast recombinase expressed in the producer cell line. To date, most HC-Ad vectors have been produced using the Cre-*loxP* recombination system of the bacteriophage P1 (Sauer et al. 1987). In this system the packaging signal of the helper virus is flanked by two *loxP* sites (Parks et al. 1996; Hardy et al. 1997). The HC-Ad vector is produced in E1-complementing cells that constitutively express the recombinase. The Cre-mediated excision is surprisingly efficient, and the contamination of vector by helper virus is reduced compared to the earlier production system described above. The yeast Flp-*frt* system has also been adapted recently for production of HC-Ad vectors (Umana et al. 2001; Ng et al. 2001) with comparable results. It is possible that for the production of HC-Ad vectors, the use of other recombination systems will be applied in the near future.

7
Vector Targeting

As an adjunct to the manipulation of viral genes and their expression in order to reduce immune responses and other toxic effects of viral infection, a significant amount of attention is being paid to the issue of re-targeting the virus such that it will only infect and/or express genes in cells in which transgene expression is intended. In theory, such changes to the tropism of the virus would improve its safety profile and could allow for systemic delivery of the therapeutic vector. We have already discussed the topic of transcriptional targeting, in which one either replaces a viral promoter(s) with a tissue-specific promoter to restrict gene expression and/or viral replication, or uses a binary system towards the same end. A related area of research, however, is receptor re-targeting, in which attempts are made to alter the cell-binding properties of the virion. This topic was reviewed by Wickham (2000), and is summarized here.

Most adenovirus serotypes share the ability to bind to a receptor on the surface of the cell called the Coxsackie and Adenovirus Receptor (CAR; Bergelson et al. 1997). Binding is mediated by a ligand on the adenovirus fiber knob, that domain of the fiber which is farthest from the surface of the virion. This initial interaction is followed by one between an RGD (arginine-glycine-aspartic acid) motif on the penton base and α_V integrins on the cell surface, resulting in the internalization of the particle (Wickham et al. 1993). Two main types of approaches have been taken to alter this pathway. The first is a biochemical approach, in which the virus is treated with a molecule that acts as a bridge between the virion and the intended receptor on the cell. The bridge can be a bifunctional antibody or a single-chain antibody that is covalently attached or genetically fused to a heterologous ligand. In this case, the combining site of the antibody is usually directed to the fiber knob, blocking its ability to recognize CAR while conferring the altered specificity. The second approach is to genetically modify the fiber, penton, or, most recently, hexon protein. In these situations, the gene encoding the viral protein is often mutated so as to remove its native receptor binding domain, and the homologous domain from a different adenovirus serotype or a heterologous ligand is inserted. This avenue of experimentation is a little trickier, as changing the structural proteins of the virus could conceivably interfere with virus assembly or stability, but has been moderately successful.

8
Summary and Conclusion

The use of adenovirus as a vector for gene transfer dates well over a decade now. Initially, clinicians felt that adenovirus might be the ideal delivery vehicle: it can be grown to high titer, infects a wide variety of dividing and nondividing cells, and can be easily genetically manipulated. The first experiences with adenovirus-mediated gene transfer demonstrated, however, that although it holds great promise, there is still much about which we need a deeper understanding. What are the functions of many of the viral genes in animal and human hosts compared to the functions that have been so well characterized in cells in culture? The absence of a good animal model of human adenovirus disease makes answering this question difficult, but progress is being made. Will HC-Ad vectors behave like wild-type virus in terms of infectivity and replication ability, and will one be able to produce large quantities of these vectors without contaminating helper virus? Technological advances along with a renewed interest in the later stages of the viral life cycle will surely assist these efforts. Will adenovirus be usable in systemic applications such as treatment of metastatic cancer or DMD? The ability to target the virus to specific cell populations, and to restrict gene expression to those cells, are important steps in that direction. Clearly as much progress as has been made to date could not have occurred without the strong basic background in adenovirus biology that has been laid by multiple investigators since the discovery of the virus half a century ago, and it is just as clear that future developments in this technology will require an ongoing commitment to elucidating the intricacies of the viral life cycle.

Acknowledgments. M.J.I. would like to acknowledge support from NIH grants HL64762 and AI52150. S.K. would like to acknowledge support by the Federal Ministry of Education and Research (FKZ: 01KS9502) and the Center for Molecular Medicine, Cologne.

References

Abrahamsen K, Kong HL, Mastrangeli A, Brough D, Lizonova A, Crystal RG, Falck-Pedersen E (1997) Construction of an adenovirus type 7a E1A-vector. J Virol 71:8946–8951

Adachi Y, Reynolds PN, Yamamoto M, Wang M, Takayama K, Matsubara S, Muramatsu T, Curiel DT (2001) A midkine promoter-based conditionally replicative adenovirus for treatment of pediatric solid tumors and bone marrow tumor purging. Cancer Res 61:7882–7888

Amalfitano A, Hauser MA, Hu H, Serra D, Begy CR, Chamberlain JS (1998) Production and characterization of improved adenovirus vectors with the E1, E2b, and E3 genes deleted. J Virol 72:926–933

Armentano D, Smith MP, Sookdeo CC, Zabner J, Perricone MA, St George JA, Wadsworth SC, Gregory RJ (1999) E4ORF3 requirement for achieving long-term transgene expression from the cytomegalovirus promoter in adenovirus vectors. J Virol 73:7031–7044

Armentano D, Zabner J, Sacks C, Sookdeo CC, Smith MP, St George JA, Wadsworth SC, Smith AE, Gregory RJ (1997) Effect of the E4 region on the persistence of transgene expression from adenovirus vectors. J Virol 71:2408–2416

Benjamin R, Helman L, Meyers P, Reaman G (2001) A phase I/II dose escalation and activity study of intravenous injections of OCaP1 for subjects with refractory osteosarcoma metastatic to lung. Hum Gene Ther 12:1591–1593

Bergelson JM, Cunningham JA, Droguett G, Kurt-Jones EA, Krithivas A, Hong JS, Horwitz MS, Crowell RL, Finberg RW (1997) Isolation of a common receptor for Coxsackie B viruses and adenoviruses 2 and 5. Science 275:1320–1323

Berkner KL (1988) Development of adenovirus vectors for the expression of heterologous genes. BioTechniques 6:616–629

Bischoff JR, Kirn DH, Williams A, Heise C, Horn S, Muna M, Ng L, Nye JA, Sampson-Johannes A, Fattaey A, McCormick F (1996) An adenovirus mutant that replicates selectively in p53-deficient human tumor cells. Science 274:373–376

Brough DE, Hsu C, Kulesa VA, Lee GM, Cantolupo LJ, Lizonova A,Kovesdi I (1997) Activation of transgene expression by early region 4 is responsible for a high level of persistent transgene expression from adenovirus vectors in vivo. J Virol 71:9206–9213

Bruder JT, Jie T, McVey DL., Kovesdi I (1997) Expression of gp19K increases the persistence of transgene expression from an adenovirus vector in the mouse lung and liver. J Virol 71:7623–7628

Chinnadurai G, Chinnadurai S, Brusca J (1979) Physical mapping of a large-plaque mutation of adenovirus type 2. J.Virol. 32:623–628

Clarke MF, Apel IJ, Benedict MA, Eipers PG, Sumantran V, Gonzalez-Garcia M, Doedens M, Fukunaga N, Davidson B, Dick JE, Minn AJ, Boise LH, Thompson CB, Wicha M, Nunez G (1995) A recombinant bcl-xs adenovirus selectively induces apoptosis in cancer cells but not in normal bone marrow cells. Proc Natl Acad Sci U S A 92:11024–11028

Clayman GL, El Naggar AK, Lippman SM., Henderson YC., Frederick M, Merritt JA, Zumstein LA, Timmons TM, Liu TJ, Ginsberg L, Roth JA, Hong WK, Bruso P, and Goepfert H (1998) Adenovirus-mediated p53 gene transfer in patients with advanced recurrent head and neck squamous cell carcinoma. J Clin Oncol 16:2221–2232

Clemens PR, Kochanek S, Sunada Y, Chan S, Chen H-H, Campbell KP, Caskey CT (1996) In vivo muscle gene transfer of full-length dystrophin with an adenoviral vector that lacks all viral genes. Gene Ther 3:965–972

Crystal RG, Harvey BG, Wisnivesky JP, O'Donoghue KA, Chu KW, Maroni J, Muscat JC, Pippo AL, Wright CE, Kaner RJ, Leopold PL, Kessler PD, Rasmussen HS, Rosengart TK, Hollmann C (2002) Analysis of risk factors for local delivery of low- and intermediate- dose adenovirus gene transfer vectors to individuals with a spectrum of comorbid conditions. Hum Gene Ther 13:65–100

Daniell E (1976) Genome structure of incomplete particles of adenovirus. J Virol 19:685–708

Danthinne X, Imperiale MJ (2000) Production of first-generation adenovirus vectors: a review. Gene Ther 7:1707–1714

Dedieu JF, Vigne E, Torrent C, Jullien C, Mahfouz I, Caillaud JM, Aubailly N, Orsini C, Guillaume JM, Opolon P, Delaere P, Perricaudet M, and Yeh P (1997) Long-term gene delivery into the livers of immunocompetent mice with E1/E4-defective adenoviruses. J Virol 71:4626–4637

Deuring R, Klotz G, Doerfler W (1981) An unusual symmetric recombinant between adenovirus type 12 DNA and human cell DNA. Proc Natl Acad Sci USA 78:3142–3146

Ding EY, Hodges BL, Hu H, McVie-Wylie AJ, Serra D, Migone FK, Pressley D, Chen YT, Amalfitano A (2001) Long-term efficacy after E1-, polymerase- adenovirus-mediated transfer of human acid-alpha-glucosidase gene into glycogen storage disease type II knockout mice. Hum Gene Ther 12:955-965

Doronin K, Toth K, Kuppuswamy M, Ward P, Tollefson AE, Wold WS (2000) Tumor-specific, replication-competent adenovirus vectors overexpressing the adenovirus death protein. J Virol 74:6147–6155

Doronin K, Kuppuswamy M, Toth K, Tollefson AE, Krajcsi P, Krougliak V, Wold WS (2001) Tissue-specific, tumor-selective, replication-competent adenovirus vector for cancer gene therapy. J Virol 75:3314–3324

Engelhardt JF, Litzky L, Wilson JM (1994a) Prolonged transgene expression in cotton rat lung with recombinant adenoviruses defective in E2a. Hum.Gene Ther. 5:1217–1229

Engelhardt JF, Ye X, Doranz B, Wilson JM (1994b) Ablation of E2A in recombinant adenoviruses improves transgene persistence and decreases inflammatory response in mouse liver. Proc.Natl.Acad.Sci.USA 91:6196–6200

Fallaux FJ, Bout A, Van Der Velde I, Van Den Wollenberg DJ, Hehir KM, Keegan J, Auger C, Cramer SJ, Van Ormondt H, Van Der Eb AJ, Valerio D Hoeben RC (1998) New helper cells and matched early region 1-deleted adenovirus vectors prevent generation of replication-competent adenoviruses. Hum Gene Ther 9:1909–1917

Fisher KJ, Choi H, Burda J, Chen S-J, Wilson JM (1996) Recombinant adenovirus deleted of all viral genes for gene therapy of cystic fibrosis. Virology 217:11–22

Ganly I, Kirn D, Eckhardt G, Rodriguez GI, Soutar DS, Otto R, Robertson AG, Park O, Gulley ML, Heise C, Von Hoff DD, Kaye SB, Eckhardt SG. (2000) A phase I study of ONYX-015, an E1B attenuated adenovirus, administered intratumorally to patients with recurrent head and neck cancer. Clin Cancer Res 6:798–806

Gaynor RB, Berk AJ (1983) Cis-acting induction of adenovirus transcription. Cell 33:683–693

Gluzman Y,Van Doren K (1983) Palindromic adenovirus type 5-simian virus 40 hybrid. J Virol 45:91–103

Grable M, Hearing P (1990) Adenovirus type 5 packaging domain is composed of a repeated element that is functionally redundant. J Virol 64:2047–2056

Grable M, Hearing P (1992) Cis and trans requirements for the selective packaging of adenovirus type 5 DNA. J Virol 66:723–731

Graham FL, Smiley J, Russell WC, Nairn R (1977) Characteristics of a human cell line transformed by DNA from human adenovirus type. J Gen Virol 36:59–74

Haecker SE, Stedman HH, Balice-Gordon RJ, Smith DB, Greelish JP, Mitchell MA, Wells A, Sweeney HL Wilson JM (1996) *In vivo* expression of full-length human dystrophin from adenoviral vectors deleted of all viral genes. Hum Gene Ther 7:1907–1914

Hallenbeck PL, Chang YN, Hay C, Golightly D, Stewart D, Lin J, Phipps S, Chiang YL (1999) A novel tumor-specific replication-restricted adenoviral vector for gene therapy of hepatocellular carcinoma. Hum Gene Ther 10:1721–1733

Hammarskjold ML, Winberg G (1980) Encapsidation of adenovirus 16 DNA is directed by a small DNA sequence at the left end of the genome. Cell 20:787–795

Hardy S, Kitamura M, Harris-Stansil T, Dai Y, Phipps ML (1997) Construction of adenovirus vectors through Cre-*lox* recombination. J Virol 71:1842–1849

Hartley JW, Heubner RJ, Rowe WP (1956) Serial propagation of adenoviruses (APC) in monkey kidney tissue cultures. Proc Soc Exp Biol Med 9:667

Harvey BG, Maroni J, O'Donoghue KA, Chu KW, Muscat JC, Pippo AL, Wright CE, Hollmann C, Wisnivesky JP, Kessler PD, Rasmussen HS, Rosengart TK, Crystal RG (2002) Safety of local delivery of low- and intermediate-dose adenovirus gene transfer vectors to individuals with a spectrum of morbid conditions. Hum Gene Ther 13:15–63

Hassell JA, Lukanidin E, Fey, G, Sambrook J (1978) The structure and expression of two defective adenovirus 2/simian virus 40 hybrids. J Mol Biol 120:209–247

Hearing P, Samulski RJ, Wishart WL, Shenk T (1987) Identification of a repeated sequence element required for efficient encapsidation of the adenovirus type 5 chromosome. J Virol 61:2555–2558

Heise C, Hermiston T, Johnson L, Brooks G, Sampson-Johannes A, Williams A, Hawkins L, Kirn D (2000) An adenovirus E1A mutant that demonstrates potent and selective systemic anti-tumoral efficacy. Nat Med 6:1134–1139

Hernandez-Alcoceba R, Pihalja M, Wicha MS, Clarke MF (2000) A novel, conditionally replicative adenovirus for the treatment of breast cancer that allows controlled replication of E1a-deleted adenoviral vectors. Hum Gene Ther 11:2009–2024

Imperiale MJ, Kao HT, Feldman LT, Nevins JR, Strickland S (1984) Common control of the heat shock gene and early adenovirus genes: evidence for a cellular E1A-like activity. Mol.Cell.Biol. 4:867–874

Jones N, Shenk T (1979) Isolation of adenovirus type 5 host range deletion mutants defective for transformation of rat embryo cells. Cell 17:683–689

Khuri FR, Nemunaitis J, Ganly I, Arseneau J, Tannock IF, Romel L, Gore M, Ironside J, MacDougall RH, Heise C, Randlev B, Gillenwater AM, Bruso P, Kaye SB, Hong WK, Kirn DH (2000) A controlled trial of intratumoral ONYX-015, a selectively-replicating adenovirus, in combination with cisplatin and 5-fluorouracil in patients with recurrent head and neck cancer. Nat Med 6:879–885

Kim IH, Jozkowicz A, Piedra PA, Oka K, Chan L. (2001) Lifetime correction of genetic deficiency in mice with a single injection of helper-dependent adenoviral vector. Proc Natl Acad Sci U S A 98:13282–13287

Klessig DF (1984) Adenovirus-Simian Virus 40 interactions. In: The Adenoviruses, ed. H.S. Ginsberg, Plenum Press, New York

Kochanek S, Clemens PR, Mitani K, Chen H-H, Chan S, Caskey CT (1996) A new adenoviral vector: Replacement of all viral coding sequences with 28 kb of DNA independently expressing both full-length dystrophin and b-galactosidase. Proc Natl Acad Sci USA 93:5731–5736

Kovesdi I, Reichel R, Nevins JR (1986) Identification of a cellular transcription factor involved in E1A trans-activation. Cell 45:219–228

Kumar-Singh R, Chamberlain JS (1996) Encapsidated adenovirus minichromosomes allow delivery and expression of a 14 kb dystrophin cDNA to muscle cells. Hum Mol Genet 5:913–921

Kumar-Singh R, Farber DB.(1998) Encapsidated adenovirus mini-chromosome-mediated delivery of genes to the retina: application to the rescue of photoreceptor degeneration. Hum Mol Genet 7:1893–1900

Kurihara T, Brough DE, Kovesdi I, Kufe DW (2000) Selectivity of a replication-competent adenovirus for human breast carcinoma cells expressing the MUC1 antigen. J Clin Invest 106:763–771

Lee MG, Abina MA, Haddada H, Perricaudet M (1995) The constitutive expression of the immunomodulatory gp19k protein in E1-, E3- adenoviral vectors strongly reduces the host cytotoxic T cell response against the vector. Gene Ther 2:256–262

Lewis Jr AM, Baum SG, Prigge KO, Rowe WP (1966) Occurrence of adenovirus-SV40 hybrids among monkey kidney cell adapted strains of adenovirus. Proc Soc Exp Biol Med 122:214–218

Lewis Jr AM, Rowe WP (1970) Isolation of two plaque variants from the adenovirus type 2-simian virus 40 hybrid population which differ in their efficiency in yielding simian virus 40. J Virol 5:413–420

Lieber A, He C, Kirillova I, Kay MA (1996) Recombinant adenoviruses with large deletions generated by cre-mediated excision exhibit biological properties compared with first-generation vectors *in vitro* and *in vivo.* J Virol. 70:8944–8960

Lieber A, He C-Y, Kay MA (1997) Adenoviral preterminal protein stabilizes mini-adenoviral genomes in vitro and in vivo. Nat Biotechnol 15:1383–1387

Lochmuller H, Jani A, Huard J, Prescott S, Simoneau M, Massie B, Karpati G, Acsadi G (1994) Emergence of early region 1-containing replication-competent adenovirus in stocks of replication-defective adenovirus recombinants (delta E1 + delta E3) during multiple passages in 293 cells. Hum Gene Ther 5:1485–1491

Lusky M, Christ M, Rittner K, Dieterle A, Dreyer D, Mourot B, Schultz H, Stoeckel F, Pavirani A, Mehtali M (1998) In vitro and in vivo biology of recombinant adenovirus vectors with E1, E1/E2A, or E1/E4 deleted. J Virol 72:2022–2032

Mansour SL, Grodzicker T, Tjian R (1985) An adenovirus vector system used to express polyoma virus tumor antigens. Proc Natl Acad Sci USA 82:1359–1363

Matsubara S, Wada Y, Gardner TA, Egawa M, Park MS, Hsieh CL, Zhau HE, Kao C, Kamidono S, Gillenwater JY, Chung LW (2001) A conditional replication-competent adenoviral vector, Ad-OC-E1a, to cotarget prostate cancer and bone stroma in an experimental model of androgen-independent prostate cancer bone metastasis. Cancer Res 61:6012–6019

Mitani K, Graham FL, Caskey CT, Kochanek S (1995) Rescue, propagation, and partial purification of a helper virus-dependent adenovirus vector. Proc Natl Acad Sci USA 92:3854–3858

Morral N, Parks RJ, Zhou H, Langston C, Schiedner G, Quinones J, Graham FL, Kochanek S Beaudet AL (1998) High doses of a helper-dependent adenoviral vector yield supraphysiological levels of alpha1-antitrypsin with negligible toxicity. Hum Gene Ther 9:2709–2716

Morris JC, Ramsey WJ, Wildner O, Muslow HA, Aguilar-Cordova E, Blaese RM (2000) A phase I study of intralesional administration of an adenovirus vector expressing the HSV-1 thymidine kinase gene (AdV.RSV-TK) in combination with escalating doses of ganciclovir in patients with cutaneous metastatic malignant melanoma. Hum Gene Ther 11:487–503

Morsy MA, Gu M, Motzel S, Zhao J, Lin J, Su Q, Allen H, Franlin L, Parks RJ, Graham FL, Kochanek S, Bett AJ, Caskey CT (1998) An adenoviral vector deleted for all viral coding sequences results in enhanced safety and extended expression of a leptin transgene. Proc Natl Acad Sci USA 95:7866–7871

Nanda D, Vogels R, Havenga M, Avezaat CJ, Bout A, Smitt PS (2001) Treatment of malignant gliomas with a replicating adenoviral vector expressing herpes simplex virus-thymidine kinase. Cancer Res 61:8743–8750

Nemunaitis J, Ganly I, Khuri F, Arseneau J, Kuhn J, McCarty T, Landers S, Maples P, Romel L, Randlev B, Reid T, Kaye S, Kirn D (2000) Selective replication and oncolysis in p53 mutant tumors with ONYX-015, an E1B-55kD gene-deleted adenovirus, in patients with advanced head and neck cancer: a phase II trial. Cancer Res 60:6359–6366

Nemunaitis J, Khuri F, Ganly I, Arseneau J, Posner M, Vokes E, Kuhn J, McCarty T, Landers S, Blackburn A, Romel L, Randlev B, Kaye S, Kirn D (2001) Phase II trial of intratumoral administration of ONYX-015, a replication-selective adenovirus, in patients with refractory head and neck cancer. J Clin Oncol 19:289–298

Ng P, Beauchamp C, Evelegh C, Parks R, Graham FL (2001) Development of a FLP/frt system for generating helper-dependent adenoviral vectors. Mol Ther 2001:3:809–815

Parks RJ, Chen L, Anton M, Sankar U, Rudnicki MA Graham FL (1996) A helper-dependent adenovirus vector system: Removal of helper virus by Cre-mediated excision of the viral packaging signal. Proc Natl Acad Sci USA 93:13565–13570

Parks RJ, Graham FL (1997) A helper-dependent system for adenovirus vector production helps define a lower limit for efficient DNA packaging. J Virol 71:3293–3298

Parks RJ, Bramson JL, Wan Y, Addison CL, Graham FL (1999) Effects of stuffer DNA on transgene expression from helper-dependent adenovirus vectors. J Virol 73:8027–8034

Pierce WE, Rosenbaum MJ, Edwards EA, Peckinpaugh RO, Jackson GG (1968) Live and inactivated adenovirus vaccines for the prevention of acute respiratory illness in naval recruits. Am J Epidemiol 87:237–246

Qiao J, Doubrovin M, Sauter BV, Huang Y, Guo ZS, Balatoni J, Akhurst T, Blasberg RG, Tjuvajev JG, Chen SH, Woo SL (2002) Tumor-specific transcriptional targeting of suicide gene therapy. Gene Ther 9:168–175

Rajagopalan S, Shah M, Luciano A, Crystal R, Nabel EG (2001) Adenovirus-mediated gene transfer of VEGF(121) improves lower-extremity endothelial function and flow reserve. Circulation 104:753–765

Rao L, Debbas M, Sabbatini P, Hockenbery D, Korsmeyer S, White E (1992) The adenovirus E1A proteins induce apoptosis, which is inhibited by the E1B 19-kDa and Bcl-2 proteins. Proc.Natl.Acad.Sci.USA 89:7742–7746

Raper SE, Yudkoff M, Chirmule N, Gao GP, Nunes F, Haskal ZJ, Furth EE, Propert KJ, Robinson MB, Magosin S, Simoes H, Speicher L, Hughes J, Tazelaar J, Wivel NA, Wilson JM, Batshaw ML (2002) A pilot study of in vivo liver-directed gene transfer with an adenoviral vector in partial ornithine transcarbamylase deficiency Hum Gene Ther 13:163–175

Rodriguez R, Schuur ER, Lim HY, Henderson GA, Simons JW, Henderson DR (1997) Prostate attenuated replication competent adenovirus (ARCA) CN706: a selective cytotoxic for prostate-specific antigen-positive prostate cancer cells. Cancer Res 57:2559–2563

Rogulski KR, Wing MS, Paielli DL, Gilbert JD, Kim JH, Freytag SO (2000) Double suicide gene therapy augments the antitumor activity of a replication-competent lytic adenovirus through enhanced cytotoxicity and radiosensitization. Hum Gene Ther 11:67–76

Rosengart TK, Lee LY, Patel SR, Kligfield PD, Okin PM, Hackett NR, Isom OW, Crystal RG (1999a) Six-month assessment of a phase I trial of angiogenic gene therapy for the treatment of coronary artery disease using direct intramyocardial administration of an adenovirus vector expressing the VEGF121 cDNA. Ann Surg 230:466–470; discussion 470–472

Rosengart TK, Lee LY, Patel SR, Sanborn TA, Parikh M, Bergman GW, Hachamovitch R, Szulc M, Kligfield PD, Okin PM, Hahn RT, Devereux RB, Post MR, Hackett NR, Foster T, Grasso TM, Lesser ML, Isom OW, Crystal RG (1999b) Angiogenesis gene therapy: phase I assessment of direct intramyocardial administration of an adenovirus vector expressing VEGF121 cDNA to individuals with clinically significant severe coronary artery disease. Circulation 100:468–474

Roth JA, Swisher SG, Merritt JA, Lawrence DD, Kemp BL, Carrasco CH, El-Naggar AK, Fossella FV, Glisson BS, Hong WK, Khurl FR, Kurie JM, Nesbitt JC, Pisters K, Putnam JB, Schrump DS., Shin DM, Walsh GL (1998) Gene therapy for non-small cell lung cancer: a preliminary report of a phase I trial of adenoviral p53 gene replacement. Semin Oncol 25:33–37

Sauer, B (1987) Functional expression of the cre-lox site-specific recombination system in the yeast Saccharomyces cerevisiae. Mol Cell Biol 7:2087–2096

Schiedner G, Morral N, Parks RJ, Wu Y, Koopmans SC, Langston C, Graham FL, Beaudet AL Kochanek S (1998) A high capacity adenovirus vector with all viral genes deleted results in improved in vivo expresssion and decreased toxicity. Nature Genet 18:180–183

Schiedner G, Hertel S, Kochanek S (2000) Efficient transformation of primary human amniocytes by E1 functions of Ad5: generation of new cell lines for adenoviral vector production. Hum Gene Ther 11:2105–2116

Schiedner G, Hertel S, Johnston M, Biermann V, Dries V, Kochanek S (2002) Variables affecting in vivo performance of high-capacity adenovirus vectors. J Virol 76:1600–1609

Shenk T, Jones N, Colby W, Fowlkes D (1980) Functional analysis of adenovirus-5 host-range deletion mutants defective for transformation of rat embryo cells. Cold Spring Harb Symp Quant Biol 44 Pt 1, 367–375

Swisher, SG, Roth JA, Nemunaitis J, Lawrence DD, Kemp BL, Carrasco CH, Connors DG, El-Naggar AK, Fossella F, Glisson BS, Hong WK, Khuri FR, Kurie JM., Lee JJ., Lee JS, Mack M, Merritt JA, Nguyen DM, Nesbitt JC, Perez-Soler R, Pisters KM,

Putnam JB Jr, Richli WR, Savin M, Waugh MK et al (1999) Adenovirus-mediated p53 gene transfer in advanced non-small-cell lung cancer. J Natl Cancer Inst 91:763–771

Tibbets C (1977) Viral DNA sequences from incomplete particles of human adenovirus type 7. Cell 12: 243–249

Trask TW, Trask RP, Aguilar-Cordova E, Shine HD, Wyde PR, Goodman JC, Hamilton WJ, Rojas-Martinez A, Chen SH, Woo SL, Grossman RG (2000) Phase I study of adenoviral delivery of the HSV-tk gene and ganciclovir administration in patients with current malignant brain tumors. Mol Ther 1:195–203

Ueda K, Iwahashi M, Nakamori M, Nakamura M, Matsuura I, Yamaue H, Tanimura H (2001) Carcinoembryonic antigen-specific suicide gene therapy of cytosine deaminase/5-fluorocytosine enhanced by the cre/loxP system in the orthotopic gastric carcinoma model. Cancer Res 61:6158–6162

Umana P, Gerdes CA, Stone D, Davis JR, Ward D, Castro MG, Lowenstein PR (2001) Efficient FLPe recombinase enables scalable production of helper-dependent adenoviral vectors with negligible helper-virus contamination. Nat Biotechnol 19:582–585

Wang Q, Greenburg G, Bunch D, Farson D, Finer MH (1997) Persistent transgene expression in mouse liver following in vivo gene transfer with a delta E1/delta E4 adenovirus vector. Gene Ther 4:393–400

White E, Faha B, Stillman B (1986) Regulation of adenovirus gene expression in human WI38 cells by an E1B-encoded tumor antigen. Mol Cell Biol 6:3763–3773

Wickham T (2000) Targeting adenovirus. Gene Ther 7:110–114

Wickham TJ, Mathias P, Cheresh DA, Nemerow GR (1993) Integrins alpha v beta 3 and alpha v beta 5 promote adenovirus internalization but not virus attachment. Cell 73:309–319

Yamada M, Lewis JA, Grodzicker T (1985) Overproduction of the protein product of a nonselected foreign gene carried by an adenovirus vector. Proc Natl Acad Sci USA 82:3567–3571

Yang Y, Nunes FA, Berencsi K, Gonczol E, Engelhardt JF, Wilson JM (1994) Inactivation of E2a in recombinant adenoviruses improves the prospect for gene therapy in cystic fibrosis. Nat.Genet. 7:362–369

Subject Index

Current Topics in Microbiology and Immunology

Volumes published since 1989 (and still available)

Vol. 230: **Kärre, Klas; Colonna, Marco (Eds.):** Specificity, Function, and Development of NK Cells. 1998. 22 figs. IX, 248 pp. ISBN 3-540-63941-1

Vol. 231: **Holzmann, Bernhard; Wagner, Hermann (Eds.):** Leukocyte Integrins in the Immune System and Malignant Disease. 1998. 40 figs. XIII, 189 pp. ISBN 3-540-63609-9

Vol. 232: **Whitton, J. Lindsay (Ed.):** Antigen Presentation. 1998. 11 figs. IX, 244 pp. ISBN 3-540-63813-X

Vol. 233/I: **Tyler, Kenneth L.; Oldstone, Michael B. A. (Eds.):** Reoviruses I. 1998. 29 figs. XVIII, 223 pp. ISBN 3-540-63946-2

Vol. 233/II: **Tyler, Kenneth L.; Oldstone, Michael B. A. (Eds.):** Reoviruses II. 1998. 45 figs. XVI, 187 pp. ISBN 3-540-63947-0

Vol. 234: **Frankel, Arthur E. (Ed.):** Clinical Applications of Immunotoxins. 1999. 16 figs. IX, 122 pp. ISBN 3-540-64097-5

Vol. 235: **Klenk, Hans-Dieter (Ed.):** Marburg and Ebola Viruses. 1999. 34 figs. XI, 225 pp. ISBN 3-540-64729-5

Vol. 236: **Kraehenbuhl, Jean-Pierre; Neutra, Marian R. (Eds.):** Defense of Mucosal Surfaces: Pathogenesis, Immunity and Vaccines. 1999. 30 figs. IX, 296 pp. ISBN 3-540-64730-9

Vol. 237: **Claesson-Welsh, Lena (Ed.):** Vascular Growth Factors and Angiogenesis. 1999. 36 figs. X, 189 pp. ISBN 3-540-64731-7

Vol. 238: **Coffman, Robert L.; Romagnani, Sergio (Eds.):** Redirection of Th1 and Th2 Responses. 1999. 6 figs. IX, 148 pp. ISBN 3-540-65048-2

Vol. 239: **Vogt, Peter K.; Jackson, Andrew O. (Eds.):** Satellites and Defective Viral RNAs. 1999. 39 figs. XVI, 179 pp. ISBN 3-540-65049-0

Vol. 240: **Hammond, John; McGarvey, Peter; Yusibov, Vidadi (Eds.):** Plant Biotechnology. 1999. 12 figs. XII, 196 pp. ISBN 3-540-65104-7

Vol. 241: **Westblom, Tore U.; Czinn, Steven J.; Nedrud, John G. (Eds.):** Gastroduodenal Disease and Helicobacter pylori. 1999. 35 figs. XI, 313 pp. ISBN 3-540-65084-9

Vol. 242: **Hagedorn, Curt H.; Rice, Charles M. (Eds.):** The Hepatitis C Viruses. 2000. 47 figs. IX, 379 pp. ISBN 3-540-65358-9

Vol. 243: **Famulok, Michael; Winnacker, Ernst-L.; Wong, Chi-Huey (Eds.):** Combinatorial Chemistry in Biology. 1999. 48 figs. IX, 189 pp. ISBN 3-540-65704-5

Vol. 244: **Daëron, Marc; Vivier, Eric (Eds.):** Immunoreceptor Tyrosine-Based Inhibition Motifs. 1999. 20 figs. VIII, 179 pp. ISBN 3-540-65789-4

Vol. 245/I: **Justement, Louis B.; Siminovitch, Katherine A. (Eds.):** Signal Transduction and the Coordination of B Lymphocyte Development and Function I. 2000. 22 figs. XVI, 274 pp. ISBN 3-540-66002-X

Vol. 245/II: **Justement, Louis B.; Siminovitch, Katherine A. (Eds.):** Signal Transduction on the Coordination of B Lymphocyte Development and Function II. 2000. 13 figs. XV, 172 pp. ISBN 3-540-66003-8

Vol. 246: **Melchers, Fritz; Potter, Michael (Eds.):** Mechanisms of B Cell Neoplasia 1998. 1999. 111 figs. XXIX, 415 pp. ISBN 3-540-65759-2

Vol. 247: **Wagner, Hermann (Ed.):** Immunobiology of Bacterial CpG-DNA. 2000. 34 figs. IX, 246 pp. ISBN 3-540-66400-9

Vol. 248: **du Pasquier, Louis; Litman, Gary W. (Eds.):** Origin and Evolution of the Vertebrate Immune System. 2000. 81 figs. IX, 324 pp. ISBN 3-540-66414-9

Vol. 249: **Jones, Peter A.; Vogt, Peter K. (Eds.):** DNA Methylation and Cancer. 2000. 16 figs. IX, 169 pp. ISBN 3-540-66608-7

Vol. 250: **Aktories, Klaus; Wilkins, Tracy, D. (Eds.):** Clostridium difficile. 2000. 20 figs. IX, 143 pp. ISBN 3-540-67291-5

Vol. 251: **Melchers, Fritz (Ed.):** Lymphoid Organogenesis. 2000. 62 figs. XII, 215 pp. ISBN 3-540-67569-8

Vol. 252: **Potter, Michael; Melchers, Fritz (Eds.):** B1 Lymphocytes in B Cell Neoplasia. 2000. XIII, 326 pp. ISBN 3-540-67567-1

Vol. 253: **Gosztonyi, Georg (Ed.):** The Mechanisms of Neuronal Damage in Virus Infections of the Nervous System. 2001. approx. XVI, 270 pp. ISBN 3-540-67617-1

Vol. 254: **Privalsky, Martin L. (Ed.):** Transcriptional Corepressors. 2001. 25 figs. XIV, 190 pp. ISBN 3-540-67569-8

Vol. 255: **Hirai, Kanji (Ed.):** Marek's Disease. 2001. 22 figs. XII, 294 pp. ISBN 3-540-67798-4

Vol. 256: **Schmaljohn, Connie S.; Nichol, Stuart T. (Eds.):** Hantaviruses. 2001, 24 figs. XI, 196 pp. ISBN 3-540-41045-7

Vol. 257: **van der Goot, Gisou (Ed.):** Pore-Forming Toxins, 2001. 19 figs. IX, 166 pp. ISBN 3-540-41386-3

Vol. 258: **Takada, Kenzo (Ed.):** Epstein-Barr Virus and Human Cancer. 2001. 38 figs. IX, 233 pp. ISBN 3-540-41506-8

Vol. 259: **Hauber, Joachim, Vogt, Peter K. (Eds.):** Nuclear Export of Viral RNAs. 2001. 19 figs. IX, 131 pp. ISBN 3-540-41278-6

Vol. 260: **Burton, Didier R. (Ed.):** Antibodies in Viral Infection. 2001. 51 figs. IX, 309 pp. ISBN 3-540-41611-0

Vol. 261: **Trono, Didier (Ed.):** Lentiviral Vectors. 2002. 32 figs. X, 258 pp. ISBN 3-540-42190-4

Vol. 262: **Oldstone, Michael B.A. (Ed.):** Arenaviruses I. 2002, 30 figs. XVIII, 197 pp. ISBN 3-540-42244-7

Vol. 263: **Oldstone, Michael B. A. (Ed.):** Arenaviruses II. 2002, 49 figs. XVIII, 268 pp. ISBN 3-540-42705-8

Vol. 264/I: **Hacker, Jörg; Kaper, James B. (Eds.):** Pathogenicity Islands and the Evolution of Microbes. 2002. 34 figs. XVIII, 232 pp. ISBN 3-540-42681-7

Vol. 264/II: **Hacker, Jörg; Kaper, James B. (Eds.):** Pathogenicity Islands and the Evolution of Microbes. 2002. 24 figs. XVIII, 228 pp. ISBN 3-540-42682-5

Vol. 265: **Dietzschold, Bernhard; Richt, Jürgen A. (Eds.):** Protective and Pathological Immune Responses in the CNS. 2002. 21 figs. X, 278 pp. ISBN 3-540-42668-X

Vol. 266: **Cooper, Koproski (Eds.):** The Interface Between Innate and Acquired Immunity, 2002, 15 figs. XIV, 116 pp. ISBN 3-540-42894-1

Vol. 267: **Mackenzie, John S.; Barrett, Alan D. T.; Deubel, Vincent (Eds.):** Japanese Encephalitis and West Nile Viruses. 2002. 66 figs. X, 418 pp. ISBN 3-540-42783-X

Vol. 268: **Zwickl, Peter; Baumeister, Wolfgang (Eds.):** The Proteasome-Ubiquitin Protein Degradation Pathway. 2002, 17 figs. X, 213 pp. ISBN 3-540-43096-2

Vol. 269: **Koszinowski, Ulrich H.; Hengel, Hartmut (Eds.):** Viral Proteins Counteracting Host Defenses. 2002, 32 figs. XII, 328 pp. ISBN 3-540-43261-2

Vol. 270: **Beutler, Bruce; Wagner, Hermann (Eds.):** Toll-Like Receptor Family Members and Their Ligands. 2002, 31 figs. X, 192 pp. ISBN 3-540-43560-3

Vol. 271: **Koehler, Theresa M. (Ed.):** Anthrax. 2002, 14 figs. X, 169 pp. ISBN 3-540-43497-6

Vol. 272: **Doerfler, Walter; Böhm, Petra (Eds.):** Adenoviruses: Model and Vectors in Virus-Host Interactions. Virion-Structure. Viral Replication and Host-Cell Interactions. 2003, 63 figs. X, 452 pp. ISBN 3-540-00154-9

MIX
Papier aus verantwortungsvollen Quellen
Paper from responsible sources
FSC® C105338

If you have any concerns about our products,
you can contact us on
ProductSafety@springernature.com

In case Publisher is established outside the EU,
the EU authorized representative is:
Springer Nature Customer Service Center GmbH
Europaplatz 3, 69115 Heidelberg, Germany

Printed by Libri Plureos GmbH
in Hamburg, Germany